Observations of ocean circulation have vastly increased during the last two decades, as a result of both international field programs and remote sensors on artificial earth satellites. Oceanographers are turning to inverse methods in order to combine these observations with numerical models of ocean circulation. Recent advances in high-performance computing enable rigorous implementation of inverse methods for massive data sets and complex models. Accordingly this book develops the fundamentals of inverse theory, emphasizing rigor over expedience.

In addition to interpolating the data and adding realism to the model solutions, inverse methods can yield estimates for unobserved flow variables, forcing fields and model parameters. Inverse formulations can resolve ill-posed modeling problems, provide assessment of oceanic observing systems and test models as formal scientific hypotheses.

Ocean models considered range from linear, finite-dimensional systems of equality and inequality constraints to the nonlinear primitive equations. Optimization methods include Kalman filters, descent algorithms and especially representer expansions. Exercises of varying difficulty rehearse technique, and supplement the central theoretical development.

This book is intended for environmental scientists and engineers interested in data analysis and quantitative forecasting, for advanced undergraduates in applied mathematics, and for graduate students in physical oceanography.

CAMBRIDGE MONOGRAPHS ON MECHANICS

Founding Editor
G. K. BATCHELOR

General Editors

S.H. DAVIS
Northwestern University

L. B. FREUND
Brown University

S. LEIBOVICH
Cornell University

V. TVERGAARD
Technical University of Denmark

INVERSE METHODS IN PHYSICAL OCEANOGRAPHY

Inverse Methods in Physical Oceanography

ANDREW F. BENNETT

College of Oceanography

Oregon State University

Published by the Press Syndicate of the University of Cambridge
The Pitt Building, Trumpington Street, Cambridge CB2 1RP
40 West 20th Street, New York, NY 10011, USA
477 Williamstown Road, Port Melbourne VIC 3207, Australia
Ruiz de Alarcón 13, 28014 Madrid, Spain
Dock House, The Waterfront, Cape Town 8001, South Africa

First published 1992
Reprinted with corrections 1999, 2003

Typeset in TEX

Library of Congress Cataloging-in-Publication Data

Bennett, Andrew F.

Inverse methods in physical oceanography / Andrew F. Bennett

p. cm. (Cambridge monographs on mechanics and applied mathematics)
Includes bibliographical references (p.) and index.

ISBN 0-521-38568-7 (hc)

1. Oceanography—Mathematics. 2. Inverse problems (Differential equations)
I. Title. II. Series.

GC10.4.M36B46 1992	92-5107
551.46'0015'1535—dc20	CIP

A catalog record for this book is available from the British Library

ISBN 0-521-38568-7 hardback

Transferred to digital printing 2003

CONTENTS

9. Primitive-equation models

PREFACE

Inverse methods combine oceanic observations with theoretical models of ocean circulation. The methods lead to

- estimates of oceanic fields from sparse data, guided by physical laws;
- estimates of meteorological forcing fields;
- estimates of parameters in the physical laws;
- designs for oceanic observing systems;
- resolution of mathematically ill-posed modeling problems; and
- tests of scientific hypotheses.

The rapid development of inverse methods in physical oceanography over the last decade has been greatly influenced by the work of meteorologists and solid-earth geophysicists; the growing interest of oceanographers in inverse methods is largely in response to the arrival of great volumes of data collected by artificial earth satellites. Collected articles may be found in conference proceedings edited by Anderson & Willebrand (1989), and by Haidvogel & Robinson (1989). Even greater volumes of data are anticipated from planned, future missions. The emergence of inverse theory as a scientific tool has provided a major stimulus to numerical modeling of ocean circulation, which has moved beyond the infancy of thought-experiments but has only tentatively entered the maturity of operational forecasting long occupied by meteorologists. The extraordinary recent gains in computer performance will enable theoretical oceanographers, who have come to inverse theory relatively lately, to consider applying the most elegant and powerful inverse methods to their most complex models. (While this preface was being written, a computer manufacturer announced a 128–gigaflops, 32–gigabyte machine.) In this regard, oceanographers are even more fortunate in that their synoptic time

scale is so much longer than the meteorologists' (months, instead of days), so powerful methods may also be brought to bear upon real-time forecasting problems.

The purpose of this book is, therefore, to introduce advanced graduate students, research scientists and that newly emerging professional, the ocean forecaster, to the *possibilities* for inverse theory, rather than the necessities which are the emphasis in Roger Daley's new and important book: *Atmospheric Data Analysis* (Daley, 1991). There the emphasis is almost entirely on objective analysis and the practical initialization problem in operational forecasting, equivalent to the first item in our list.

It is assumed here that the reader has a basic knowledge of geophysical fluid dynamics and ocean modeling. The texts by Pedlosky (1987) and by Pickard & Emery (1990) are excellent references for oceanographic theory and practice; the latter also provides sketches of the major ocean current systems. It is also assumed that the reader is confident in linear algebra and advanced calculus. The material is based largely on lectures given to advanced graduate students in physical oceanography and to advanced undergraduates in applied mathematics. Some tools from functional analyses are exploited (embedding theorems, reproducing-kernel Hilbert spaces and *a priori* inequalities), but only in context and with no formality. In particular, the need for such tools is explained in simple terms. Some of these tools, while being powerful, can also be crude to the point of being misleading. The major analytical tool used throughout is the calculus of variations. An excellent and readily accessible discussion may be found in Courant & Hilbert (1953). A very brief introduction is offered at the end of the fifth chapter here.

This monograph does not review every contribution to the subject. About 50% of the contents derive from the work of the author and collaborators. It is hoped that the result will serve as a coherent presentation of the essential ideas. At this fundamental level, inverse methods in physical oceanography are not so different from those used in geophysics, as described in *Geophysical Data Analysis: Discrete Inverse Theory* (Menke, 1984) and in *Inverse Problem Theory* (Tarantola, 1987). A brief but useful theoretical review for oceanographers (Miller, 1987) may be noted. However, these outstanding texts do not provide for the requirements of oceanographers, that is, detailed analysis of the application of such methods to nonlinear, time-dependent, dynamical models of stratified ocean circulation;

detailed discussion of data requirements, including initial conditions and boundary conditions; a theory for the design of observing systems, and interpretations of simple data-assimilation techniques in terms of inverse methods. Close attention is given to many real inverse calculations drawn from the recent oceanographic literature.

The theoretical discussion is supported with exercises. These do not usually appear in monographs, so some justification is warranted. A number of the exercises merely invite the reader to derive results given in the text; had the full derivations been included, the size of this work would have increased enormously. Other exercises pursue lines of analysis which, while being of interest, were not considered essential. It is hoped that the exercises will encourage the habit of exploring inverse methods in simple settings using elementary analytical techniques, before applying the methods to complex ocean models using sophisticated numerical techniques.

To the extent that this monograph contains essentials, it should be useful for some time. The case studies will quickly become dated, as inverse calculations are made with increasingly more general models and more significant data sets.

The chapters may be summarized as follows:

1. Finite-dimensional inverse theory

Applications of finite-dimensional inverse theory to ocean modeling are discussed, with the objective of introducing some fundamental concepts in a relatively simple setting. These concepts are listed in the introduction to the chapter. Much of the theoretical material is well covered elsewhere, so the discussion of the concepts is brief and directed toward subsequent applications in more complex settings.

2. Smoothing of observations

Two fundamental techniques for estimating fields are discussed here. Both involve least-squares principles.

The first, the well-known "optimal interpolation" or "objective analysis," is based on the Gauss-Markov theorem. The essential concept is unbiased linear regression with minimal error variance. Measurements are expressed as linear functionals acting upon the oceanic field of interest. The necessary prior statistical information will be identified. Simplifying assumptions, such as statistical homogeneity or isotropy, will be discussed along with current practice in estimating the necessary statistics.

The second method, spline smoothing, is less well known to oceanographers but is indispensable later. The powerful constructs known as reproducing kernels and representers are introduced. The remarkable, virtual one-to-one relation between spline smoothing and Gauss-Markov smoothing is delineated.

As a refrain, Wunsch's finite-dimensional inverse methods, described in the preceding chapter using the simple language of linear algebra, are reinterpreted using both linear regression and the theory of representers.

This chapter is also brief, with preparation for subsequent chapters in mind rather than comprehensive presentation.

3. Data assimilation

Oceanographers have already met with some success in ocean forecasting when accurate initial conditions have been available. The quality of these conditions owes much to the smoothing of significant amounts of good initial data. These data may also be used to revise the covariance fields used in the smoothing; the resulting strategy (the Kalman filter) is the optimal "sequential" data-assimilation scheme, in the sense that the growth rate of the forecast error variance is the least. The Kalman filter is an intricate algorithm. It has proved to be so demanding of computing resources (time and memory) that it has, so far, been applied with any success only to simplified ocean models, such as linear models of the equatorial wave guide.

4. The spatial structure of the Kalman filter

An oft-quoted application of the Kalman filter involves forecasting the flight of a point projectile. Accuracy aside, there is no inherent significance in the cross-covariances of the six degrees of freedom which describe the flight. Ocean circulation, on the other hand, is conceived as the flow of a continuum possessing infinitely many degrees of freedom. For example, the spatial distributions of momentum and pressure may be represented as infinite Fourier series. A least-squares best fit to a model and to some data may be achieved with a series which, while being square-summable, might be divergent at isolated points such as observing sites. The corresponding fields of momentum and pressure would be "delta-correlated" in space, or have even worse behavior.

An analysis of the Kalman filter reveals the relationship of the physical realizability of the forecasts to the internal scales of the fore-

casting model, and to the assumed statistical scales of the model errors. The intuitive sequential assimilation technique known as "nudging" is shown to be a special and typically pathological case of the Kalman filter; the demonstration here uses hypothetical along-track altimetric data.

5. Generalized inverses of dynamical models

Most oceanic data assimilation is hindcasting (after an observing program), so the demand for a sequential assimilation scheme may be relaxed in favor of a best fit over the duration of the program. The fit may include the dynamics (simultaneously with the prior forcing estimate), the prior initial and boundary estimates, and the data. The approach is a natural extension of the inverse formulation in Chapter 1. To a mathematician, a conventional "forward" integration of a model, together with its initial conditions and boundary conditions, amounts to inverting the operator which they define. If the model is well posed then the inverse is non-singular. Specifying additional data would overdetermine the problem, rendering it ill-posed. The operator would then be singular, but would possess a generalized inverse in the least-squares sense. As Reid (1968) remarks, mathematicians must be possessed of a genetic urge to reinvent this least-squares inverse, seemingly without end. The first discussion by a meteorologist or an oceanographer appears to be Sasaki (1955).

The residual of the least-squares best fit, expressed here as the minimal value of a quadratic penalty functional, is the χ^2-variable with one degree of freedom per observation. The proof assumes that the prior estimates for the means and the covariances of the model errors, initial errors, and data errors are all correct. Thus two simple concepts have been extended and combined. The first is the concept of a model as a system of equations and side conditions. The second is the concept of model testing, by simple comparison with data. An inverse formulation combines these two into a null hypothesis concerning the first two moments of the model errors and data errors. Repeated inversions with different forcing estimates, initial values and data allow testing of the hypothesis, by comparing the theoretical and sample distributions of the minimal penalty functional.

For an inverse estimate of ocean circulation to be of greatest use, it should be supported by estimates of the error statistics for the best fit, that is, by estimates of the posterior error covariances. Several methods are available for constructing the generalized inverse

and the posterior error statistics. Each method has its adherents. The case is made here for the use of the so-called representer fields. It is argued that they reveal the structure of the inverse, they lead to the best-conditioned algorithm, and they lead to the most efficient computation of posterior error covariances. Representer methods are widely used in solid-earth geophysics.

6. Antenna analysis

A benefit of constructing the inverse of a model and a data set is the opportunity to analyze the efficiency of the "antenna" which collected the data. Here, an antenna might be an array of current meters, or a ship-of-opportunity schedule, or an acoustic tomography web. If the model and antenna are linear, then the efficiency is independent of the actual data values, and may be analyzed prior to an observing program. That is, inverse methods lead to prior experimental design. The design will be sound if the error statistics for the model and the data have been correctly estimated. Several real examples of antennae are analyzed in detail.

7. Nonlinear quasi-geostrophic models

After listing five ways in which an inverse problem can be nonlinear, the chapter focuses on just one of the ways: through advection of the ocean circulation fields, by the circulation itself. The simplest example, based on a nonlinear quasi-geostrophic model in a doubly periodic domain, is examined in detail. The inverse is constructed using an iterative technique, which reduces the nonlinear inverse problem to a sequence of linear inverse problems. It can be proved that the sequence of inverse estimates is bounded, and therefore has limit points. Computations indicate the existence of actual limits, if the Reynolds number is sufficiently small. Posterior error statistics are derived.

8. Open-ocean modeling: quasi-geostrophy

Satellite-borne observing systems are becoming increasingly accurate, and computers become ever faster, yet realistic models of the world ocean circulation are still beyond our reach. Subsurface data are only available in limited regions, while mesoscale ocean circulation (which meteorologists would describe as synoptic-scale circulation) cannot be resolved by present global numerical models running on readily accessible computers. Thus open-ocean modeling remains

an important activity. However, boundary data are invariably unreliable or fragmentary. To make matters worse, seemingly well-chosen boundary conditions, even for quasi-geostrophic models, lead to ill-posed problems. We must rely on inverse methods for correcting boundary errors, by exploiting good data collected in the interior of the open-ocean region. Inverse methods also offer some hope for resolving ill-posedness.

Tests of inverse methods must distinguish between shortcomings of the methods and inadequacies of the test data. Really good oceanographic data sets are unavailable, so oceanographers must turn to meteorologists who have data that are incomparably better in both quality and quantity. A regional quasi-geostrophic model for tropical cyclone steering flows is therefore formulated as an inverse problem, and tested with real data. A discussion of the transient-tracer problem follows naturally from the material in this chapter.

9. Primitive-equation models

Open boundary conditions for primitive-equation models are notoriously ill-posed. Some leading meteorologists have declared the problem to be insoluble (Robert & Yakimiw, 1986; Yakimiw & Robert, 1990); they conclude that boundaries for limited-area numerical weather prediction models must be pushed so far out that they have no influence upon the region of interest. This strategy is not available to oceanographers; the size of the model region is restricted by the availability of initial data, especially below the surface. Recourse must be made to inverse formulations, which either resolve conflicts arising from too many boundary conditions or else remove uncertainties arising from too few boundary conditions. Of course, inverse formulations using good data would also lead to more realistic hindcasts, but ill-posedness is the motivating problem in this chapter. The ideal algorithm for solving inverse problems is characterized.

10. Outstanding problems

The β-spiral and finite-dimensional inverse calculations of general circulation described in Chapter 1 are all based on linear dynamical models. Inversion of steady, nonlinear circulation models remains as a major technical challenge. Even if the dynamics of the model are linear, the dependence of the solution upon parameters such as eddy diffusivities is highly nonlinear. Experience in engineering control theory suggests that parameter estimation can be highly unstable.

Theories of error statistics for parameter estimates are linearizations of doubtful utility.

Simple variational inverse methods for ocean models are inadequate if non-smooth circulations are admissible, or if the dynamics are non-smooth owing to branching processes, for example. The processes may represent unresolved scales of motion such as convective mixing. Regardless of the choice of inverse method, real oceanic inverse problems will require massive computing resources, which will only be available from parallel computers. Inverse methods must be adapted to these new architectures.

Finally, and probably the most challenging problem of all, theoretical oceanographers must persuade experimental oceanographers to collect statistics of dynamical errors, so that least-squares inverse calculations may be correctly weighted.

Acknowledgement

Figures 1.2.1 and 1.2.2 are reprinted with the permission of Professor F. Schott and Pergamon Press, 1977. Figures 1.3.2 and 1.3.3 are reprinted with the permission of the American Association for the Advancement of Science, 1977. Figures 1.5.1, 1.5.2, 1.5.3, 1.5.4, 1.6.1 and 1.6.2 are reprinted with the permission of the American Geophysical Union, 1978. Figures 1.7.1, 1.7.2, 1.7.3, 1.8.1, 1.8.2, 1.8.3 and 1.8.4 are reprinted with the permission of the American Geophysical Union, 1987. Figures 2.2.2 and 2.2.3 are reprinted with the permission of the American Meteorological Society, 1976. Figure 2.2.4 is reprinted with the permission of the American Meteorological Society, 1986. Figures 3.2.1 and 3.2.2 are reprinted with the permission of the American Meteorological Society, 1987. Figures 3.2.3 and 3.2.4 are reprinted with the permission of the American Meteorological Society, 1988. Figures 3.5.1, 3.5.2, 3.5.3 and 3.5.4 are reprinted with the permission of the American Meteorological Society, 1989. Figure 6.3.5 is reprinted with the permission of the American Meteorological Society, 1983. Figures 6.3.6, 6.3.7, 6.3.8 and 6.3.9 are reprinted with the permission of the American Geophysical Union, 1990. Figures 6.3.10 and 6.3.11 are reprinted with the permission of Pergamon Press PLC, 1985. Figure 6.3.12 is reprinted with the permission of the American Meteorological Society, 1984. Figure 6.3.13 is reprinted with

the permission of Dr. P. McIntosh, 1984. Figure 6.3.14 is reprinted with the permission of the American Meteorological Society, 1987. Figures 6.3.15, 6.3.16, 6.3.17 and 6.3.18 are reprinted with the permission of the American Meteorological Society, 1990. Figures 7.6.1, 7.6.2, 7.6.3, 7.6.4, 7.6.5, 7.6.6, 7.6.7, 7.6.8 and 7.6.9 are reprinted with the permission of the American Meteorological Society, 1992. Figures 7.6.10 and 7.6.11 were prepared by Mr. Edward Zaron, who also provided thoughtful comments. Figures 8.5.2, 8.5.3 and 8.6.1 are reprinted with the permission of *Tellus*, 1988. Figure 9.1.1 is reprinted with the permission of Dr. T. Whitworth and Elsevier Science Publishers, 1990. Figure 10.1.1 is reprinted with the permission of the American Meteorological Society, 1986. Figures 10.2.1, 10.2.2, 10.2.3 and 10.2.4 are reprinted with the permission of the American Geophysical Union, 1991. Figure 10.3.1 is reprinted with the permission of Dr. M. Ghil and Springer-Verlag GmbH & Co. KG, 1987. Figure 10.3.2 is reprinted with the permission of the American Geophysical Union, 1989. Figure 10.4.2 is reprinted with the permission of the American Meteorological Society, 1992.

To my editors, Peter-John Leone and Alan Harvey, and to my production manager, Louise Calabro Gruendel, all at Cambridge University Press: sincere thanks for all their help.

I am grateful to Carl Wunsch for suggesting that I write this book, and for his continuing encouragement. My own interest in inverse methods dates from his celebrated Woods Hole seminar in 1977, on the circulation of the Western North Atlantic. Much of my subsequent education has been stimulated by, I hope, endless arguments with my collaborators Peter McIntosh, Paul Budgell and Robert Miller. Thoughtful comments by Paul Budgell, Michele Rienecker, Leonard Walstad and Carl Wunsch on a draft manuscript were very helpful and greatly appreciated. Special thanks are owed to Bill McMechan for his careful typing, infinite patience and cheerful visits during which heaps of scrawl were exchanged for immaculate text. David Reinart did all the art work.

At various stages of this labor I have been fortunate to have the support of the National Science Foundation (OCE 8800004), the Office of Naval Research (N 00014–90–J–1142) and the National Aeronautics and Space Administration (NAGW–1305, NAGW–2715).

Finally, I wish to thank my wife Elaine for her unfailing support.

Preface to the Second Printing

The potential for inverse theory is now becoming realized: global ocean tides (Egbert, Bennett & Foreman, 1994; Egbert & Bennett, 1996; Egbert, 1998); global numerical weather prediction (Bennett, Chua & Leslie, 1996, 1997), and the tropical Pacific ocean-atmosphere (Bennett et al., 1998). These applications involve models having as many as 10^9 computational degrees of freedom in space and time, and incorporate as many as 10^5 observations. Much of this progress owes to the parallel computing power mentioned in Chapter 10, but an equally significant development has been the indirect use of representers (Egbert et al., 1994). The representer coefficients \hat{b} may be found without explicit construction of the representers themselves. That is, (5.4.8) may be solved iteratively, given an algorithm that calculates \mathbf{Pb} for any \mathbf{b}. As pointed out by Gary Egbert, one backward and one forward integration suffice. The coefficients \hat{b} having been found, the coupling of the Euler-Lagrange equations (5.3.2) – (5.3.5) is broken, thanks to (5.4.22). A similar approach has been advocated by Amodei (1995). In essence, the penalty functional is being minimized not by a search in the state or control space, but in the space of observable degrees of freedom or 'data space'.

It would be impossible to review here the literature of the 1990's on data assimilation in meteorology and oceanography. Survey articles include Courtier et al. (1993) and Anderson, Sheinbaum & Haines (1996); for collections of papers see Malanotte-Rizzoli (1996) and Ghil et al. (1997). Two major new texts are: Parker (1994) and Wunsch (1996).

Many thanks are owed to my secretary Florence Beyer, to the College artist David Reinert, and to Alan Harvey of Cambridge University Press for assembling the second printing.

1

FINITE-DIMENSIONAL INVERSE THEORY

1.1 Introduction

This is an outline of some applications of finite-dimensional inverse theory to ocean modeling. The objective is not to offer a comprehensive discussion of every application and its consequences; rather it is to introduce several concepts in a relatively simple setting:

- an incomplete ocean model, based on physical laws but possessing multiple solutions;
- measurements of quantities not included in the original model but related through additional physical laws;
- inequality constraints on the model fields or the data;
- prior estimates of errors in the physical laws and the data; and
- analysis of the level of information in the system of physical laws, measurements and inequalities.

Much of this material is well covered in mathematical texts, geophysical monographs and scientific review articles. Thus the presentation is brief and directed towards subsequent application of these concepts in more complex settings. However actual oceanographic studies are discussed, and tutorial problems are posed.

1.2 The β-spiral

A major objective of physical oceanography in the 1970's was the exploration of the dynamics of mesoscale eddies and their influence

on large-scale ocean circulation (Robinson, 1983). It was therefore
something of a surprise when, in 1977, Stommel & Schott showed
that the vertical structure of large-scale horizontal velocity fields could
be explained using simple equations expressing geostrophy and mass
balance. A compelling aspect of their study was the use of data in
order to complete and then test their calculations. The original paper
(Stommel & Schott, 1977) is somewhat cryptic. A more measured
presentation may be found in the review by Olbers, Wenzel & Wille-
brand (1985); the latter approach will be followed here.

Consider, then, the following system of equations:

$$-\overline{\rho}fv = -p_x\,,\qquad\qquad(1.2.1)$$

$$\overline{\rho}fu = -p_y\,,\qquad\qquad(1.2.2)$$

$$0 = -p_z - \rho g\,,\qquad\qquad(1.2.3)$$

$$\nabla\cdot\mathbf{u} + w_z = 0\,.\qquad\qquad(1.2.4)$$

These are familiar to oceanographers (Pedlosky, 1987, Chapters 2
and 3). The first two express β-plane geostrophy of a Boussinesq
liquid, the third expresses the hydrostatic balance for dynamical per-
turbations in pressure and density while the fourth expresses con-
servation of volume, or mass, for a Boussinesq liquid. The common
notation is: f for the Coriolis parameter; u, v and w for eastward,
northward and upward velocities; x, y and z for eastward, northward
and upward Cartesian coordinates; $\overline{\rho} = \overline{\rho}(z)$ and $\rho = \rho(x,y,z)$ for
mean and perturbation densities, p for perturbation pressure, g for
the gravitational acceleration and subscripts for partial derivatives.
Vector notation will be reserved for horizontal fields and operators:
$\mathbf{x} = (x,y)$, $\mathbf{u} = (u,v)$, $\nabla = \left(\frac{\partial}{\partial x}, \frac{\partial}{\partial y}\right)$. These notations will be used
throughout this book.

Some well-known relations may be derived from this system.
First, there are the vertically integrated thermal wind relations

$$u(\mathbf{x}, z) = u_0(\mathbf{x}) + \left(\frac{g}{f\overline{\rho}}\right)\int_{z_0}^{z} \rho_y(\mathbf{x}, \zeta)d\zeta\,,\qquad(1.2.5)$$

$$v(\mathbf{x}, z) = v_0(\mathbf{x}) - \left(\frac{g}{f\overline{\rho}}\right)\int_{z_0}^{z} \rho_x(\mathbf{x}, \zeta)d\zeta\,,\qquad(1.2.6)$$

or simply

$$\mathbf{u}(\mathbf{x}, z) = \mathbf{u}_0(\mathbf{x}) + \mathbf{u}'(\mathbf{x}, z)\,,\qquad\qquad(1.2.7)$$

where \mathbf{u}_0 is the horizontal velocity at depth z_0, while \mathbf{u}' is determined by the density field. Second, there is the Sverdrup interior vorticity balance:

$$w_z = \frac{\beta v}{f}, \qquad (1.2.8)$$

where β is the northward gradient of the Coriolis parameter: $f = f(y) = f_* + \beta(y - y_*)$ for y close to some local latitude y_*. These relations would not determine the velocity field (\mathbf{u}, w), even if the density field ρ were known. However, the indeterminacy is only that of unknown fields at a single depth, namely $\mathbf{u}(\mathbf{x}, z_0)$ and $w(\mathbf{x}, z_0)$. Stommel & Schott showed that these unknown reference fields may be estimated by assuming the availability of measurements of some conservative tracer ϕ, satisfying the steady-state conservation law

$$\mathbf{u} \cdot \nabla\phi + w\phi_z = 0. \qquad (1.2.9)$$

The tracer ϕ might be salinity S or potential temperature θ (Pedlosky, 1987, Chapter 1) or some function of them both: $F = F(S, \theta)$, such as potential density. Combining the vertical derivative of (1.2.9) with (1.2.7) and (1.2.8) yields

$$\left(\mathbf{u} \cdot \nabla + w\frac{\partial}{\partial z}\right)(f\phi_z) = (g/\overline{\rho})\, J, \qquad (1.2.10)$$

where $J(\rho, \phi) = \rho_x\phi_y - \rho_y\phi_x$. The quantity $f\phi_z$ is a potential vorticity which would be conserved if ρ were itself conserved: see Eliassen & Kleinschmidt (1957) for a comprehensive discussion. The tracer equation (1.2.9) may be used again to eliminate the vertical velocity w:

$$\mathbf{u} \cdot \mathbf{a} = (g/\overline{\rho})\, J(\rho, \phi), \qquad (1.2.11)$$

where the vector \mathbf{a} is given by

$$\mathbf{a}(\mathbf{x}, z) = \nabla(f\phi_z) - \frac{\nabla\phi}{\phi_z}\, f\phi_{zz}. \qquad (1.2.12)$$

Finally, using the integrated thermal wind relations (1.2.7) yields

$$\mathbf{u}_0 \cdot \mathbf{a} = c, \qquad (1.2.13)$$

where

$$c\left(\mathbf{x}, z\right) = -\mathbf{u}' \cdot \mathbf{a} + \left(g/\overline{\rho}\right) J(\rho, \phi). \tag{1.2.14}$$

Now \mathbf{a} and c depend upon g, $\overline{\rho}$, f, $\boldsymbol{\nabla}\rho$, $\boldsymbol{\nabla}\phi$, ϕ_z and ϕ_{zz} and so are measurable, in principle, using closely spaced hydrographic stations (e.g., Pickard & Emery, 1990). Consequently, the horizontal reference velocity \mathbf{u}_0 may be measured indirectly, using the hydrographic data. Since (1.2.13) holds at all levels, any two levels may be chosen to determine the two components u_0 and v_0. Then the vertical velocity w may be determined using the tracer equation (1.2.9). All that would remain would be the choice of a pair of levels, and provided the model physics are realistic, the solution for \mathbf{u}_0 should be independent of the choice. However, (1.2.13) is not an exact relation, being derived from the approximate dynamical laws (1.2.1)–(1.2.4), (1.2.9), and the data will contain experimental errors. Consequently the estimate of \mathbf{u}_0 from (1.2.13) should be a best fit using data from at least two levels.

Suppose N levels are chosen ($N \geq 2$). Let $c_n = c\left(\mathbf{x}, z_n\right)$ and $\mathbf{a}_n = \mathbf{a}(\mathbf{x}, z_n)$ for $1 \leq n \leq N$. Then a simple least-squares best fit minimizes

$$R^2 \equiv \sum_{n=1}^{N} R_n^2 \equiv \sum_{n=1}^{N} (c_n - \mathbf{u}_0 \cdot \mathbf{a}_n)^2, \tag{1.2.15}$$

where R_n is the residual at the nth level and R is the root-mean-square (rms) total error. It is a trivial exercise to show that R^2 is least if \mathbf{u}_0 satisfies a simple linear system:

$$\mathbf{M}\mathbf{u}_0 = \mathbf{d}, \tag{1.2.16}$$

where the 2×2 symmetric, non-negative matrix \mathbf{M} depends upon the components of \mathbf{a}_n, while \mathbf{d} depends upon \mathbf{a}_n and c. So (1.2.16) would seem to yield an estimate of the reference velocity \mathbf{u}_0. However, adding levels works only if the coefficient vector \mathbf{a} appearing in the left-hand side of (1.2.13) is significantly depth-dependent. Otherwise there is effectively only one indirect observation of the vector \mathbf{u}_0 at each station \mathbf{x}. If \mathbf{a} or c varies with depth, then (1.2.11) immediately implies that the total velocity vector \mathbf{u} also depends on depth, that is, the large-scale ocean current performs a vertical spiral at each station.

Before proceeding to examine Stommel & Schott's results, it is necessary to explain their terse formulation. Assuming that ρ itself suffices as a near-conservative tracer and, neglecting $(\rho\rho_{zz}/\rho_z^2)$,

reduces (1.2.10) to

$$uf\rho_{zx} + v(f\rho_{zy} + \beta\rho_z) \cong 0. \tag{1.2.17}$$

Let $z = h(\mathbf{x}, \rho)$ denote a surface of constant density. On this surface,

$$\nabla\rho + \rho_z\nabla h = 0. \tag{1.2.18}$$

Once more neglecting second vertical derivatives of ρ, and combining (1.2.17), (1.2.18) leads to

$$uh_{xz} + v\left(h_y - \frac{\beta z}{f}\right)_z = 0, \tag{1.2.19}$$

which is Stommel & Schott's equation (1). The field h_{xz}, for example, may be evaluated as $\rho_z\, h_{x\rho}$. Again, this holds for the depth h of each isopycnal surface, and holds at each station \mathbf{x}.

Hydrographic data from *Atlantis* Stations 5754–5764 at 28°N, *Discovery* Stations 3594–3607 at 24°N and *Discovery* Stations 3627–3641 at 32°N were used by Stommel & Schott to estimate \mathbf{u}_0 for a reference level at $z_0 = 1000$ m depth. They obtained $u_0 = 0.0034$ $\pm\, .0003$ m s^{-1}, $v_0 = 0.006 \pm .00013$ m s^{-1} at 28°N, 36°W, the errors being the standard errors for the linear fit.

The spiral for the horizontal velocity \mathbf{u} at 28°N, 36°W is shown in Fig. 1.2.1. A profile for the vertical velocity w was also obtained; a negative value for w was found at the surface, consistent with the local value of the wind stress curl (Pedlosky, 1987, Chapter 4). Stommel & Schott then "tested" their circulation model by calculating the minimal value of the squared residual R^2 for incorrect values of the Coriolis parameter f appearing in (1.2.19). They normalized R with respect to the inhomogeneous term in their equation for \mathbf{u}_0, that is, with respect to the analogue of c in (1.2.13): see Fig. 1.2.2. The residual finds a clear minimum for the correct value of f at 28°N, indicating that the "the density field in the ocean senses its latitude."

The β-spiral theory includes two of the concepts essential to inverse methods. First, there is the incomplete set of physical laws (1.2.1)–(1.2.4) or their rearrangement as (1.2.7) and (1.2.8), exposing the indeterminate reference velocity. Second, there is the indirect measurement of an additional quantity, in this case a conserved tracer.

Figure 1.2.1. The β-spiral in horizontal velocity $\mathbf{u} = \mathbf{u}(z)$ at 28°N, 36°W.
Depths in 100 meters; error bars at the origin (after Stommel & Schott, 1977).

Figure 1.2.2. Normalized least-squares residual for 28°N, 36°W. Density data
computed for various values of latitude (after Stommel & Schott, 1977). Note
that the minimal square value is about 0.02, so the minimal rms value is about
0.14.

What are most conspicuously absent are prior estimates of the errors in the final dynamical equation (1.2.13), and of the errors in the hydrographic data. Were the variances of these errors known to vary with depth, it would be appropriate to seek a non-uniformly weighted least-squares fit, instead of the uniformly weighted fit (1.2.15). Moreover, comparison of the residual with prior estimates would, in effect, test the several hypotheses tacitly included in the β-spiral theory. For example, the minimal normalized rms residual at 28°N is, judging from Fig. 1.2.2, about 0.14. Such a value is somewhat larger than the order of the Rossby number or scale estimate for the mesoscale variability neglected in (1.2.1)–(1.2.2). The value is comparable with estimates of relative data error. Stommel & Schott concluded their study by remarking that the simple dynamics could be augmented by including diffusion in the tracer conservation law, the eddy diffusivity being chosen by the least-squares approach. The suggestion has been explored at length, and the reader is referred to Olbers *et al.* (1985) for a very detailed analysis of the North Atlantic circulation along these lines. A careful discussion of β-spiral concepts may be found in Stommel (1987).

Exercise 1.2.1 Make scale estimates of the errors in the approximate dynamics of the β-spiral. □

Exercise 1.2.2 Derive the forms of matrix **M** and vector **d** in (1.2.16). Verify that **M** is symmetric and non-negative. Under what conditions is **M** singular? □

1.3 Wunsch's method: a first analysis

At the same time that Stommel & Schott were estimating reference velocities locally from hydrographic data, Wunsch (1977) showed that reference velocities could be estimated simultaneously around a closed path in the ocean. His investigation was prompted by Worthington's inability to find a uniform depth of no motion in the North Atlantic consistent with geostrophy and the conservation of total heat and salt at various levels (Worthington, 1976).

Again the unknown reference velocity is \mathbf{u}_0 in (1.2.7), the thermal wind \mathbf{u}' being known from hydrographic data. By construction, $\mathbf{u}' = \mathbf{0}$ at the reference level z_0 which in general may be chosen to depend upon position: $z_0 = z_0(\mathbf{x})$. Wunsch chose the reference level to be the ocean bottom at $z_0(\mathbf{x}) = H(\mathbf{x})$, so $\mathbf{u}_0(\mathbf{x})$ was defined to

Figure 1.3.1. Subdivision of water column into layers defined by temperature ranges (after Worthington, 1976).

be the bottom velocity. Guided by Worthington's classical water mass classifications, Wunsch divided the water column everywhere into a number of layers (see Fig. 1.3.1) defined by temperature ranges. These layers need not have the same thickness at each station. The stations used by Wunsch, together with the U.S. coastline, form a closed path in the Western North Atlantic (Fig. 1.3.2). Let v now denote the outward velocity across the sections; thus $v = \mathbf{u} \cdot \mathbf{n}$ where \mathbf{n} is the outward unit normal. In particular, let $v' = \mathbf{u}' \cdot \mathbf{n}$ be the outward thermal wind, and let $b = \mathbf{u}_0 \cdot \mathbf{n}$ be the outward velocity at the bottom. Let $v'_n(z)$ and b_n denote the thermal wind estimate and unknown bottom velocity midway between the nth station pair, where $1 \le n \le N$, and let \bar{v}'_{mn} denote the average value of v'_n in the mth layer of the water column, where $1 \le m \le M$. Wunsch chose the Mth layer to be the total water column, thus the Mth tracer is the total mass of the column. The assumptions of tracer conservation within each layer are

$$\sum_{n=1}^{N}(\bar{v}'_{mn} + b_n)\Delta z_{mn}\Delta x_n = 0, \quad (1 \le m \le M), \qquad (1.3.1)$$

where Δz_{mn} is the thickness of the mth layer at the nth station pair, while Δx_n is the separation of the nth station pair. This system of M equations for the N unknowns b_n, $1 \le n \le N$, may be written in matrix notation as

$$\mathbf{Ab} = \mathbf{c}, \qquad (1.3.2)$$

where \mathbf{A} is an $M \times N$ matrix with elements $A_{mn} = \Delta z_{mn}\Delta x_n$, while \mathbf{c} is a column vector of length M with elements $c_m = -\sum_{n=1}^{N} \bar{v}'_{mn} A_{mn}$.

Figure 1.3.2. Locations of *Atlantis* 215 Stations used by Wunsch (after Wunsch, 1977).

Wunsch chose $M = 5$ layers defined by the four temperature ranges $12°$ to $17°$C, $7°$ to $12°$C, $4°$ to $7°$C, $2.5°$ to $4°$C, and by the total mass. The hydrographic data were comprised of $N = 43$ station pairs. Thus (1.3.2) represents 5 equations for the 43 unknown bottom velocities b_j, and would seem to be an underdetermined system having many solutions.

Somewhat arbitrarily, Wunsch chose the smallest solution, that is, the vector **b** of smallest length $(\mathbf{b}^*\mathbf{b})^{1/2}$ where superscript $*$ denotes the transpose. Minimization of the length of **b**, subject to the linear constraint (1.3.2), may be achieved by minimizing the quadratic form \mathcal{J}_1 defined by

$$\mathcal{J}_1 = \mathbf{b}^*\mathbf{b} + 2\boldsymbol{\lambda}^*(\mathbf{Ab} - \mathbf{c}) = \mathcal{J}_1[\mathbf{b}, \boldsymbol{\lambda}]. \qquad (1.3.3)$$

In (1.3.3) the constraint (1.3.2) has been appended to the squared

length using an unknown Lagrange multiplier $\boldsymbol{\lambda}$, which is a column vector of length M. The factor of 2 is included for later convenience. It is a simple exercise to show that \mathcal{J}_1, as a function of \mathbf{b} and $\boldsymbol{\lambda}$, is least if

$$\mathbf{b} + \mathbf{A}^*\boldsymbol{\lambda} = 0 \tag{1.3.4}$$

and, as required, (1.3.2) holds. Further simple manipulations yield the sought-after smallest solution:

$$\mathbf{b} = \mathbf{A}^*(\mathbf{A}\mathbf{A}^*)^{-1}\mathbf{c}, \tag{1.3.5}$$

which clearly satisfies (1.3.2). Note that the symmetric matrix $\mathbf{A}\mathbf{A}^*$ has dimension $M \times M$, and is non-negative. However, it may be singular, and so (1.3.5) is at best a formal solution. The difficulty may be overcome by admitting errors in the hydrography and conservation laws, that is, by not seeking exact satisfaction of (1.3.2). This may be accomplished by replacing (1.3.3) with the quadratic form

$$\mathcal{J}_2 = W_1\mathbf{b}^*\mathbf{b} + W_2(\mathbf{A}\mathbf{b} - \mathbf{c})^*(\mathbf{A}\mathbf{b} - \mathbf{c}) = \mathcal{J}_2[\mathbf{b}], \tag{1.3.6}$$

where W_1 and W_2 are positive weights. The *strong constraint* (1.3.2), appended to the penalty function (1.3.3) using a Lagrange multiplier, has been replaced by a *weak constraint* which may be satisfied, at the expense of minimizing the length of \mathbf{b}, by increasing the ratio W_2/W_1. It is readily shown that (1.3.6) is minimized by

$$\mathbf{b} = (\mathbf{A}^*\mathbf{A} + (W_1/W_2)\mathbf{I})^{-1}\mathbf{A}^*\mathbf{c}, \tag{1.3.7}$$

or, equivalently,

$$\mathbf{b} = \mathbf{A}^*(\mathbf{A}\mathbf{A}^* + (W_1/W_2)\mathbf{I})^{-1}\mathbf{c}, \tag{1.3.8}$$

where \mathbf{I} is the $N \times N$ unit matrix in (1.3.7), but is the $M \times M$ matrix in (1.3.8). Since the weights are positive, the matrix inverses in (1.3.7) and (1.3.8) are well defined, so a unique solution has been obtained.

It is instructive to examine (1.3.8) in its limits. If the hydrographic data and conservation laws are perfect, then $(W_2/W_1) \to \infty$, and (1.3.5) and (1.3.2) are recovered. If the data are worthless or the laws invalid, then $(W_2/W_1) \to 0$ and so $\mathbf{b} \to \mathbf{0}$. That is, no

information has been provided about **b**, beyond the first guess that it is small. Note that only W_2/W_1 matters; clearly minimization of \mathcal{J}_2 is equivalent to minimization of \mathcal{J}_2/W_1.

Consideration of the choice of weights leads to inspection of the coefficient matrix **A**. As defined, $A_{nj} \propto \Delta x_j$, which is larger for larger station spacing. In effect, widely spaced pairs are being more heavily weighted, which seems an unreasonable choice. To remedy this, Wunsch introduced an $N \times N$ diagonal weight matrix **W** defined by $W_{ij} = \delta_{ij}\Delta x_j^{-1}$, where δ_{ij} is the Kronecker delta. A more reasonably scaled set of coefficients is provided by $\mathbf{A}' = \mathbf{AW}$; then (1.3.2) becomes $\mathbf{A}'\mathbf{b}' = \mathbf{c}$, where $\mathbf{b} = \mathbf{Wb}'$. The weakly constrained smallest solution becomes

$$\mathbf{b} = \mathbf{WWA}^*(\mathbf{AWWA}^* + (W_1/W_2)\mathbf{I})^{-1}\mathbf{c}, \qquad (1.3.9)$$

which, in general, differs from (1.3.8). Wunsch went further, after recognizing that the northward transport through the Florida Strait subsection had been previously estimated at about 33 Sverdrup (Sv) or 33×10^6 m^3 s^{-1}, which is equivalent to a uniform velocity of 0.29 m s^{-1} throughout the subsection. Accordingly he replaced $\mathbf{b}^*\mathbf{b}$ in the quadratic forms by

$$(\mathbf{b} - \mathbf{b}_0)^*(\mathbf{b} - \mathbf{b}_0) = \sum_{n=1}^{N-k} b_n^2 + \sum_{n=N-k+1}^{N} (b_n - 0.29\Delta x_n)^2, \quad (1.3.10)$$

where station pairs $N - k + 1$ to N are in the Florida Strait. That is, he sought to minimize the departure from a *first-guess* bottom velocity distribution of 0.29 m s^{-1} in the strait, and zero elsewhere: $\mathbf{b}_0^* = (0,\ldots,0,0.29,\ldots 0.29)$. Wunsch chose the weight ratio W_2/W_1 somewhat arbitrarily to be 0.08% of the trace of $\mathbf{A}^*\mathbf{WWA}$, on the grounds that the choice led to stable results.

Some of Wunsch's results are shown in Fig. 1.3.3 and Table 1.3.1. It is clear that significant corrections have been made to the first-guess bottom velocity \mathbf{b}_0, and that there is nothing like a single level of no motion. Wunsch described this solution as illustrative rather than definitive. The first-guess \mathbf{b}_0 yielded a worst-case transport imbalance of 2 Sv in the layer 12°–17°C. The imbalance is no smaller after the least-squares correction $(\mathbf{b}-\mathbf{b}_0)$ is added. This result is a consequence of admitting data errors or dynamical errors $(W_2/W_1 > 0)$. As an explanation, ageostrophic effects are to be expected in sections which include current shears as large as 3×10^{-5} s^{-1}.

Figure 1.3.3. (a) Geostrophic velocity contours (in units of 10^{-2} m s^{-1}) computed with a level of no motion at the bottom. Flow violates mass and heat transport requirements. It is the first guess used to obtain flow (b), except that 0.29 m s^{-1} was first added to the Fort Pierce section velocities. Positive values are to the north. (b) Geostrophic velocity contours computed from the flow of (a) plus the barotropic velocity found from (1.2.9). Actual bottom speed is denoted by an underlined number. Conservation requirements are now met to the accuracy stated in Table 1.3.1 (after Wunsch, 1977).

This preliminary presentation of Wunsch's method further illustrates two of the important concepts in inverse methods: a system of incomplete physical laws, and measurements of related fields. The need to admit errors in order to stabilize results has become apparent, as has the impact of weights and also first guesses for the solution. The method has made use of simple algebraic minimization and manipulation. However, it will be seen that linear least-squares problems, such as those just described, are amenable to general solution methods which reveal the structure of the problem and the range of solutions.

Table 1.3.1						
First-guess and inverse transports through the three sections in Sverdrups (10^6 m^3 s^{-1}), rounded to the nearest Sverdrup. Negative values are to the south.						
Water mass	Fort Pierce		Antilles		Cape Henry	
	F	I	F	I	F	I
Total mass	33	34	25	7	68	42
2.5° to 4°C	0	0	−2	−10	4	−9
4° to 7°C	0	0	3	1	8	2
7° to 12°C	5	5	10	6	13	10
12° to 17°C	7	7	14	7	19	16
Notes: F = First-guess; I = Inverse						

Exercise 1.3.1 Derive (1.3.4) and (1.3.5). □

Exercise 1.3.2 Derive (1.3.7) by completing the square in (1.3.6); deduce (1.3.8). □

Exercise 1.3.3 Derive (1.3.9) and find conditions on **A** for (1.3.9) to agree with (1.3.8). □

Exercise 1.3.4 Show that introducing a first-guess b_0 as in (1.3.10) does in general alter the minimizing solution **b**. □

Exercise 1.3.5 Find the minimum of **b*b** subject to (1.3.2) by parametric differentiation instead of using Lagrange multipliers as in (1.3.3). Interpret the Lagrange multipliers geometrically. □

1.4 The general problem of linear algebra: the Moore-Penrose inverse and its singular-value decomposition

The β-spiral theory leads to a system of linear equations (1.2.13) for the unknown reference velocity. When more than two levels of hydrographic data are used, the system is apparently overdetermined and a least-squares best fit may be sought; that is, unknown misfits or residuals (1.2.15) may be introduced. Then the system is apparently underdetermined, and the solution which minimizes the length of the

residuals is sought. However, that solution for the reference velocity must satisfy a linear system (1.2.16), for which the coefficient matrix \mathbf{M} may be singular. Thus there may be *no* solution for the reference velocity. A similar situation arose in Wunsch's problem. The formal solution (1.3.5) to the apparently undetermined system (1.3.2) may not be well defined. The general problem of linear algebra addresses these situations. The following is a brief outline following Lanczos (1961), using the notation of Wunsch (1978).

Consider the general real linear system

$$\mathbf{Ab} = \mathbf{c}, \qquad\qquad (1.4.1) = (1.3.2)$$

where \mathbf{A} is a known $M \times N$ matrix, \mathbf{b} is an unknown $N \times 1$ column vector, and \mathbf{c} is a known $M \times 1$ column vector. All solutions for \mathbf{b} are sought. First, some abstract results may be stated. Let $\mathbf{r} = \mathbf{c} - \mathbf{Ab}$ be the residual vector.

i) Definition: A "symmetric g-inverse" of \mathbf{A} is any matrix \mathbf{G} satisfying

$$\mathbf{AGA} = \mathbf{A}, \ (\mathbf{AG})^* = \mathbf{AG}. \qquad\qquad (1.4.2)$$

Theorem: \mathbf{b} minimizes $\|\mathbf{r}\|^2 = \mathbf{r}^*\mathbf{r}$ if and only if $\mathbf{b} = \mathbf{Gc}$, where \mathbf{G} is some symmetric g-inverse of \mathbf{A}.

Proof: Show

$$\|\mathbf{r}\|^2 = \|\mathbf{AGc} - \mathbf{c}\|^2 + \|\mathbf{A}(\mathbf{b} - \mathbf{Gc})\|^2 \qquad\qquad (1.4.3)$$

for any \mathbf{b} satisfying (1.4.1), and any symmetric g-inverse \mathbf{G}.

ii) Definition: A "Moore-Penrose" inverse of \mathbf{A} is any matrix \mathbf{M} satisfying

$$\mathbf{AMA} = \mathbf{A}, \qquad (\mathbf{AM})^* = \mathbf{AM} \qquad\qquad (1.4.4)$$

and

$$\mathbf{MAM} = \mathbf{M}, \qquad (\mathbf{MA})^* = \mathbf{MA}. \qquad\qquad (1.4.5)$$

Note that a Moore-Penrose inverse of \mathbf{A} is also a symmetric g-inverse of \mathbf{A}.

Theorem: \mathbf{b} is the shortest vector (least $\|\mathbf{b}\|^2$) minimizing $\|\mathbf{r}\|^2$ if and only if $\mathbf{b} = \mathbf{Mc}$, where \mathbf{M} is some Moore-Penrose inverse of \mathbf{A}.

Proof: Show

$$\|\mathbf{Gc}\|^2 = \|\mathbf{Mc}\|^2 + \|\mathbf{Gc} - \mathbf{Mc}\|^2 \qquad (1.4.6)$$

for any \mathbf{c}, any symmetric g-inverse \mathbf{G} and any Moore-Penrose inverse \mathbf{M} of \mathbf{A}.

iii) Theorem: A matrix \mathbf{A} can have only one Moore-Penrose inverse \mathbf{M}.

iv) Theorem: If \mathbf{A} is non-singular, $\mathbf{M} = \mathbf{A}^{-1}$.

v) Remark: The minimal residual \mathbf{r} in (*i*) and (*ii*) may be zero.

Were the Moore-Penrose inverse to exist, it would provide precisely the desired solution for the general problem (1.4.1). It remains to construct this inverse for any coefficient matrix \mathbf{A}. The construction is provided by the singular value decomposition (SVD), which now follows. For definiteness, suppose $M \leq N$, that is, (1.4.1) is apparently underdetermined. Consider the dual eigenvalue problem

$$\mathbf{Av} = \lambda \mathbf{u}, \qquad (1.4.7)$$

$$\mathbf{A}^*\mathbf{u} = \lambda \mathbf{v}. \qquad (1.4.8)$$

Recall that \mathbf{A} is $M \times N$, so \mathbf{A}^* is $N \times M$; \mathbf{v} is $N \times 1$ and \mathbf{u} is $M \times 1$. Now any solution \mathbf{u} satisfies

$$\mathbf{AA}^*\mathbf{u} = \lambda^2 \mathbf{u}. \qquad (1.4.9)$$

The $M \times M$ matrix \mathbf{AA}^* is symmetric and non-negative, so there are M real singular values $\lambda_1 \geq \lambda_2 \geq \ldots \geq \lambda_M \geq 0$, and M associated real orthonormal left singular vectors $\mathbf{u}_1, \ldots, \mathbf{u}_M$ (e.g., Noble & Daniel, 1988). The latter are the columns of an $M \times M$ unitary matrix \mathbf{U}.

Now suppose $\lambda_1 \geq \lambda_2 \geq \ldots \lambda_K > 0$ but $\lambda_{K+1} = \ldots = \lambda_M = 0$. The K right singular vectors

$$\mathbf{v}_m = \lambda_m^{-1} \mathbf{A}^* \mathbf{u}_m \qquad (1.4.10)$$

are orthonormal, $(1 \leq m \leq K)$, and by the Gram-Schmidt process may be augmented to a total of N orthonormal vectors which are the

columns of an $N \times N$ unitary matrix \mathbf{V}. The supplementary vectors $\mathbf{v}_{K+1}, \ldots, \mathbf{v}_N$ are necessarily right null vectors of \mathbf{A}. Hence (1.4.7), which holds for all the \mathbf{v}_m, may be expressed in matrix form:

$$\mathbf{AV} = \mathbf{U}\Lambda, \tag{1.4.11}$$

where Λ is the $M \times N$ matrix

$$\Lambda = (\mathbf{L}|\mathbf{Z}), \tag{1.4.12}$$

in which \mathbf{L} is the $M \times M$ diagonal matrix with diagonal elements $(\lambda_1, \ldots, \lambda_K, 0, \ldots, 0)$ and \mathbf{Z} is an $M \times (N - M)$ matrix of zeros. Rearrangement of (1.4.11) yields

$$\mathbf{A} = \mathbf{U}\Lambda\mathbf{V}^*, \tag{1.4.13}$$

which is the SVD of \mathbf{A}. It is readily shown that

$$\mathbf{A}^+ = \mathbf{V}\Lambda^+\mathbf{U}^* \tag{1.4.14}$$

is the Moore-Penrose inverse of \mathbf{A}, where

$$\Lambda^+ = \left(\frac{\mathbf{L}^+}{\mathbf{Z}^*}\right), \tag{1.4.15}$$

in which \mathbf{L}^+ is the $M \times M$ diagonal matrix with diagonal elements $(\lambda_1^{-1}, \ldots, \lambda_K^{-1}, 0, \ldots, 0)$. Note that Λ^+ and \mathbf{L}^+ are the Moore-Penrose inverses of Λ and \mathbf{L} respectively. Finally, the sought-after smallest best-fit to (1.4.1) is

$$\mathbf{b} = \mathbf{V}\Lambda^+\mathbf{U}^*\mathbf{c}. \tag{1.4.16}$$

It is instructive to rederive (1.4.16) by a singular value expansion. The vectors \mathbf{b} and \mathbf{c} and the residual \mathbf{r} have the representations

$$\mathbf{b} = \sum_{n=1}^{N} \beta_n \mathbf{v}_n \,, \quad \mathbf{c} = \sum_{m=1}^{M} \gamma_m \mathbf{u}_m \,, \quad \mathbf{r} = \sum_{m=1}^{M} \rho_m \mathbf{u}_m \,, \tag{1.4.17}$$

where $\beta_n = \mathbf{v}_n^* \mathbf{b}$, $\gamma_m = \mathbf{u}_m^* \mathbf{c}$ and $\rho_m = \mathbf{u}_m^* \mathbf{r}$. Again, the β_n and ρ_m are unknown; the γ_m are known. With these representations, the equation $\mathbf{r} = \mathbf{c} - \mathbf{A}\mathbf{b}$ becomes

$$\sum_{m=1}^{M} \rho_m \mathbf{u}_m = \sum_{m=1}^{M} \gamma_m \mathbf{u}_m - \sum_{m=1}^{K} \beta_m \lambda_m \mathbf{u}_m , \qquad (1.4.18)$$

where $K \leq M \leq N$.

Hence $\beta_n = (\gamma_n - \rho_n)\lambda_n^{-1}$ for $1 \leq n \leq K$, and $\rho_n = \gamma_n$ for $K+1 \leq n \leq M$, while $\beta_{K+1}, \ldots, \beta_N$ and ρ_1, \ldots, ρ_K are indeterminate. The residual \mathbf{r} is smallest if $\rho_1 = \ldots = \rho_K = 0$, and the solution is smallest if $\beta_{K+1} = \ldots = \beta_N = 0$. In summary, the minimal residual is

$$\mathbf{r} = \sum_{m=K+1}^{M} (\mathbf{u}_m^* \mathbf{c})\mathbf{u}_m , \qquad (1.4.19)$$

and the smallest solution is

$$\mathbf{b} = \sum_{m=1}^{K} (\mathbf{u}_m^* \mathbf{c})\lambda_m^{-1}\mathbf{v}_m , \qquad (1.4.20)$$

which is in agreement with (1.4.16). Thus the minimal residual \mathbf{r} is the projection of \mathbf{c} on the left null vectors of \mathbf{A}, while the smallest solution belongs to the complement of the right null space of \mathbf{A}.

Exercise 1.4.1 Paraphrase the above analysis for the case $M \geq N$. □

Exercise 1.4.2 Let \mathbf{b}_1 be the *smallest* best fit to (1.4.1), that is, the first guess for \mathbf{b}_1 is zero. Let \mathbf{b}_2 be the best fit to (1.4.1), closest to a non-zero first-guess \mathbf{b}_0. Show that \mathbf{b}_2 and \mathbf{b}_1 differ by the projection of \mathbf{b}_0 onto the right null space of \mathbf{A}. □

Exercise 1.4.3 Show that the solution \mathbf{b} does not differ from the first-guess \mathbf{b}_0 if the first-guess residual $\mathbf{c} - \mathbf{A}\mathbf{b}_0$ belongs to the left null space of \mathbf{A}. In a practical sense, such a residual may be described as unobservable. □

Exercise 1.4.4 The quadratic form \mathcal{J}_2 in (1.3.6) is minimized by \mathbf{b} given in (1.3.7). Use the SVD representation $\mathbf{A} = \mathbf{U}\mathbf{\Lambda}\mathbf{V}^*$ to show that (1.3.7) is equivalent to

$$\mathbf{b} = \mathbf{V}(\mathbf{\Lambda}^* \mathbf{\Lambda} + \sigma^2 \mathbf{I})^{-1} \mathbf{\Lambda}^* \mathbf{U}^* \mathbf{c} , \qquad (1.4.21)$$

where $\sigma^2 = (W_1/W_2)$. Compare the respective contributions of small
eigenvalues to (1.4.21) and to (1.4.16), which is the solution of (1.4.1)
obtained using the Moore-Penrose inverse. Once again, the vector
(1.4.21) minimizes a weighted sum of $\|\mathbf{b}\|^2$ *and* $\|\mathbf{Ab} - \mathbf{c}\|^2$, while
(1.4.16) is the smallest vector which minimizes $\|\mathbf{Ab} - \mathbf{c}\|^2$. □

Exercise 1.4.5 Consider two choices of reference level for the ther-
mal wind, denoted level 1 and level 2. Let \mathbf{v}_1' and \mathbf{v}_2' be the thermal
wind fields relative to their respective reference levels. Their differ-
ence is independent of depth. Let \mathbf{b}_1 and \mathbf{b}_2 be estimates of the
total velocities at the respective reference levels, the estimates being
obtained by generalized inversion of Wunsch's equation (1.3.2). (Here
vectors denote N-tuples, where there are N station pairs.) Then the
estimates of total velocities at all levels are $\mathbf{v}_1 = \mathbf{b}_1 + \mathbf{v}_1'$, $\mathbf{v}_2 = \mathbf{b}_2 + \mathbf{v}_2'$
respectively. In general, $\mathbf{v}_1 \neq \mathbf{v}_2$. Show, however, that if \mathbf{v}_1' and \mathbf{v}_2' are
smoothed at each level by projecting onto the row of $\mathbf{A}^+\mathbf{A}$ appropriate
to that level, then the eventual solutions for the total velocities will be
the same. On the other hand, the total velocities are still dependent in
general upon non-zero first guesses for the velocities at the reference
levels. Recall that the inverse estimate for the reference velocity
depends upon the thermal wind. □

The SVD may be computed using subroutines in the LINPACK
library (Dongarra *et al.*, 1979). Introductory information and advice
may be found in Coleman & Van Loan (1988).

1.5 Wunsch's method revisited

The SVD leads naturally to the concept of resolution. Suppose,
in the language of Wunsch's ocean circulation problem, that \mathbf{b} is the
true bottom velocity. Then \mathbf{c} $(= \mathbf{Ab})$ is the indirect observation of \mathbf{b}.
The smallest, best-fit estimate for \mathbf{b} is

$$\mathbf{b}_{\text{SVD}} = \mathbf{V}\Lambda^+\mathbf{U}^*\mathbf{c}, \tag{1.5.1}$$

where $\mathbf{A} = \mathbf{U}\Lambda\mathbf{V}^*$. Since many of the entries of Λ^+ are zero, (1.5.1)
may be compressed:

$$\mathbf{b}_{\text{SVD}} = \hat{\mathbf{V}}\hat{\Lambda}^+\hat{\mathbf{U}}^*\mathbf{c}, \tag{1.5.2}$$

where $\hat{\mathbf{V}}$ is the $N \times K$ matrix with columns $\mathbf{v}_1, \ldots, \mathbf{v}_K$; $\hat{\Lambda}^+$ is the
$K \times K$ diagonal matrix with diagonal components $(\lambda_1^{-1}, \ldots, \lambda_K^{-1})$,

and $\hat{\mathbf{U}}$ is the $M \times K$ matrix with columns $\mathbf{u}_1, \ldots, \mathbf{u}_K$. Similarly, $\mathbf{A} = \hat{\mathbf{U}}\hat{\mathbf{\Lambda}}\hat{\mathbf{V}}^*$. Thus (1.5.2) may be expressed as

$$\mathbf{b}_{\text{SVD}} = \hat{\mathbf{V}}\hat{\mathbf{\Lambda}}^+\hat{\mathbf{U}}^*\hat{\mathbf{U}}\hat{\mathbf{\Lambda}}\hat{\mathbf{V}}^*\mathbf{b} \qquad (1.5.3)$$

$$= \hat{\mathbf{V}}\hat{\mathbf{V}}^*\mathbf{b} . \qquad (1.5.4)$$

Since $K \leq N \leq M$ here, $\hat{\mathbf{V}}$ is not unitary, and so $\mathbf{b}_{\text{SVD}} \neq \mathbf{b}$. However, (1.5.4) yields the part of \mathbf{b} that can be resolved with the observations, and so $\hat{\mathbf{V}}\hat{\mathbf{V}}^*$ is the resolution matrix. For example, if \mathbf{b} has a unit value for velocity between the jth station pair but is zero elsewhere: $\mathbf{b} = \mathbf{e}_j$, then $\mathbf{e}_{j_{\text{SVD}}} = \hat{\mathbf{V}}\hat{\mathbf{V}}^*\mathbf{e}_j$ is the jth column of the resolution matrix. The column defines the jth impulse-response. Of course, if $K = M$ then the resolution matrix is the identity, and $\mathbf{e}_{j_{\text{SVD}}} = \mathbf{e}_j$.

The SVD effectively replaces the observations \mathbf{c} with

$$\mathbf{c}_{\text{SVD}} = \hat{\mathbf{U}}\hat{\mathbf{U}}^*\mathbf{c} . \qquad (1.5.5)$$

If $K < M$, then $\hat{\mathbf{U}}\hat{\mathbf{U}}^*$ is not the identity, so the observations are "blurred." In particular, suppose \mathbf{c} consists of a unit value in the kth layer and is zero elsewhere: $\mathbf{c} = \mathbf{e}_k$. If all the components of $\hat{\mathbf{U}}\hat{\mathbf{U}}^*\mathbf{e}_k$ (the kth column of $\hat{\mathbf{U}}\hat{\mathbf{U}}^*$) are small, then the kth observation \mathbf{e}_k has little influence.

Suppose finally that the components of the observation \mathbf{c} are uncorrelated, zero-mean random variables with the same variance σ^2: $\overline{\mathbf{c}} = 0$, $\overline{\mathbf{c}\mathbf{c}^*} = \sigma^2\mathbf{I}$. Then

$$\overline{\mathbf{b}_{\text{SVD}}\mathbf{b}_{\text{SVD}}^*} = \sigma^2\hat{\mathbf{V}}(\hat{\mathbf{\Lambda}}^+)^2\hat{\mathbf{V}}^* , \qquad (1.5.6)$$

which shows that random noise is amplified by small singular values.

The preliminary investigation of North Atlantic hydrographic data by Wunsch (1977) was expanded greatly in a subsequent article (Wunsch, 1978). Some of the preliminary details were altered. The number of usable station pairs was increased to $N = 44$, the roles of the symbols M and N were reversed (leading to the choice made here!), and the $N \times N$ weight matrix \mathbf{W}, where $W_{ij} = \delta_{ij}\Delta x_j^{-1}$, was replaced with $\mathbf{W}^{-1/2}$, where $W_{ij} = \delta_{ij}\Delta z_{jM}\Delta x_j$ (Δz_{jM} being the total depth of the jth station pair). This choice of weight reduces the

Figure 1.5.1. (a) The five singular values λ_l for Wunsch's first problem (arising from the sections shown in Fig. 1.3.2). (b) The first five of the left singular vectors \mathbf{u}_l, in the *observation* space. (c) The five right singular vectors \mathbf{v}_l, related to λ_l and \mathbf{u}_l by (1.3.10), in the *solution* space. There are another 39 orthonormal \mathbf{v}_l, each corresponding to $\lambda_l = 0$, defining the (right) null space of \mathbf{A} (after Wunsch, 1978).

undesirable tendency of the previous choice to yield a solution with large amplitudes between deep, widely spaced stations. The SVD of the modified weighted matrix $\mathbf{A}' = \mathbf{AW}^{-1/2}$ is shown in Fig. 1.5.1. Recall that $M = 5$. The singular values range from the order of unity

Figure 1.5.2. The bottom velocity vector **b** according to (1.4.20), truncated at $K = 3$, 4 and 5. Implausibly large velocities are obtained for $K = 5$ (after Wunsch, 1978).

to 10^{-5}. Also shown are the $M \times 1$ singular vectors $\{\mathbf{u}_1, \ldots, \mathbf{u}_M\}$, which Wunsch named the "observation space," and the $N \times 1$ singular vectors $\{\mathbf{v}_1, \ldots, \mathbf{v}_M\}$, which Wunsch named the "solution space." The latter are related to the \mathbf{u}_m by (1.4.10).

In the solution space, the finest structure is found in \mathbf{v}_4 and \mathbf{v}_5. Hence (see 1.4.20) the solution will be noisy unless the observations are nearly orthogonal to the corresponding vectors in the observation space: $\mathbf{u}_m^* \mathbf{c} = O(\lambda_m)$, $m = 4, 5$. Such is not the case for the North Atlantic data. Fig. 1.5.2 shows the solution (1.4.20) truncated successively at $K = 3$, 4 and 5. Implausibly large bottom velocities are obtained for $K = 5$.

The 26th column of the resolution matrix $\hat{\mathbf{V}}\hat{\mathbf{V}}^*$ is shown in Fig. 1.5.3. This is the solution obtained with the SVD if the true solution were unity at station pair 26 (the western edge of the Cape Henry section–see Fig. 1.3.1) and zero elsewhere. The SVD solution is spread

Figure 1.5.3. Column 26 of the resolution matrix $\widehat{\mathbf{V}}\widehat{\mathbf{V}}^*$. This is the solution obtained with the SVD if the true solution were unity at station pair 26 (the western edge of the Cape Henry section—see Fig. 1.3.1) and zero elsewhere. The SVD solution is spread over all the sections (after Wunsch, 1978).

Figure 1.5.4. The solution variance for $\sigma^2 = 1$ and $K = 5$, given by (1.5.6). Retaining all five singular values optimizes the resolution but amplifies the noise, since several singular values are relatively small (after Wunsch, 1978).

across all three sections. The "blurring" matrix $\widehat{\mathbf{U}}\widehat{\mathbf{U}}^*$ was constructed assuming $K = 4$, that is, the smallest singular value λ_5 was set to zero. The small values in the first column imply that the observations from the first layer ($12°–17°$C) do not contribute to the SVD solution. The solution variance (1.5.6) for $\sigma^2 = 1$ and $K = 5$ is shown in Fig. 1.5.4. Retaining all 5 singular values optimizes the resolution but amplifies the noise, since several singular values are relatively small.

The SVD exemplifies the concept of structural analysis of an inverse problem, in order to reveal the information content in the dynamics and data.

1.6 Remarks

The close correspondence between Wunsch's method and the β-spiral theory was pointed out by Davis (1978). Both approaches assume the integrated thermal wind relation (1.2.7), and both provide estimates for the reference velocity \mathbf{u}_0. The β-spiral theory assumes the existence of a steady, conserved tracer ϕ satisfying (1.2.9). It follows that for incompressible flow,

$$\boldsymbol{\nabla} \cdot \int_{h_l}^{h_{l+1}} \mathbf{u}\phi dz = 0, \qquad (1.6.1)$$

where h_l and h_{l+1} are any two members of a sequence of material surfaces. A tracer-weighted average velocity $\overline{\mathbf{u}}_{l+\frac{1}{2}}$ may be defined such that (1.6.1) becomes, after cancelling a layer-representative tracer value $\overline{\phi}_{l+\frac{1}{2}}$,

$$\boldsymbol{\nabla} \cdot \{\Delta h_{l+\frac{1}{2}} \overline{\mathbf{u}}_{l+\frac{1}{2}}\} = 0, \qquad (1.6.2)$$

where $\Delta h_{l+\frac{1}{2}} = h_{l+1} - h_l$. Integrating (1.6.2) around a closed, piecewise smooth path C yields

$$\int_C \mathbf{n} \cdot \overline{\mathbf{u}}_{l+\frac{1}{2}} \Delta h_{l+\frac{1}{2}} ds = 0, \qquad (1.6.3)$$

where \mathbf{n} is the outward normal on the path and s is arc length along the path. This result indicates that the average outward thermal wind \overline{v}' appearing in Wunsch's equation (1.3.1) should be tracer-weighted. Large offsets in tracer values should be removed when calculating such averages; departures from the offsets are also conserved.

The integral along the path in (1.6.3) is approximated by a sum over M station pairs with spacing Δs_m, implying that the horizontal length scales of the unknown reference velocity, and that of the thermal wind, exceed the station spacing. A finite number N of interface pairs or slabs is chosen, implying that the hydrographic data are so noisy that thinner slabs could not be used. Recall that the tracer-weighted vertical averaging should suppress noise. The underdetermining choice $M > N$ implies that the data are too noisy to allow unambiguous estimation of the bottom velocity field.

A β-spiral calculation requires horizontal gradients of tracer and density. These are assumed to have non-vanishing means in a region

of sufficient size such that local smoothing will suppress noise in the form of mesoscale variability. Then (1.2.13) is used at many levels to estimate u_0, in order to suppress noise in the vertical structure of the data.

Davis concluded that the underdetermined and overdetermined problems arising from Wunsch's method and β-spiral theory are consequences of tacit assumptions about noise levels and scales. He suggested that a more orderly approach based on Gauss-Markov smoothing (see Chapter 2) may be appropriate, with explicit and quantitative recognition of noise structure. He outlined such an approach; a fundamentally similar analysis of North Atlantic hydrographic data and current meter data may be found in Mercier (1989).

Some applications of inverse methods to ocean circulation problems will now be briefly reviewed. Wunsch (1978) examined a large, high-quality hydrographic data set collected in the North Atlantic, west of 50°W. The data had been collected in different years, and in different seasons. Accordingly there were gross inconsistencies in the full data set which were resolvable using inverse methods, but the resulting solutions for the circulation were implausible. A reduced, more consistent data set was identified, consisting of five conserved fields (total mass and four temperature ranges) plus seven regions bounded by coastline or hydrographic sections. Thus the SVD was of dimension 35 × 35. Various solutions were obtained using 15, 20 and 25 singular values, a relative data noise variance of $\sigma^2 = .0005$ or $.001$ (the former corresponding empirically to rms solution errors of about 2 Sv), and two choices of first guess for the reference velocity. The results were relatively insensitive to the level of SVD truncation and choice for σ^2; the choice of first guess was far more influential. To use Wunsch's nomenclature, Model I had as a first guess a reference velocity vanishing at the bottom everywhere save in the Florida Current, where, as discussed earlier, a bottom velocity of 29 cm s^{-1} was assumed, equivalent to a Gulf Stream transport of 28 Sv. Model II had a first-guess reference velocity vanishing at the shallower of 2000 dbar or the bottom, save in the Florida Current, where the first guess coincided with that of Model I. Contours of the total transport for the solutions are shown in Figs. 1.6.1 and 1.6.2. Wunsch warned against giving too much credit to the details in these subjective contour plots, since for both models almost all of the diagonal values of the resolution matrices were very small. The solution for Model I was distinguished by a large Gulf Stream

Figure 1.6.1. Circulation diagram resulting from Model I for total transports. Points A–E have been used as arbitrary demarcation points for computing transports. Parenthetical numbers are nominal "Gulf Stream" values (after Wunsch, 1978).

transport of 124 Sv, including recirculation, compared with 80 Sv for the solution of Model II. The latter was distinguished by an onshore recirculation as well as an offshore recirculation. Both solutions included a southern gyre not found in hydrographic data alone, but found here as a consequence of the non-vanishing bottom velocity.

Roemmich (1980) used Wunsch's inverse method to estimate northward heat flux across several subtropical sections of the North Atlantic (24°, 35° and 48°N). The first guesses for the reference velocity were levels of no motion at 1000 m and at 4000 m. The former choice led to a net southward total mass transport, the latter to a net northward transport. To the extent that there should be virtually no net transport, these guesses were considered as appropriate alternatives. The quality and quantity of the density data supported

Figure 1.6.2. Circulation diagram for Model II for total transports (analogous to Fig. 1.6.1) (after Wunsch, 1978).

the definition of as many as 14 layers. However, some layers ventilated between the 36°N and 48°N sections, and so density conservation could not be assumed for every layer at every section. The SVD indicated that the 24°N section was the best resolved. Indeed the resolution at 36°N and 48°N was so poor that in some inversions Roemmich imposed heat fluxes through those sections, the fluxes having been estimated from flux budgets at the ocean surface (Hastenrath, 1980). The inverse solutions at 24°N had northward heat fluxes ranging from 8.2×10^{14} W to 9.8×10^{14} W, exclusive of Ekman fluxes. Imposing Hastenrath's estimate of 1.55×10^{15} W (including Ekman fluxes) as a constraint at 24°N led to a highly inconsistent system giving rise to unacceptably large northward mass transports.

Fu (1981) employed Wunsch's method to study the general circulation of the South Atlantic, using IGY and *Meteor* hydrographic

data principally along sections at 8°S, 16°S, 24°S and 32°S. The inferred sections of total velocity had a highly non-uniform structure in the horizontal, and it was concluded that "... the classical picture of the layered structure of the circulation in the South Atlantic is inconsistent with the conservation of mass and salt." However, the zonally averaged or meridional circulation was "remarkably similar to the classical picture of Wüst (1935)—a northward transport of upper waters (20 Sv) above the Upper Circumpolar Water and a comparable transport of deep waters." These conclusions were not sensitive to the reference levels chosen by Fu: 2000 m or 4000 m. Meridional heat flux was found to be northward, in close agreement with direct estimates of Bennett (1978) and Hastenrath (1980). To the extent that Bennett also used hydrographic data but estimated reference velocities from the wind stress curl, the agreement simply indicates the insignificant role the barotropic circulation plays in the heat flux budget of the South Atlantic, due to the relatively small horizontal variations in depth-averaged temperature. The general agreement with Hastenrath is the more striking since the latter employed surface flux budgets alone.

Fu also sought solutions obeying the inequality constraint of a southward meridional heat flux (solution methods will be discussed in §1.7). The resulting circulation was found to be unrealistic, with "large bottom velocities and shears, too small a transport of surface water and too large a transport of bottom water."

Roemmich (1980) and Fu (1981) both derived "smoothed solutions," obtained by assuming horizontal reference levels and by smoothing the thermal wind field with the matrix $\mathbf{A}^{+}\mathbf{A} = \hat{\mathbf{V}}\hat{\mathbf{V}}^{*}$ (see Exercise 1.4.5). Such a solution is independent of choice of reference level, and is also smooth in appearance. This solution, it was argued, is more likely to be representative of the long-term time-averaged circulation than are the quasi-synoptic solutions derived from individual unsmoothed hydrographic sections.

1.7 Water mass analysis

The preceding finite-dimensional inverse problems have involved strong constraints appended to least-squares expressions using Lagrange multipliers, and weak constraints appended as weighted squared residuals. Inequality constraints are now considered, with simple mixing of water masses as an example.

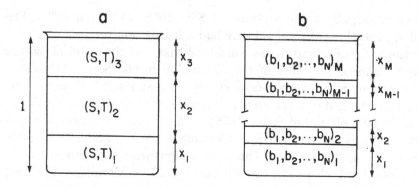

Figure 1.7.1. Simple mixing. (a) The three layers having volume fractions x_1, x_2 and x_3 are mixed without change of volume. The salinities and temperatures of the layers are averaged by simple mixing according to (1.7.1). (b) In general, there may be M volume fractions x_m, $1 \le m \le M$, and N water properties b_n, $1 \le n \le N$, such as salinity, temperature, silicates, etc.

Simple mixing is defined as the ideal process illustrated in Fig. 1.7.1. A beaker of unit volume contains three separate or unmixed samples of seawater, having volume fractions x_1, x_2 and x_3. Their respective salinities and temperatures are S_i and T_i: $i = 1, 2, 3$. After simple mixing the salinity and temperature of the mixture are S and T, which, together with the statement of conservation of volume, are given by

$$S = \sum_{i=1}^{3} x_i S_i \,, \quad T = \sum_{i=1}^{3} x_i T_i \,, \quad 1 = \sum_{i=1}^{3} x_i \,. \qquad (1.7.1)$$

If the source salinities and temperatures are known, then (1.7.1) comprises three equations for the three unknown x_i. There will be a unique solution provided the coefficient vectors in (1.7.1) are not coplanar.

In practice there will be imprecision in the measurement of S and T for the water sample or mixture, and in the definition of the S_i and T_i for the water masses. Additional well-measured, simple-mixing tracers should be used; candidates might be dissolved oxygen, nutrients, inert gases, radioisotopes or silicates. More than three water masses may be present. So let x_m be the fraction of the mth water mass: $m = 1, \ldots, M$; let b_n be the measured value of the nth water property in the mixture: $n = 1, \ldots, N$ (with the total volume $b_{N+1} = 1$); let A_{nm} be the value of the nth property in the mth water

mass (with $A_{N+1,m} = 1$ for $m = 1, \ldots, M$); then for simple mixing:

$$\mathbf{Ax} = \mathbf{b}, \tag{1.7.2}$$

where \mathbf{A} is the $(N + 1) \times M$ matrix with columns given by water mass definitions, \mathbf{x} is the M-dimensional column vector of water mass fractions and \mathbf{b} is the $(N + 1)$-dimensional column vector of mixture measurements. Not all exact or approximate solutions of (1.7.2) are acceptable; only non-negative fractions are allowed:

$$x_m \geq 0, \qquad (1 \leq m \leq M). \tag{1.7.3}$$

Thus the problem of finding the water mass fractions may be cast as a minimization problem with inequality constraints. A slightly more general formulation involving K linear constraints is as follows: Minimize

$$\mathcal{J}[\mathbf{x}] = (\mathbf{Ax} - \mathbf{b})^*(\mathbf{Ax} - \mathbf{b}) \tag{1.7.4}$$

subject to

$$\mathbf{Fx} \geq \mathbf{h}, \tag{1.7.5}$$

where $\mathbf{A}, \mathbf{x}, \mathbf{b}, \mathbf{F}$ and \mathbf{h} have dimensions $L \times M$, $M \times 1$, $L \times 1$, $K \times M$ and $K \times 1$ respectively (here $L = N + 1$, $K = M$). The casual notation in (1.7.5) is expressed more precisely as

$$\mathbf{e}_k^* \mathbf{Fx} \geq h_k, \qquad (1 \leq k \leq K), \tag{1.7.6}$$

where \mathbf{e}_k is the kth unit vector. The Karush-Kuhn-Tucker theorem states that \mathbf{x} minimizes (1.7.4), subject to (1.7.5), if there is a K-dimensional vector \mathbf{v} such that

$$\begin{aligned} &i) && \mathbf{v} \geq 0, &&&(1.7.7)\\ &ii) && \mathbf{F}^*\mathbf{v} = \nabla \mathcal{J}[\mathbf{x}], &&&(1.7.8) \end{aligned}$$

and

$$iii) \qquad v_k(\mathbf{e}_k^* \mathbf{Fx} - h_k) = 0, \qquad (1 \leq k \leq K). \tag{1.7.9}$$

The proof may be found in, for example, Peressini, Sullivan & Uhl (1988). Condition (i) means simply that $v_k \geq 0$ for $1 \leq k \leq K$. Condition (ii) relates the hyperplanes $\mathbf{Fx} = \mathbf{h}$ to the contours of

constant J. The columns of \mathbf{F}^* are the normals to the hyperplanes, so (ii) states that at the solution \mathbf{x}, there is a linear combination of the K normals which is equal to the gradient of J. Condition (iii) implies that if $v_k > 0$ then the solution must lie on the kth hyperplane, but if $v_k = 0$ then the solution may be off the hyperplane. The theorem cannot hold if the inequality constraints are *infeasible* or inconsistent. On the other hand it provides conditions which are not only sufficient but also necessary ("if and only if") when the inequality constraints are *superconsistent*, that is, they can be satisfied with strict inequality holding for each component of (1.7.5).

A simple example illustrates the theorem. Let $J = x^2 + y^2 + z^2$; let the constraints be: $x \geq -1$, $y \geq -2$ and $z = 1$. Then \mathbf{F} is the unit matrix, while $\mathbf{h} = (-1, -2, 1)^*$. Consider the vector $\mathbf{v} = (0, 0, 2)^*$. This satisfies conditions (i), (ii) and (iii) for $\mathbf{x} = (0, 0, 1)^*$, which is readily seen to be the solution.

The theorem becomes clearer if the components of \mathbf{v} (possibly after a reordering) are partitioned as follows:

$$\mathbf{v} = \begin{pmatrix} \mathbf{v}_E & > & 0 \\ \mathbf{v}_S & = & 0 \end{pmatrix} \quad \text{where} \quad \begin{pmatrix} \mathbf{F}_E \mathbf{x} & = & \mathbf{h}_E \\ \mathbf{F}_S \mathbf{x} & > & \mathbf{h}_S \end{pmatrix}. \qquad (1.7.10)$$

The subscript E denotes constraints of equality, that is, the solution \mathbf{x} lies on those hyperplanes, while S denotes constraints which are *slack*: \mathbf{x} lies off those hyperplanes or inside the *feasible region*. Now suppose $\delta\mathbf{x}$ is a small perturbation of the solution. The perturbed solution satisfies the constraints provided that

$$\delta\mathbf{x}^* \mathbf{F}_E^* \geq 0. \qquad (1.7.11)$$

The corresponding change in J is determined by condition (ii) of the theorem:

$$\delta J = \delta\mathbf{x}^* \nabla J = \delta\mathbf{x}^* \mathbf{F}^* \mathbf{v} = \delta\mathbf{x}^* \mathbf{F}_E^* \mathbf{v}_E \qquad (1.7.12)$$

since $\mathbf{v}_S = \mathbf{0}$. By virtue of (1.7.11), J must increase.

The following example shows that the theorem leads naturally to an algorithm for finding the solution. Minimize $J = x^2 + y^2$, subject to $x + 2y \geq 2$, $x \geq 0$. In this case

$$\mathbf{F} = \begin{pmatrix} 1 & 2 \\ 1 & 0 \end{pmatrix}, \quad \mathbf{h} = \begin{pmatrix} 2 \\ 0 \end{pmatrix}. \qquad (1.7.13)$$

Conditions (i), (ii) and (iii) become

$$\left.\begin{array}{c} v_1 \geq 0 \\ v_2 \geq 0 \end{array}\right\}, \qquad (1.7.14)$$

$$\left.\begin{array}{r} v_1 + v_2 = 2x \\ 2v_1 \quad\;\; = 2y \end{array}\right\}, \qquad (1.7.15)$$

$$\left.\begin{array}{l} v_1(x + 2y - 2) = 0 \\ v_2 x \qquad\qquad = 0 \end{array}\right\}. \qquad (1.7.16)$$

The guess $x = 0$ leads to $v_1 = -v_2$, $v_1 = v_2 = 0$, $y = 0$, which is not feasible. Hence $x > 0$, thus $v_2 = 0$, $v_1 = 2x$, $v_1 = y$, $2x = y$; $v_1 > 0$, $x + 2y - 2 = 0$ so $x = 2/5$, $y = 4/5$. Also $v_1 = 4/5$. Condition (ii) is readily verified. An orderly guessing procedure with FORTRAN code may be found in Lawson & Hanson (1974). Several proprietary subroutine libraries include variants of this procedure.

The Karush-Kuhn-Tucker theorem was applied to a water mass analysis by Mackas, Denman & Bennett (1987), based on a number of hydrographic sections across the continental shelf of Vancouver Island, British Columbia: see Fig. 1.7.2. Six tracers were measured: temperature, salinity, nitrate, phosphate, silicate and oxygen. It was hypothesized that the water was a mixture of five types: relatively fresh surface flow from Juan de Fuca Strait; offshore water representative of the surface on which sigma-t = 25.3 in summer (depth range 50–80 m); subarctic water representative of the sigma-t = 26.6 surface (depth 125–175 m) seaward of the shelf break; California Undercurrent water represented by the sigma-t = 26.6 surface (depth 200–300 m), and coastal deep water typical of sigma-t = 27.0 surface (depth 450–500 m). For a discussion of the density anomaly sigma-t (= density − 1000 kg m^{-3}), see Pickard & Emery (1990). Quantitative water mass or source definitions, that is, the first 6 rows of the matrix \mathbf{A}, are given in Table 1.7.1. In summary, the data consist of concentrations of $M = 6$ tracers measured on the 5 sections, while the desired results are the concentrations of $N = 5$ source types calculated on the 5 sections.

The uncertainty in measurement varied from tracer to tracer: salinity can be measured with greater precision than can silicates or nitrates, for example. The uncertainty in source definition varied from source to source due to measurement uncertainty, and also due to the real variability within the somewhat arbitrarily defined sources.

Figure 1.7.2. Map of the British Columbia coast showing the locations of the five lines of stations (A–E) and the assumed large-scale movements of the five source water types. See Table 1.7.1 for tracer characteristics of the various sources (after Mackas *et al.*, 1987).

Consequently the source fractions were estimated in a weighted least-squares sense, taking into account the different units of measurement for each tracer:

$$\mathcal{J}_Q = (\mathbf{Ax} - \mathbf{b})^*\mathbf{Q}^{-1}(\mathbf{Ax} - \mathbf{b}), \qquad (1.7.17)$$

where \mathbf{Q} is the "dispersion matrix." The choice of \mathbf{Q} encapsulates the scientific content of a water mass analysis. Mackas *et al.* assumed that measurement errors for different tracers were independent, thus errors in \mathbf{b} contribute only to the diagonal of \mathbf{Q}. From an examination of the variability in regions dominated by single sources, Mackas *et al.* concluded that in coastal waters the errors in \mathbf{A} are dominated by definition variability rather than measurement error, although this

Table 1.7.1

Tracer values for the source water types used in the British Columbia coastal analysis (after Mackas *et al.*, 1987).

	Source				
	Juan de Fuca	Offshore	Subarctic	C.U.	Coastal deep
Temperature	9.84	8.75	4.70	6.90	5.10
Salinity	31.55	32.62	33.57	33.90	34.10
Nitrate	24.68	7.63	29.00	33.40	40.00
Phosphate	2.19	1.11	2.30	2.65	3.20
Silicate	41.76	12.55	48.99	52.00	90.00
Oxygen	5.26	6.20	4.65	2.10	0.50

Notes: C.U. = California Undercurrent; temperature = $^\circ$C; nitrate, phosphate and silicate = μmol; oxygen = mL L^{-1}.

may not be the case in mid-ocean. For simplicity, the definition errors were assumed to be the same for all sources, and given by the average over all sources. That is,

$$Q_{ij} = \overline{b_i' b_j'} + M^{-1} \sum_{m=1}^{M} A_{im}'' A_{jm}'', \qquad (1 \le i, \ j \le N), \qquad (1.7.18)$$

where the prime and double prime denote measurement error and definition error, respectively, while the overbar denotes an average over calibration data. See Table 1.7.2. For i or j equal to $N + 1$, the coefficients Q_{ij} were set to zero, except for $Q_{N+1,N+1}$, which was chosen to be sufficiently large so that the normalizing condition $\sum_{m=1}^{M} x_i = 1$ was satisfied to within 1%.

The quadratic form \mathcal{J}_Q may be diagonalized, that is, the misfit vector may be rotated to achieve statistically independent components and redefined tracers. Since \mathbf{Q} is symmetric and non-negative, it has the SVD

$$\mathbf{Q} = \mathbf{U} \Lambda \mathbf{U}^*, \qquad (1.7.19)$$

Table 1.7.2

Estimate of dispersion matrix \mathbf{Q} describing expected error in the tracer variables used in the British Columbia coastal analysis. Units are products of those listed in Table 1.7.1 (after Mackas et al., 1987).

	(1)	(2)	(3)	(4)	(5)	(6)
(1) Temperature	0.05	0	−0.033	−0.003	0	0
(2) Salinity		0.005	−0.003	0	0	0
(3) Nitrate			2.50	+0.067	+0.716	−0.083
(4) Phosphate				0.033	+0.083	−0.001
(5) Silicate					3.33	0
(6) Dissolved O$_2$						0.050

where the diagonal elements of Λ are non-negative, thus \mathbf{Q} has a well-defined square root:

$$\mathbf{Q}^{1/2} = \mathbf{U}\Lambda^{1/2}\mathbf{U}^*. \tag{1.7.20}$$

Then \mathcal{J}_Q becomes

$$\mathcal{J}_Q = (\mathbf{Gx} - \mathbf{d})^*(\mathbf{Gx} - \mathbf{d}), \tag{1.7.21}$$

where $\mathbf{G} = \mathbf{Q}^{-1/2}\mathbf{A}$ and $\mathbf{d} = \mathbf{Q}^{-1/2}\mathbf{b}$.

Mackas et al. minimized (1.7.21) subject to $\mathbf{x} \geq \mathbf{0}$; the results are shown in Fig. 1.7.3. Current meter measurements collected off the

Figure 1.7.3a. Source composition solutions showing concentration (0.0–1.0) of the five source water types versus horizontal position and depth for the two southernmost British Columbia sections. Sample locations are indicated by horizontal tick marks, where the least-squares solution gave a good fit to the sample, and by solid circles where the fit was poor (usually near the surface and presumably due to nonconservative behavior of the tracers). Contours indicate the estimated 0.1, 0.25, 0.5, 0.75 and 0.9 source concentration isopleths (solid black indicates a local concentration > 0.9). Note the large amount of California Undercurrent water on the southern part of the shelf, and the movement of coastal deep water up the Juan de Fuca submarine canyon (station 19 on line A and station 78 on line B). Subarctic water is present only in very low concentrations on these two lines, while the major transition in upper-layer source types (Juan de Fuca versus offshore) occurs slightly inshore from the shelf break (after Mackas et al., 1987).

Vancouver Island shelf suggest strong upwelling; the density and low dissolved oxygen content of the upwelled water are indicative of California Undercurrent water from 200–300 m depth. Such upwelling of relative deep water would bring nutrients and enhanced productivity, but it would also flush the deeper parts of the shelf with deep water depleted in potential zooplankton and phytoplankton seed populations. The contours of source fraction shown in Fig. 1.7.3 indicate that the California Undercurrent water concentration is, as expected, highest in the southernmost sections. The core of subarctic water over the shelf break, most extensive and most highly concentrated in the northern sections, was unexpected. Thus delivery of this water is apparently from the north, rather than zonal convergence as had been hypothesized. The distribution of shallow, relatively low-salinity water did not clearly identify Juan de Fuca Strait as the source; fresh water discharge from Queen Charlotte Sound could be advected southward by the Vancouver Island Coastal Current.

1.8 Steady tracers

Both the β-spiral and Wunsch's method use salinity and temperature data in the form of geostrophically balanced vertical shear (thermal wind). The horizontal reference velocities are estimated subject to layer-by-layer conservation of mass locally (§1.2), or in a region (§1.3). The concentrations of other dissolved substances such as nitrates, silicates and oxygen were used in §1.7 to identify the components in a mixture of water masses, that is, the source waters were identified from tracer concentrations. The latter information also provides constraints on inverse estimates of reference velocities, once consistency with the conservation laws for the tracers is required. Assuming that the concentration C of a particular tracer is steady over the time scales of interest, conservation of the tracer may be expressed as

Figure 1.7.3b. Source composition solutions along the three northernmost British Columbia sections. Sample locations and contouring as in Fig. 1.7.3a. Note the reduced amounts of California Undercurrent and coastal deep water and the stronger core of subarctic water along the upper continental slope. Note also the progressive reduction from south (line C) to north (line E) in the amount of Juan de Fuca water, and the greatly increased vertical and horizontal spread of offshore water (after Mackas et al., 1987).

$$\mathbf{u} \cdot \nabla C + w \frac{\partial C}{\partial z} = \nabla \cdot (K_H \nabla C) + \frac{\partial}{\partial z} \left(K_V \frac{\partial C}{\partial z} \right) + Q, \quad (1.8.1)$$

where $C = C(\mathbf{x}, z)$, while $Q = Q(\mathbf{x}, z)$ is the steady distribution of sources and sinks for the tracer. A solenoidal advecting velocity field may be assumed:

$$\nabla \cdot \mathbf{u} + \frac{\partial w}{\partial z} = 0, \quad (1.8.2)$$

thereby expressing conservation of mass, with sufficient accuracy.

The diffusivities K_H and K_V may be replaced with a diffusivity tensor, but the quantity and quality of presently available data do not support such generality. Indeed, the second spatial derivatives of concentration appearing in (1.8.1) cannot be reliably estimated; the only resolution supported by the data is that obtained by averaging (1.8.1) over a rectangular volume oriented along the coordinate axes. Then (1.8.1) becomes

$$(\Delta x)^{-1} \left[u\, C \right]_W^E + (\Delta y)^{-1} \left[v\, C \right]_S^N + (\Delta z)^{-1} \left[w\, C \right]_L^U$$
$$= (\Delta x)^{-1} \left[K_H \frac{\partial C}{\partial x} \right]_W^E + (\Delta y)^{-1} \left[K_H \frac{\partial C}{\partial y} \right]_S^N$$
$$+ (\Delta z)^{-1} \left[K_V \frac{\partial C}{\partial z} \right]_L^U + Q, \quad (1.8.3)$$

where [] denotes the difference between values on opposite faces, averaged over those faces; E, W, N, S, L and U denote the East, West, North, South, Lower and Upper faces; Δx, Δy and Δz are the lengths of the edges of the rectangular volume, while Q now denotes the volume average of the source strength. The first derivatives of concentration appearing in (1.8.3) may be estimated using centered differences, for example.

No matter how the concentration gradients are estimated, (1.8.3) is a linear relation amongst unknown local values of the velocity and the diffusivities, with measured tracer concentrations, gradients and source strengths entering as coefficients. A collection of such volumes yields a set of linear equations for the locally averaged velocities and diffusivities, given adequate data to evaluate C, Q and the gradients

of C. If use is made of the thermal wind, then the unknowns are reduced to the horizontal reference velocities, vertical velocities and diffusivities. In any event, the various conservation laws for mass and tracer concentrations may be arranged in the general linear form (1.3.2), where **b** is now a vector of unknown horizontal velocities (or reference velocities), vertical velocities and diffusivities. Accordingly, the methods of §1.3 may be used to find the smallest solution vector yielding the best fit. The solutions for the diffusivities could be negative and thus unacceptable on physical grounds. This possibility may be avoided by appending linear inequality constraints, as in §1.7. In the event that the constrained linear system has solutions, that is, it is a feasible system, it may be desirable to find the solution which maximizes, say, the meridional heat flux. The latter is a linear combination of the unknown velocities and diffusivities, with combination coefficients depending on temperature measurements. The problem may be viewed as a search for the maximum heat flux consistent with the observed tracer concentrations, while obeying positivity constraints on the diffusivities.

A linear combination of variables is a trivial example of a *convex function* of the variables, so the problem just described is an example of *convex programming*. The solubility conditions for such a problem constitute the general form of the Karush-Kuhn-Tucker theorem (Peressini *et al.*, 1988). The standard computational method for the preceding *linear programming problem* (the Simplex Algorithm) is widely available in subroutine libraries. The algorithm restricts its search to the vertices of the region bounded by the defining hyperplanes, since the solution must be at a vertex. Much more efficient algorithms have been devised recently; these search within the interior of the region (Noble & Daniel, 1988). The properties of these newer algorithms are not completely understood.

Estimates of North Atlantic circulation using steady tracer data have been made by Wunsch & Minster (1982) and Wunsch (1984). Both studies considered only the zonally averaged circulation and concentration fields; in each case it was further assumed that the stationary eddy fluxes (zonal correlations of meridional velocity and concentration) could be represented by diffusive fluxes, that is,

$$\overline{vC} = \overline{v}\,\overline{C} + \overline{v\prime C\prime} \tag{1.8.4}$$

$$= \overline{v}\,\overline{C} + K_H \frac{\partial \overline{C}}{\partial y}, \tag{1.8.5}$$

where the overbar denotes a zonal average, and the prime denotes a departure from the zonal average. Such a simple parameterization of the stationary flux is especially suspect when the zonal section spans the entire horizontal gyre, including a western boundary current. The least unsatisfactory estimate of circulation may be one in which the diffusivity K_H is small, that is, the mean meridional circulation $(\overline{v}, \overline{w})$ alone is consistent with the zonally averaged tracer concentration, and with any other constraints. Thermal wind data were not used in the 1982 study, but were employed in the second study. The latter also made use of isopycnal coordinates, thus the volumes were bounded above and below by surfaces of constant potential density, while the unknown variables were isopycnal and diapycnal velocities and diffusivities. The 1984 study also incorporated bounds on certain velocities arising from current meter and subsurface drifter data, bounds on regional tracer fluxes, and bounds on reference velocities. Wunsch (1984) described this use of a broad range of oceanographic data as "eclectic modeling." Sensitivities of the results to the values of the bounds were determined using the *dual variables* of the linear programming problems (Peressini *et al.*, 1988). The vector **v** in (1.7.7) is the dual variable for the quadratic programming problem of minimizing (1.7.4), subject to the linear inequality constraint (1.7.5). Notice that (1.7.8), by virtue of (1.7.10), relates changes in the bounds vector **h** to changes in the quadratic function \mathcal{J} according to

$$\delta \mathcal{J} = \delta \mathbf{h}_E^* \, \mathbf{v}_E . \qquad (1.8.6)$$

The subscript E indicates those components of (1.7.5) which are satisfied as equalities at the solution **x**.

In a subsequent study with North Atlantic data of higher quality, Wunsch (1988b) used a box model in the meridional plane in order to infer circulation from tracer data and dynamical constraints. A source-sink term was allowed in the O_2 budget, representing the oxygen utilization ratio (OUR). The dissolved gas oxidizes sinking particulate carbon. However, circulations meeting the dynamical constraints led to improbably large values for the OUR. Arguing that the latter were present in the model solution in order to balance downward advection of oxygen-rich shallow water, Wunsch abandoned the two-dimensional model in favor of one having three meridional stacks of boxes. Westward advection was imposed as a further constraint. A solution with an entirely advective O_2 budget was obtained; the OUR values were negligible.

Figure 1.8.1. Map of the East Atlantic showing *Meteor* and GEOSECS stations used by Schlitzer (1987). Crossed solid dots are *Meteor*, cruise 56, leg 5 (1981); crosses are GEOSECS (1972–1973). Water depth in the hatched areas is less than 4000 m. The sills between the major basins of the East Atlantic are indicated by heavy lines; the box columns of Schlitzer's model are divided along these lines (after Schlitzer, 1987).

An attractive example of steady, two-dimensional tracer analysis is the box model of East Atlantic Deep Water developed by Schlitzer (1987). The East Atlantic basin deeper than 4000 m is shown in Fig. 1.8.1. The deep water is renewed partially from above, partially from the north (Iceland-Scotland Overflow Water: ISOW), and partially from the south (Antarctic Bottom Water: AABW), but principally from the west, with deep water entering through the

Figure 1.8.2. Schematic diagram of boxes in Schlitzer's model of the deep East Atlantic. The upper and lower surfaces are isopycnals; the vertical divisions along the sills correspond to the heavy lines in Fig. 1.8.1. The Romanche fracture zone (RFZ) inflow is directed into boxes 5 and 6 from the west; the Vema fracture zone (VFZ) inflow is directed into box 9 from the west (after Schlitzer, 1987).

Romanche and Vema fracture zones (RFZ, VFZ) in the Mid-Atlantic Ridge. Schlitzer's formal objective was an estimate of the renewal rate for the basin, using data for the radioactive isotope ^{14}C. For a discussion of the formation and decay of this tracer, plus its utility for estimating fluxes, see Broecker & Peng (1982, Ch. 5 and definitions on p. 681). Other tracers used were: potential temperature, silicate and ΣCO_2 (total carbon). The potential temperature data were obtained from the atlas of Levitus (1982); the other three data sets were obtained on a *Meteor* cruise and from GEOSECS stations (see Broecker & Peng, 1982). The deep East Atlantic basin in Fig. 1.8.1 was modeled using the 11 boxes shown in Fig. 1.8.2. Their upper and lower surfaces were defined by uniform values of σ_4, the potential density relative to a depth of 4000 m.

The model distinguished between mean flow and turbulent mixing (parameterized by eddy viscosities), but did not use thermal wind data. Eastward flow into boxes 5 and 6 was allowed, representing renewal through the RFZ and VFZ. A linear decay term was included in the volume-integrated conservation law for ^{14}C (the tracer has a half-life of 5600 years). A distributed source in that equation

represented dissolution of sinking particulate ^{14}C or, in the case of the deepest boxes, ^{14}C arising from sediments. Fluxes of geothermal heat were allowed in the energy budget for those boxes: see Schlitzer (1987) for details of the tracer data, including concentrations at inflow. The tracer fields were almost uniform in the deepest boxes. The acceptable errors for the conservation laws in each box were imposed as linear inequality constraints. Further, the acceptable error for each column of the boxes, and the error for the composite region, were also imposed as constraints. Otherwise the combined errors could have been unacceptably large, with the error in each box achieving a limiting value. Additional constraints included: inflow only through the fracture zones; VFZ inflow at most 25% of RFZ inflow; ISOW and AABW inflow at most 10% of the combined FZ inflows; upward vertical fluxes through the top surfaces of the Angola Basin, the Cape Verde/Canary Basin and the Iberian/West European Basin, with upper limits proportional to the respective surface areas. Downward flow across the top surface of the Sierra Leone/Guinea Basin was allowed, but at a rate not exceeding the RFZ inflow. Finally, differences between diffusivities in adjacent boxes were bounded.

Schlitzer (1987) found the system of 70 model parameters subject to 110 constraints to be feasible, provided diapycnal mixing was allowed. Solutions having minimal and maximal inflows through the RFZ were obtained, as were solutions minimizing the sums of the diffusivities, solutions maximizing and minimizing the ΣCO_2 and silicate sources, and others. Several common features emerged. The minimal value of the inflow through the fracture zones (Fig. 1.8.3) mainly depends upon the ^{14}C balances of the entire region and of the Iberian/West European box column. Removing all ^{14}C constraints (in fact Schlitzer worked with the anomaly Δ^{14}C: see Broecker and Peng, 1982) led to considerably less FZ inflow. The minimal value for the FZ inflow was also sensitive to the constraints on the flow rates across the upper surfaces of the boxes. Those flow rates attained their upper bounds. Indeed no FZ inflow was needed to satisfy the ^{14}C budget, if arbitrary upward or downward flows were allowed across the upper surfaces of the top boxes. Diffusivities were found to increase with depth (Fig. 1.8.4). The energy source for this deep mechanical mixing can only be speculated. The differences in tracer concentrations between the deepest boxes and the boxes immediately above were comparable with Schlitzer's estimates of measurement error. It must be concluded that fluxes inferred from the large diapycnal

Figure 1.8.3. Transports in the solution which minimize the combined RFZ and VFZ inflows. Their separate values are: RFZ, 2.6 Sv; VFZ, 0 Sv. The ISOW and AABW contributions are negligible (after Schlitzer, 1987).

diffusivities shown in Fig. 1.8.4, and from the insignificant diapycnal concentration differences, are suspect. Deep isopycnal flows were typically polewards, but varied considerably from solution to solution.

In summary, the tracer data employed by Schlitzer left the flow field and diffusivities well underdetermined, with a large null space of "free solutions." Linear inequalities were used to constrain the null space, which was then explored by seeking solutions yielding maxima or minima for certain variables, or for linear combinations of variables of interest. While closed circulation cells within the basin could be found consistent with the ^{14}C budget, plausible constraints indicated that there could be significant renewal through the fracture zones. Schlitzer concluded with a recommendation for current metering in those passages.

1.9 A smoothed inversion

The singular value decomposition (SVD) leads, in general, to smooth solutions for linear least-squares problems. Smoothing is effected by discarding projections onto the finely structured singular vectors, which are in general the vectors associated with small singular values. Provost & Salmon (1986) chose not to make a SVD, in their

Figure 1.8.4. Diffusivities in the solution which minimize FZ inflow. The isopycnal diffusivities have units of 10^3 m^2 s-1, the diapycnal diffusivities have units of 10^{-4} m^2 s^{-1}.

inverse estimation of geostrophic velocities using hydrographic data from the Labrador Sea. However, they were able to control the smoothness of their inverse estimate by augmenting their quadratic form with terms penalizing the Laplacians of the flow variables. That is, if $\mathcal{J}_0 = \mathcal{J}_0[\mathbf{u}, \ldots]$ is a weighted integral of squares of residuals in the thermal wind relations (which relate the vertical shear \mathbf{u}_z to observations of the horizontal density gradient, $\nabla \rho$) and residuals in the continuity equation, in the linearized conservation law for potential density, and in the hydrostatic balance, then Provost & Salmon minimized

$$\mathcal{J}_\theta[\mathbf{u}, \ldots] = \mathcal{J}_0[\mathbf{u}, \ldots] + \theta \int dz \int d\mathbf{x} \left| \nabla^2 \mathbf{u} \right|^2 + \ldots, \qquad (1.9.1)$$

where ∇^2 is the horizontal Laplacian operator. The diaerases indicate other circulation variables. The augmented penalty functional \mathcal{J}_θ is highly sensitive to the roughness of \mathbf{u}, if $\theta > 0$. A mixed finite-element, global-polynomial representation yielded an approximation to (1.9.1) as a quadratic form of finite but very large dimension. With 600 nodes in the horizontal plane, and 5 vertical modes, the dimension was 3000×3000. The large linear system for the minimum of \mathcal{J}_θ was then solved iteratively. Satisfactorily smooth estimates of the

circulation in the Labrador Sea were obtained by experimenting with the weight θ.

The selection of a criterion of smoothness in generalized inverse estimates of ocean circulation is a fundamental issue, which reflects in detail the prior knowledge of the circulation. The next chapter outlines two seemingly disparate but in fact closely related techniques for smoothing observations, without regard to dynamical or other constraints. In subsequent chapters these techniques will be incorporated into the formulation of generalized inverses for circulation models.

2

THE SMOOTHING OF OBSERVATIONS

2.1 Introduction

This chapter provides a brief treatment of two smoothing techniques: Gauss-Markov smoothing, also known to meteorologists and oceanographers as "objective analysis," and spline smoothing. The close relationship between these two methods will be demonstrated, and some oceanographic examples will be discussed. These smoothing techniques do not by themselves belong in a work on inverse methods, but the concepts are essential in subsequent development of inverse methods for infinite-dimensional systems. The partial differential equations which model ocean circulation are examples of such systems. The treatment here is influenced strongly by Wahba & Wendelberger (1980), and also McIntosh (1990).

2.2 Gauss-Markov smoothing

Standard references for this method are Gandin (1965) and Bretherton, Davis & Fandry (1976). A recent text is Thiébaux & Pedder (1987). The following presentation is slightly non-standard in procedure and notation, for later convenience.

A typical application is as follows. An estimate of the planar field or "map" of streamfunction $\psi = \psi(\mathbf{x})$ is required, given some data d_i, $1 \leq i \leq N$. The physical nature of the data will be described later. A simple linear interpolation is provided by

$$\psi_L(\mathbf{x}) = \overline{\psi}(\mathbf{x}) + \mathbf{w}^*(\mathbf{x})(\mathbf{d} - \overline{\mathbf{d}}), \qquad (2.2.1)$$

where the overbar denotes an expected or ensemble mean value, \mathbf{d} is

the data vector and $\mathbf{w}(\mathbf{x})$ is a prescribed vector field. The estimate (2.2.1) is *unbiased*:

$$\overline{\psi}_L(\mathbf{x}) = \overline{\psi}(\mathbf{x}). \qquad (2.2.2)$$

The variance of the error in the estimate is therefore

$$\overline{e^2}(\mathbf{x}) \equiv \overline{[\psi(\mathbf{x}) - \psi_L(\mathbf{x})]^2}. \qquad (2.2.3)$$

Substitution of (2.2.1) yields

$$\overline{e^2} = \overline{(\psi - \overline{\psi})^2} + (\mathbf{w} - \mathbf{c}^{-1}\mathbf{q})^*\mathbf{c}(\mathbf{w} - \mathbf{c}^{-1}\mathbf{q}) - \mathbf{q}^*\mathbf{c}^{-1}\mathbf{q}, \qquad (2.2.4)$$

where \mathbf{q} and \mathbf{c} are second-order moments of ψ and \mathbf{d}:

$$\mathbf{q} = \overline{(\psi - \overline{\psi})(\mathbf{d} - \overline{\mathbf{d}})}, \qquad (2.2.5)$$

and

$$\mathbf{c} = \overline{(\mathbf{d} - \overline{\mathbf{d}})(\mathbf{d} - \overline{\mathbf{d}})^*}. \qquad (2.2.6)$$

Note that $\mathbf{q} = \mathbf{q}(\mathbf{x})$ while \mathbf{c} is independent of \mathbf{x}. Assuming that all first- and second-order moments are known, only \mathbf{w} in (2.2.4) is disposable. The choice which minimizes $\overline{e^2}$ is $\mathbf{w} = \mathbf{c}^{-1}\mathbf{q}$; that is, the minimum-variance or best linear unbiased estimate ("BLUE") for $\psi(\mathbf{x})$ is

$$\psi_L = \overline{\psi} + \mathbf{q}^*\mathbf{c}^{-1}(\mathbf{d} - \overline{\mathbf{d}}). \qquad (2.2.7)$$

The minimum error variance $\overline{e^2}(\mathbf{x})$ is given by

$$\overline{e^2} = \overline{(\psi - \overline{\psi})^2} - \mathbf{q}^*\mathbf{c}^{-1}\mathbf{q}. \qquad (2.2.8)$$

The use of the linear estimate (2.2.1) is sensible only if the data are linearly related to the field, that is,

$$d_i = \mathcal{L}_i[\psi(\mathbf{x})] + \epsilon_i. \qquad (2.2.9)$$

In (2.2.9), \mathcal{L}_i is a linear functional: it acts on a function to yield a scalar, in a linear fashion. The error in this measurement process is represented by ϵ_i. The following are examples of \mathcal{L}_i having integral representations.

i) Direct measurement of streamfunction:

$$d_i = \psi(\mathbf{X}_i) + \epsilon_i \,. \qquad (2.2.10a)$$

The integral representation is

$$\mathcal{L}_i\,[\psi] = \int \psi(\mathbf{x})\,\delta(\mathbf{x} - \mathbf{X}_i)\,d\mathbf{x} \qquad (2.2.10b)$$

where δ is the Dirac delta function.

ii) Measurement of a velocity component:

$$d_i = u(\mathbf{X}_i) + \epsilon_i = -\frac{\partial \psi}{\partial y}(\mathbf{X}_i) + \epsilon_i \,. \qquad (2.2.11a)$$

The integral representation is

$$\mathcal{L}_i\,[\psi] = -\frac{\partial}{\partial y} \int \psi(\mathbf{x})\,\delta(\mathbf{x} - \mathbf{X}_i)\,d\mathbf{x} \,. \qquad (2.2.11b)$$

iii) Acoustic tomography (reciprocal shooting path):

$$d_i = \int_{AB} \mathbf{u}(\mathbf{X}(s)) \cdot d\mathbf{s} + \epsilon_i \,. \qquad (2.2.12a)$$

See Fig. (2.2.1). The integral representation is

$$\mathcal{L}_i\,[\psi] = \int_{AB} \frac{\partial}{\partial n} \int \psi(\mathbf{x})\,\delta(\mathbf{x} - \mathbf{X}(s))\,d\mathbf{x}\,ds \qquad (2.2.12b)$$

The important emerging technology of ocean acoustic tomography is reviewed in Worcester, Cornuelle & Spindel (1991).

It is clear that the evaluation functional (2.2.10*b*) is fundamentally related to the other functionals: this will be exploited later. The errors ϵ_i may be due to instrumental or theoretical imprecision. An example of the latter might be the use of scaled pressure as a streamfunction. For example, if d_i is dynamic height at \mathbf{X}_i, then ϵ_i in (2.2.10*a*) includes errors due to measurement of dynamic height. It

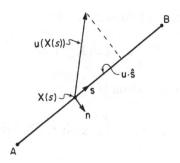

Figure 2.2.1. Acoustic tomography: reciprocal shooting. Differential acoustic travel times AB–BA yield the average fluid velocity in direction \overline{AB}. That is, the path integral (2.2.12a) is obtained.

also includes errors in equating streamfunction and dynamic height, such as variability in the Coriolis parameter or ageostrophic effects (Pickard & Emery, 1990).

The statistics of \mathbf{d} may be expressed in terms of those of $\psi(\mathbf{x})$, the measurement error vector $\boldsymbol{\epsilon}$ and the functional vector $\boldsymbol{\mathcal{L}}$, using (2.2.9):

$$\mathbf{c} \equiv \overline{(\mathbf{d} - \overline{\mathbf{d}})(\mathbf{d} - \overline{\mathbf{d}})^*}$$
$$= \boldsymbol{\mathcal{L}}_{(\mathbf{x_1})}\boldsymbol{\mathcal{L}}^*_{(\mathbf{x_2})}\left[C(\mathbf{x_1}, \mathbf{x_2})\right] + \overline{\boldsymbol{\epsilon}\boldsymbol{\epsilon}^*}, \qquad (2.2.13)$$

where it has been assumed that the measurement errors are not correlated with the streamfunction field, and have zero mean. The argument for the functionals in (2.2.13) is the covariance C for the field $\psi(\mathbf{x})$, and is defined by

$$C(\mathbf{x_1}, \mathbf{x_2}) = \overline{(\psi(\mathbf{x_1}) - \overline{\psi}(\mathbf{x_1}))(\psi(\mathbf{x_2}) - \overline{\psi}(\mathbf{x_2}))}. \qquad (2.2.14)$$

In (2.2.13), $\boldsymbol{\mathcal{L}}_{(\mathbf{x_1})}$ acts on the $\mathbf{x_1}$-dependence of C while $\boldsymbol{\mathcal{L}}_{(\mathbf{x_2})}$ acts on the $\mathbf{x_2}$-dependence. Similarly,

$$q(\mathbf{x_1}) \equiv \overline{(\psi(\mathbf{x_1}) - \overline{\psi}(\mathbf{x_1}))(\mathbf{d} - \overline{\mathbf{d}})}$$
$$= \boldsymbol{\mathcal{L}}_{(\mathbf{x_2})}[C(\mathbf{x_1}, \mathbf{x_2})]. \qquad (2.2.15)$$

With these expressions for \mathbf{c} and q, the estimate (2.2.7) becomes

$$\psi_L(\mathbf{x_1}) = \overline{\psi}(\mathbf{x_1}) + \boldsymbol{\mathcal{L}}^*_{(\mathbf{x_2})}[C(\mathbf{x_1}, \mathbf{x_2})]\{\boldsymbol{\mathcal{L}}_{(\mathbf{x_3})}\boldsymbol{\mathcal{L}}^*_{(\mathbf{x_4})}[C(\mathbf{x_3}, \mathbf{x_4})]$$
$$+ \overline{\boldsymbol{\epsilon}\boldsymbol{\epsilon}^*}\}^{-1}(\mathbf{d} - \overline{\mathbf{d}}), \qquad (2.2.16)$$

while the error covariance becomes

$$\overline{(\psi(\mathbf{x}_1) - \psi_L(\mathbf{x}_1))\,(\psi(\mathbf{x}_2) - \psi_L(\mathbf{x}_2))} = C(\mathbf{x}_1, \mathbf{x}_2)$$
$$- \mathcal{L}^*_{(\mathbf{x}_3)} [C(\mathbf{x}_1, \mathbf{x}_3)] \left\{ \mathcal{L}_{(\mathbf{x}_4)} \mathcal{L}^*_{(\mathbf{x}_5)} [C(\mathbf{x}_4, \mathbf{x}_5)] \right.$$
$$\left. + \overline{\epsilon \epsilon^*} \right\}^{-1} \mathcal{L}_{(\mathbf{x}_6)} [C(\mathbf{x}_6, \mathbf{x}_2)] . \tag{2.2.17}$$

These expressions require, in addition to the data \mathbf{d}, only knowledge of the measurement functionals \mathcal{L} together with the statistics $\overline{\mathbf{d}}$, $\overline{\psi}(\mathbf{x})$ and $C(\mathbf{x}_1, \mathbf{x}_2)$. The elaborate general expression (2.2.16) is simplest if the \mathcal{L} are direct-measurement functionals (the case (i) considered above). Then (2.2.16) reduces to

$$\psi_L(\mathbf{x}) = \overline{\psi}(\mathbf{x}) + \sum_{i=1}^{N} C(\mathbf{x}, \mathbf{X}_i) \sum_{j=1}^{N} F_{ij}(d_j - \overline{d}_j), \tag{2.2.18}$$

where F_{ij} are components of a matrix \mathbf{F}, which is the inverse of

$$c_{ij} \equiv C(\mathbf{X}_i, \mathbf{X}_j) + \overline{\epsilon_i \epsilon_j} . \tag{2.2.19}$$

Examination of (2.2.18) shows that the spatial pattern of the estimate $\psi_L(\mathbf{x})$ is a linear combination of the "field-to-measurement point" covariances $C(\mathbf{x}, \mathbf{X}_i)$. The coefficients in the linear combination are themselves linear combinations of the data $\mathbf{d} - \overline{\mathbf{d}}$. Notice that while the estimate $\psi_L(\mathbf{x})$ is unbiased (see (2.2.2)), it does not have the same covariance as ψ. Also, if the data are worthless ($\overline{\epsilon \epsilon^*} \to \infty$) then (2.2.18) reduces to $\psi_L(\mathbf{x}) = \overline{\psi}(\mathbf{x})$: the only rational estimate is the mean. If, on the other hand, the data are perfect: $\overline{\epsilon \epsilon^*} = \mathbf{0}$, then it is readily seen that $\psi_L(\mathbf{X}_i) = \psi(\mathbf{X}_i)$, $1 \le i \le N$. That is, the estimator coincides with the perfect data at the measurement sites \mathbf{X}_i, and interpolates between the sites. Otherwise the estimate $\psi_L(\mathbf{X}_i)$ at best comes close to the data, and smooths in between. Values of $\psi_L(\mathbf{x})$ for \mathbf{x} far from the \mathbf{X}_i (extrapolation) tend to be unreliable in either case.

An application of Gauss-Markov smoothing to mid-ocean data is provided by McWilliams (1976), who estimated streamfunction field using dynamic height relative to 1500 m and deep float velocities at 1500 m. The data were collected in the western North Atlantic

during the Mid-Ocean Dynamics Experiment. More precisely, the
time-dependent data were of the form

$$d_{\Delta D}(\mathbf{x}, z, t) = \Delta D_{1500}(\mathbf{x}, z, t)$$

$$= \mathcal{L}_{\Delta D}\big[\psi(\mathbf{x}, z, t)\big] + \epsilon_{\Delta D}(\mathbf{x}, z, t)$$

$$= \psi(\mathbf{x}, z, t) - \psi(\mathbf{x}, 1500\text{m}, t)$$

$$+ \,\epsilon_{\Delta D}(\mathbf{x}, z, t)\,, \qquad (2.2.20)$$

$$d_u(\mathbf{x}, 1500\text{m}, t) = u(\mathbf{x}, 1500\text{m}, t)$$

$$= -(g/f_0)\frac{\partial}{\partial y}\psi(\mathbf{x},\ 1500\text{m}, t)$$

$$+\,\epsilon_u(\mathbf{x}, t), \qquad (2.2.21)$$

$$d_v(\mathbf{x}, 1500\text{m}, t) = v(\mathbf{x}, 1500\text{m}, t)$$

$$= (g/f_0)\frac{\partial}{\partial x}\,\psi(\mathbf{x}, 1500\text{m}, t)$$

$$+\,\epsilon_v(\mathbf{x}, t)\,. \qquad (2.2.22)$$

The data were collected at many stations and times, and expressed
in consistent units. Note the use of the local value f_0 for the Coriolis
parameter. McWilliams proceeded in two steps. First the float data
were smoothed to yield maps of $\psi(\mathbf{x}, 1500\text{m}, t)$. Second, the dynamic
height data were smoothed to get $\psi(\mathbf{x}, z, t) - \psi(\mathbf{x}, 1500\text{m}, t)$. The
smoothed fields were then added to obtain $\psi(\mathbf{x}, z, t)$. Thus the total
streamfunction field was not obtained by direct regression on the float
data. To the extent that the field and float data were correlated, this
was a suboptimal smoothing. Covariance functions were specified
for streamfunction at each level, and for velocities at 1500 m. The
streamfunction covariance, for example, was taken to be

$$C(r) \equiv \overline{\psi(\mathbf{x}, z, t)'\psi(\mathbf{x} + \mathbf{r}, z, t)'}$$

$$= \overline{\psi'^2}(1 - \epsilon_*^2)(1 - \gamma^2 r^2)e^{-\delta^2 r^2/2}, \qquad (2.2.23)$$

where ψ' denotes a random fluctuation: $\psi' = \psi - \overline{\psi}$; \mathbf{r} is a horizontal
separation vector with magnitude r, while ϵ_*, γ and δ are the depth-
dependent parameters listed in Table 2.2.1. Their values were ob-
tained by fitting (2.2.23) to prior data. Note that horizontal isotropy
has been assumed: $C = C(r)$. This is not plausible owing to the

β-effect alone, but the alternative would be the introduction of many more parameters into the form of C. Estimation of these parameters would not have been feasible, given the limited information available for the region at that time. The velocity covariances were expressed as longitudinal and transverse covariances, which are related to the streamfunction covariance in isotropic turbulence (Batchelor, 1960).

Table 2.2.1				
Parameters in isotropic covariances for $\Delta D_{1500}(\mathbf{x}, z, t)$ (anomaly of dynamic height relative to 1500 m) for $z = 150$ m, 500 m and and 1000 m. The covariances have the same form as (2.2.23). At 1500 m, the variable is the streamfunction $\psi(\mathbf{x}, 1500, t)$; at that level, the units for the second column are accordingly cm^4 s^{-2} (after McWilliams, 1976).				
Depth (m)	$\overline{(\Delta D'_{1500})^2}$ (cm^2)	ϵ_*^2	$10^7\gamma$ (cm^{-1})	$10^7\delta$ (cm^{-1})
150	29.30	0.10	0.84	1.24
500	11.26	0.14	0.85	1.24
1000	0.66	0.27	0.89	1.29
1500	1.5	0.2	0.001	2.0

According to (2.2.23), $C(r) \to 0$ as $r \to \infty$, and C changes sign at $r = \gamma^{-1}$. Also, for small r, $C(r) \cong \overline{\psi'^2}(1 - \epsilon_*^2) < \overline{\psi'^2}$ since, in practice, $C(r)$ cannot be reliably estimated at very small lag. McWilliams chose to equate $\epsilon_*^2\mathbf{I}$ and $\overline{\epsilon\epsilon^*}$, thereby attributing the drop in covariance at small separations to measurement errors. Smoothed streamfunction maps at 150 m and 1500 m are shown in Fig. 2.2.2. It is emphasized that the spatial patterns in these figures are linear combinations of the covariance $C(r)$, where r separates field positions and data locations. Relative error maps are shown in Fig. 2.2.3. These maps are calculated using (2.2.8), and normalized by the field variance. The error is generally greatest at points farthest from the data locations.

McWilliams examined the sensitivity of his maps to changes in the estimates for the measurement noise ϵ_*^2 and the decorrelation lengths γ^{-1}, δ^{-1} appearing in (2.2.23). Halving the noise variance had little impact on the smoothed streamfunctions. Multiplying the decorrelation lengths by 1.5 (with the ϵ_*^2 fixed at their original values) led to "somewhat smoother patterns." Both increases and decreases

Figure 2.2.2. The total geostrophic streamfunction ψ at (a) 150 m depth and (b) 1500 m depth, at 30-day and 10-day separations, respectively. Tic marks on the sides indicate interpolated grid intervals (15 km), the small cross lies in the map center at (69°40'W, 28°N). The asterisks in (b) indicate float locations at these times (after McWilliams, 1976).

in eddy amplitude resulted from these changes, but the qualitative features of the maps were "not greatly altered."

An application of Gauss-Markov smoothing to coastal waters may be found in Denman & Freeland (1985). The data were collected over the continental shelf off Vancouver Island, British Columbia. The fields were plainly not only anisotropic but also inhomogeneous. That

Figure 2.2.3. Patterns of relative error variance for dynamic height and stream-function (after McWilliams, 1976).

is, the statistics clearly depended upon spatial location. In particular Denman & Freeland were obliged to estimate the mean streamfunction $\overline{\psi}(\mathbf{x}, z, t)$ as a linear function of \mathbf{x}. However, they did assume that the horizontal covariance for the streamfunction was isotropic: $C = C(r)$. Denman & Freeland examined various candidates for $C(r)$, the eventual choice being smooth at the origin and having a positive Bessel transform. These realizability conditions for $C(r)$ are discussed in, for example, Yaglom (1962). A statistical test of geostrophy was made by comparing the longitudinal and transverse velocity covariances,

derived from dynamic height covariances, with those obtained directly from current meter moorings. Good agreement was found.

The extensive high-quality hydrographic data collected during the POLYMODE Local Dynamics Experiment in 1978 enabled Taft *et al.* (1986) to estimate anisotropic covariances for what was essentially streamfunction at 600 m: see Fig. 2.2.4. A NE-SW orientation clearly dominated during the two-month period of the hydrographic survey. Nevertheless, Shen *et al.* (1986) adopted the isotropic covariance (2.2.23) for smoothing purposes, arguing that less anisotropy was present for the 15-month duration of the current meter deployment during the experiment and on the grounds of "the insensitivity of objective maps to anisotropy in the covariance functions."

2.3 Smoothing splines

Consider once again the typical problem of constructing a map of streamfunction $\psi(\mathbf{x})$, given the data \mathbf{d} and linear measurement functionals \mathcal{L}. In order to state a smoothing criterion, an inner product notation must be introduced. For definiteness, assume $\mathbf{x} = (x, y)$ is planar. A sequence of inner products $\langle \ , \ \rangle_m$ may be defined by

$$\langle f, g \rangle_0 = \int f(\mathbf{x}) g(\mathbf{x}) \, d\mathbf{x}, \tag{2.3.1}$$

$$\langle f, g \rangle_1 = \langle f, g \rangle_0 + a_1 \int \boldsymbol{\nabla} f \cdot \boldsymbol{\nabla} g \, d\mathbf{x}, \tag{2.3.2}$$

$$\langle f, g \rangle_2 = \langle f, g \rangle_1 + a_2 \int \left\{ \frac{\partial^2 f}{\partial x^2} \frac{\partial^2 g}{\partial x^2} + 2 \frac{\partial^2 f}{\partial x \partial y} \frac{\partial^2 g}{\partial x \partial y} + \frac{\partial^2 f}{\partial y^2} \frac{\partial^2 g}{\partial y^2} \right\} d\mathbf{x}, \tag{2.3.3}$$

and in general,

$$\langle f, g \rangle_m = \langle f, g \rangle_{m-1} + a_m \int \sum_{j=0}^{m} \frac{m!}{j!(m-j)!} \frac{\partial^m f}{\partial x^{m-j} \partial y^j} \frac{\partial^m g}{\partial x^{m-j} \partial y^j} d\mathbf{x}, \tag{2.3.4}$$

where f and g are functions of \mathbf{x} which are sufficiently well behaved that the various integrals are well defined. The parameters a_m are all positive. Note also the use of binomial coefficients; other choices are possible and may be appropriate in certain circumstances. Each of these inner products defines a norm $\| \ \|_m$: for sufficiently well behaved functions f, $\|f\|_m = \langle f, f \rangle_m^{1/2}$. The spline smoothing criterion may

Figure 2.2.4. Two-dimensional spatial-lag covariance functions: (a) pressure of σ_θ surface with a mean pressure of 601 dbar; and (b) salinity on a σ_θ surface with a mean pressure of 304 dbar (after Taft et al., 1986).

now be stated. Let $\psi_0(\mathbf{x})$ be a first-guess smooth field, and let ψ_S denote the mth-order smoothing spline estimate of $\psi(\mathbf{x})$. Then ψ_S minimizes the quadratic functional

$$\mathcal{J}_m[\psi_S] = \mu \|\psi_S - \psi_0\|_m^2 + (\mathbf{d} - \mathcal{L}[\psi_S])^* \mathbf{w}(\mathbf{d} - \mathcal{L}[\psi_S]). \qquad (2.3.5)$$

That is, $\mathcal{L}[\psi_S]$ is close to \mathbf{d}, ψ_S is close to ψ_0, $\nabla\psi_S$ is close to $\nabla\psi_0$, and $(\partial^m\psi_S/\partial x^{m-j}\partial y^j)$ is close to $(\partial^m\psi_0/\partial x^{m-j}\partial y^j)$ etc., all in a weighted least-squares sense. The parameter μ weights the fit of ψ_S to ψ_0 against the fit to the data; the parameters $(a_m)^{1/(2m)}$ are length scales for $\psi_S - \psi_0$, while \mathbf{w} is a symmetric and positive-definite matrix of data weights. For simplicity, assume that the range of integration in the inner product includes the entire plane. Assume also that the fields f, g, ψ_0, ψ_S, etc. vanish so rapidly as $|\mathbf{x}| \to \infty$ that the integrals in the inner product are finite.

The calculus of variations (e.g., Courant & Hilbert, 1953) may be used to show that the minimum of \mathcal{J}_m satisfies

$$\mu\, P_m\big(-\nabla^2_{(\mathbf{x}_1)}\big)\,\big(\psi_S(\mathbf{x}_1) - \psi_0(\mathbf{x}_1)\big)$$
$$= \mathcal{L}^*_{(\mathbf{x}_2)}\big[\delta(\mathbf{x}_1 - \mathbf{x}_2)\big]\, \mathbf{w}\big(\mathbf{d} - \mathcal{L}[\psi_S]\big), \qquad (2.3.6)$$

where $P_m(z) = \sum_{j=0}^m a_j z^j$ and $\nabla^2_{(\mathbf{x}_1)}$ is the Laplacian with respect to \mathbf{x}_1. Note the appearance of the sought-after field $\psi_S(\mathbf{x}_1)$ on the left-hand side of (2.3.6), and also its as-yet-unknown measured values $\mathcal{L}[\psi_S]$ on the right-hand side. A formal procedure for solving (2.3.6) is as follows. First, let $B_m(\mathbf{x}_1, \mathbf{x}_2)$ satisfy the scalar equation

$$P_m\big(-\nabla^2_{(\mathbf{x}_1)}\big)\, B_m(\mathbf{x}_1, \mathbf{x}_2) = \delta(\mathbf{x}_1 - \mathbf{x}_2). \qquad (2.3.7)$$

Then $\mathbf{r}^m(\mathbf{x}_1) = \mathcal{L}_{(\mathbf{x}_2)}[B_m(\mathbf{x}_1, \mathbf{x}_2)]$ satisfies the vector equation

$$P_m\big(-\nabla^2_{(\mathbf{x}_1)}\big)\mathbf{r}^m(\mathbf{x}_1) = \mathcal{L}_{(\mathbf{x}_2)}[\delta(\mathbf{x}_1 - \mathbf{x}_2)]. \qquad (2.3.8)$$

It follows that the solution of (2.3.6) is

$$\psi_S(\mathbf{x}_1) = \psi_0(\mathbf{x}_1) + \hat{\mathbf{b}}^*\, \mathbf{r}^m(\mathbf{x}_1), \qquad (2.3.9)$$

provided that the coefficient vector \mathbf{b} satisfies the linear system

$$\big(\mathcal{L}_{(\mathbf{x}_1)}[\mathbf{r}^m(\mathbf{x}_1)^*] + \mu\,\mathbf{w}^{-1}\big)\hat{\mathbf{b}} = \mathbf{d} - \mathcal{L}[\psi_0]. \qquad (2.3.10)$$

Exercise 2.3.1 Derive the partial differential equation (2.3.6), and the finite-dimensional equation (2.3.10). Show that the scalar $B_m(\mathbf{x}_1, \mathbf{x}_2)$ is symmetric in \mathbf{x}_1 and \mathbf{x}_2, and that the coefficient matrix in (2.3.10) is symmetric. □

The "influence function" $B_m(\mathbf{x}_1, \mathbf{x}_2)$ has been used to unravel the appearance of ψ_S on both sides of (2.3.6), leading to (2.3.9) and (2.3.10). Not every streamfunction $\psi(\mathbf{x}_1)$ is expressible in the form (2.3.9), thus in choosing $\psi(\mathbf{x}_1)$ to have this form, a class of fields has been discarded. The following geometrical construction of ψ_S makes clear the nature of the discard and emphasizes the collapsing of the infinite number of degrees of freedom in (2.3.5) to the finite number of degrees of freedom in (2.3.9). The value of the latter is the number N of measurements.

The geometrical construction of ψ_S uses representers, which may be derived from the reproducing kernel (rk) for the inner product. The rk is defined as follows: $B_m(\mathbf{x}_1, \mathbf{x}_2)$ is a rk for $\langle \ , \ \rangle_m$ if, for any field ψ and any point \mathbf{x}_1,

$$\langle \psi(\mathbf{x}_2), B_m(\mathbf{x}_1, \mathbf{x}_2) \rangle_m = \psi(\mathbf{x}_1) \, . \tag{2.3.11}$$

It is understood that in the inner product, the integration is with respect to \mathbf{x}_2.

Exercise 2.3.2 Show that B_m is symmetric and non-negative. □

The representer $r_n^m(\mathbf{x})$ for the linear measurement functional $\mathcal{L}_n[\]$ is constructed by measuring the rk with respect to either position:

$$r_n^m(\mathbf{x}_1) = \mathcal{L}_{n_{(\mathbf{x}_2)}}[B_m(\mathbf{x}_1, \mathbf{x}_2)] \, . \tag{2.3.12}$$

Note that r_n^m is simply a field, like ψ. It has the property that for any field ϕ,

$$\langle \phi(\mathbf{x}), r_n^m(\mathbf{x}) \rangle_m = \mathcal{L}_n[\phi] \, . \tag{2.3.13}$$

That is, the field r_n^m "represents" the linear measurement functional \mathcal{L}_n. Note that the N representers r_n^m, $(1 \leq n \leq N)$, are defined in terms of a specific inner product $\langle \ , \ \rangle_m$. The superscripts and subscripts "m" will now be dropped, for clarity.

Using representers for the measurement functionals, the quadratic functional \mathcal{J} may be expressed entirely in terms of inner products:

$$\mathcal{J}[\psi_S] = \mu \langle \psi_S - \psi_0, \psi_S - \psi_0 \rangle + (\mathbf{d} - \langle \psi_S, \mathbf{r} \rangle)^* \mathbf{w} (\mathbf{d} - \langle \psi_S, \mathbf{r} \rangle), \tag{2.3.14}$$

where **r** is the column vector field with components r_n, $(1 \leq n \leq N)$.

Minimization of the quadratic functional \mathcal{J}, which has functions as its arguments, is now easily reduced to the finite-dimensional task of minimizing a quadratic form having N-dimensional vectors as its arguments. To proceed, observe that any field $\psi_S(\mathbf{x})$ whatsoever can be expressed as

$$\psi_S(\mathbf{x}) = \psi_0(\mathbf{x}) + \sum_{n=1}^{N} b_n r_n(\mathbf{x}) + g(\mathbf{x}), \qquad (2.3.15)$$

where the field g is orthogonal to the representers:

$$\langle g, r_n \rangle = 0, \quad (1 \leq n \leq N). \qquad (2.3.16)$$

(If ψ_S and ψ_0 are given, then (2.3.16) determines the N representer coefficients b_1, \ldots, b_N.) As a consequence of (2.3.13) and (2.3.16), the field $g(\mathbf{x})$ may be regarded as *unobservable*: $\mathcal{L}[g] = \mathbf{0}$. Substituting (2.3.15) into (2.3.14), and using (2.3.16), yields

$$\mathcal{J} = \mu(\mathbf{b}^* \mathbf{R} \mathbf{b} + \|g\|^2) + (\mathbf{h} - \mathbf{R}\mathbf{b})^* \mathbf{w}(\mathbf{h} - \mathbf{R}\mathbf{b}), \qquad (2.3.17)$$

where $\mathbf{b}^* = (b_1, \ldots, b_N)$, while $\mathbf{h} = \mathbf{d} - \mathbf{d}_0$ and $\mathbf{d}_0 = \mathcal{L}[\psi_0]$. The $N \times N$ non-negative symmetric representer matrix **R** has elements

$$R_{np} = \langle r_n, r_p \rangle, \quad (1 \leq n, \ p \leq N). \qquad (2.3.18)$$

<u>Exercise 2.3.3</u> Show that B has the "reproducing property":

$$\langle B(\mathbf{x}_1, \mathbf{x}_2), \ B(\mathbf{x}_2, \mathbf{x}_3) \rangle = B(\mathbf{x}_1, \mathbf{x}_3),$$

and hence the representer matrix is obtained by measuring the rk with respect to both positions:

$$\mathbf{R} \equiv \langle \mathbf{r}, \mathbf{r}^* \rangle = \langle \mathcal{L}_{(\mathbf{x}_1)}[B(\mathbf{x}_1, \mathbf{x}_2)], \ \mathcal{L}^*_{(\mathbf{x}_3)}[B(\mathbf{x}_2, \mathbf{x}_3)] \rangle$$
$$= \mathcal{L}_{(\mathbf{x}_1)} \mathcal{L}^*_{(\mathbf{x}_3)}[B(\mathbf{x}_1, \mathbf{x}_3)]. \qquad (2.3.19)$$

□

The first step in minimizing \mathcal{J} is clear. Choose

$$g(\mathbf{x}) \equiv 0. \qquad (2.3.20)$$

That is, discard the unobservable field. Then \mathcal{J} is reduced to a quadratic form for **b**:

$$\mathcal{J}(\mathbf{b}) = \mathbf{b}^*\mathbf{RwPb} - \mathbf{h}^*\mathbf{wRb} - \mathbf{b}^*\mathbf{Rwh} + \mathbf{h}^*\mathbf{wh}, \qquad (2.3.21)$$

where **P** is the non-singular matrix $\mathbf{R} + \mu\mathbf{w}^{-1}$. It may be shown that (2.3.21) is least if $\mathbf{Pb} - \mathbf{h} = \boldsymbol{\theta}$ where $\boldsymbol{\theta}$ is any null vector of **Rw**. If **R** is positive-definite, then

$$\mathbf{b} = \hat{\mathbf{b}} = \mathbf{P}^{-1}\mathbf{h}, \qquad (2.3.22)$$

and the smoothing field is

$$\psi_S(\mathbf{x}) = \psi_0(\mathbf{x}) + \mathbf{r}(\mathbf{x})^*(\mathbf{R} + \mu\mathbf{w}^{-1})^{-1}(\mathbf{d} - \mathbf{d}_0), \qquad (2.3.23)$$

where **r** is the N-dimensional column vector field with components $r_n = \mathcal{L}_n[B]$. Comparison of the mth-order spline smoothing estimate (2.3.23) with the BLUE (2.2.16) shows a one-to-one correspondence:

(2.3.23)		(2.2.16)	
$\psi_0(\mathbf{x})$	first guess	$\overline{\psi}(\mathbf{x})$	mean
\mathbf{d}_0	measurements of first guess	$\overline{\mathbf{d}}$	measurements of mean
$B(\mathbf{x}_1, \mathbf{x}_2)$	reproducing kernel	$C(\mathbf{x}_1, \mathbf{x}_2)$	covariance
$\mu\mathbf{w}^{-1}$	inverse of data weight	$\overline{\epsilon\epsilon^*}$	measurement error covariance

An immediate practical advantage of this remarkable correspondence is that software developed originally for Gauss-Markov smoothing may be used, without modification, for spline smoothing.

It remains to construct the reproducing kernel $B(\mathbf{x}_1, \mathbf{x}_2)$. The property (2.3.11) may be written as

$$\langle \psi(\mathbf{x}_2), B(\mathbf{x}_1, \mathbf{x}_2) \rangle = \int \psi(\mathbf{x}_2)\delta(\mathbf{x}_1 - \mathbf{x}_2) \, d\mathbf{x}_2. \qquad (2.3.24)$$

Now the inner product, which is a sum of integrals, may be rearranged using integration by parts so that ψ appears only in undifferentiated form. It is assumed for the moment that the range of integration includes the entire plane, and that ψ and $B \to 0$ sufficiently rapidly as $\mathbf{x} \to \infty$. Since (2.3.24) holds for all ψ, it follows that $B(\mathbf{x}_1, \mathbf{x}_2)$ satisfies the partial differential equation

$$\sum_{j=0}^{m} a_j \left(-\nabla^2_{(\mathbf{x}_2)}\right)^j B(\mathbf{x}_1, \mathbf{x}_2) = \delta(\mathbf{x}_1 - \mathbf{x}_2).\qquad(2.3.25)$$

Thus the rk is the "influence function" introduced previously. The appearance of simple powers of the Laplacian operator $\nabla^2 = \frac{\partial^2}{\partial x^2} + \frac{\partial^2}{\partial y^2}$ in (2.3.25) motivates the choice of numerical coefficients in the definition (2.3.4) for the underlying inner product. On the other hand, anisotropic scaling may be desirable in certain circumstances.

The structure of the solution to (2.3.25) in free space is most easily seen by a Fourier integral transformation. The rk is translation-invariant:

$$B(\mathbf{x}_1, \mathbf{x}_2) = B(\mathbf{x}_1 - \mathbf{x}_2)\qquad(2.3.26)$$

and has the transform

$$\widetilde{B}(\mathbf{k}) = \int B(\mathbf{z}) e^{i\mathbf{k}\cdot\mathbf{z}} d\mathbf{z} = \left(\sum_{j=0}^{m} a_j k^{2j}\right)^{-1},\qquad(2.3.27)$$

with the inverse relation

$$B(\mathbf{z}) = (2\pi)^{-2} \int \widetilde{B}(\mathbf{k}) e^{-i\mathbf{k}\cdot\mathbf{z}} d\mathbf{k},\qquad(2.3.28)$$

where $k = |\mathbf{k}|$. As a consequence of the isotropy of the scaling in $\langle \ , \ \rangle$, it is clear from (2.3.27), (2.3.28) that B is also isotropic: $B(\mathbf{z}) = B(|\mathbf{z}|)$. Identifying the rk B with the covariance C of the field ψ, it may be seen that $2\pi k \widetilde{B}$ is the power spectrum for the field ψ.

For simplicity, consider a simple example: $a_0 = 1$, $a_1 = a_2 = \ldots = a_{m-1} = 0$, and $a_m = l^{2m}$ for some length scale l. Then $\widetilde{B}(k) = [1 + (kl)^{2m}]^{-1}$. Defining the half-power point $k_{1/2}$ by $\widetilde{B}(k_{1/2}) = 1/2$ yields $k_{1/2} = l^{-1}$, which is independent of m. However, as $k \to \infty$,

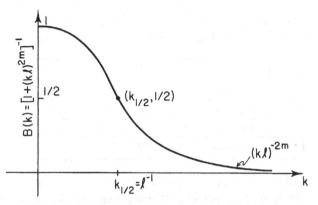

Figure 2.3.1. Fourier transform of reproducing kernel $B(x)$ associated with inner product $\langle \ , \ \rangle$ in the plane, when coefficients are $a_0 = 1$, $a_1 = a_2 = \ldots = a_{m-1} = 0$, $a_m = l^2$. See (2.3.4), (2.3.6), (2.3.27). The half-power point is at $k = k_{1/2} = l^{-1}$, the high-wavenumber asymptote is kl^{-2m}.

$\widetilde{B} \sim (kl)^{-2m}$: the decay rate is controlled by m. See Fig. 2.3.1. In this simple case the norm is

$$\|\psi\|^2 = \int \left\{ \psi^2 + l^{2m} \sum_{j=0}^{m} \frac{m!}{(m-j)!j!} \left(\frac{\partial^m \psi}{\partial x^{m-j} \partial y^j} \right)^2 \right\} dx, \quad (2.3.29)$$

which, by Parseval's theorem, is equivalent to

$$\|\psi\|^2 = \int [1 + (kl)^{2m}] |\widetilde{\psi}(\mathbf{k})|^2 \, d\mathbf{k}. \quad (2.3.30)$$

So raising the order m of the inner product and the norm forces $\widetilde{\psi}(\mathbf{k})$ to decay more rapidly as $\mathbf{k} \to \infty$. Indeed, if the integral in (2.3.30) is well defined, then necessarily

$$|\widetilde{\psi}(\mathbf{k})| = o(k^{-m-1}) \qquad \text{as } k \to \infty \quad (2.3.31)$$

Thus the high-wavenumber components of ψ have been heavily penalized by the norm $\| \ \|$: a smooth field is being sought.

The local spatial behavior of $\psi_S - \psi_0$ is determined by the representers r_n. This may be seen by assuming that \mathcal{L}_n is the direct measurement or evaluation functional at \mathbf{X}_n, so $r_n(\mathbf{x}) = B(\mathbf{x}, \mathbf{X}_n) = B(|\mathbf{x} - \mathbf{X}_n|)$. The behavior of $B(z)$ for small z is governed by the

behavior of $\widetilde{B}(k)$ for large k. Once again $\widetilde{B}(k) = O(k^{-2m})$ as $k \to \infty$. Consequently the following limiting behavior for $B(z)$ may be inferred:

m	$B(z)$
0	constant $\times\ \delta(z)$
1	constant $\times\ ln\ z$
2	constant $+$ constant $\times\ z^2 ln\ z$
m	constant $+$ constant $\times\ z^{2(m-1)} ln\ z$

Note the singular behavior as $z \to 0$, for $m = 0, 1$. For $m \geq 2$ the behavior is continuous, in fact B is at least continuously differentiable. Should the linear measurement functional \mathcal{L}_n involve integration along a track or over a local region, the behavior of r_n may be somewhat more regular than that of B. However, the general principle is that to ensure smoothness, the smoothing index m must be chosen to be sufficiently large. If, for example, it is required that ψ_S should be associated with a continuous vorticity field $\omega_S = \nabla^2 \psi_S$, then ψ_S will require at least continuous second derivatives. This will be the case if $m \geq 4$.

In summary, the rk $B(\mathbf{x}_1, \mathbf{x}_2)$ is uniquely determined by the inner product $\langle \ , \ \rangle$. Choosing the coefficients a_1, \ldots, a_m in the inner product is equivalent to choosing the scales in the streamfunction covariance $C(\mathbf{x}_1, \mathbf{x}_2)$, while selecting the smoothing index m corresponds roughly to choosing the analytical form for C. It has already been remarked above that both B and C are non-negative and symmetric. Any B can be used to define a physically realizable stochastic process $\psi(\mathbf{x})$, for example a multivariate normal random field with covariance $C = B$. However, it may be seen that not every covariance C can be used to define an inner product for which C is the rk. A natural choice would be

$$\langle f, g \rangle = \int d\mathbf{x}_1 \int d\mathbf{x}_2 f(\mathbf{x}_1) W(\mathbf{x}_1, \mathbf{x}_2)\, g(\mathbf{x}_2), \qquad (2.3.32)$$

where W is the inverse of C, that is,

$$\int W(\mathbf{x}_1, \mathbf{x}_2)\, C(\mathbf{x}_2, \mathbf{x}_3)\, d\mathbf{x}_2 = \delta(\mathbf{x}_1 - \mathbf{x}_3). \qquad (2.3.33)$$

However, consider the isotropic covariance

$$C(z) = C_0 e^{-z^2/l^2}. \qquad (2.3.34)$$

Then $\widetilde{W}(\mathbf{k}) \propto e^{k^2 l^2}$, which is not the transform of any ordinary function or even of a generalized function (Lighthill, 1958). The correspondence between covariances and rk's has been contemplated at length (Weinert, 1982; Wahba, 1990).

The only relevant, real example of spline smoothing appears to be due to Wahba & Wendelberger (1980), who considered atmospheric geopotential heights at 500 mbar. However, oceanographic applications of Gauss-Markov smoothing, using power spectra with polynomial decay, may be interpreted as spline smoothing at some order.

<u>Exercise 2.3.4</u> Consider the smoothed inverse of Provost & Salmon (1986), as outlined in §1.9. After separating the horizontal and vertical structure using vertical modes, the thermal wind data reduce to measurements of velocity, at points in the horizontal plane. The unsmoothed penalty functional \mathcal{J}_0 appearing in (1.9.1) involves only first derivatives of the velocity fields, thus the differential order of \mathcal{J}_θ is determined by the horizontal Laplacians of the velocity field when θ is positive. Show that the smoothed inverse estimate for, say, the zonal velocity component u obeys a biharmonic equation having the essential form

$$\theta \nabla^2 \nabla^2 u = K\Big(u(\mathbf{x}) - U\Big)\,\delta(\mathbf{x} - \mathbf{X}) + \dots, \qquad (2.3.35)$$

where K is a constant, and U is the velocity datum at the point \mathbf{X}. Discuss the smoothness of the solution of (2.3.35) in the neighborhood of \mathbf{X}. □

2.4 Wunsch's method as smoothing

Let us return to the problem of estimating the M-dimensional column vector \mathbf{b} of unknown bottom velocities, given the N-dimensional column data vector \mathbf{d}, where $\mathbf{Ab} = \mathbf{d} + \boldsymbol{\epsilon}$, $\boldsymbol{\epsilon}$ being the vector of measurement errors. The measurement error process is indirect; bottom velocity is inferred from tracer conservation laws and direct measurements of thermal wind. A BLUE may be sought in the form

$$\mathbf{b}_L = \overline{\mathbf{b}} + \mathbf{F}(\mathbf{d} - \overline{\mathbf{d}}) \qquad (2.4.1)$$

for some $M \times N$ matrix \mathbf{F}, with $\overline{\mathbf{d}} = \mathbf{A}\overline{\mathbf{b}}$. Defining the error variance to be $\overline{e_L^2} = \overline{(\mathbf{b} - \mathbf{b}_L)^*(\mathbf{b} - \mathbf{b}_L)}$, the estimate which minimizes $\overline{e_L^2}$ is

$$\mathbf{b}_L = \overline{\mathbf{b}} + \mathbf{C}\mathbf{A}^*(\mathbf{A}\mathbf{C}\mathbf{A}^* + \overline{\epsilon\epsilon^*})^{-1}(\mathbf{d} - \overline{\mathbf{d}}), \qquad (2.4.2)$$

where $\mathbf{C} = \overline{(\mathbf{b} - \overline{\mathbf{b}})(\mathbf{b} - \overline{\mathbf{b}})^*}$. However, (2.4.2) also minimizes the quadratic form

$$\mathcal{J} = (\mathbf{b} - \overline{\mathbf{b}})^*\mathbf{C}^{-1}(\mathbf{b} - \overline{\mathbf{b}}) + (\mathbf{A}\mathbf{b} - \mathbf{d})^*\overline{(\epsilon\epsilon^*)}^{-1}(\mathbf{A}\mathbf{b} - \mathbf{d}). \quad (2.4.3)$$

That is, the linear regression or smoothing formula (2.4.2) also yields the bottom velocity vector \mathbf{b} which is closest to the first-guess $\overline{\mathbf{b}}$, and has the smallest measurement residual $\mathbf{d} - \mathbf{A}\mathbf{b}$, with respective weights \mathbf{C}^{-1} and $\overline{(\epsilon\epsilon^*)}^{-1}$. In this latter context, size is simply measured by Euclidean length, without reference to an ensemble average.

Alternatively, an inner product may be defined by $\langle \mathbf{a}, \mathbf{b} \rangle \equiv \mathbf{a}^*\mathbf{C}^{-1}\mathbf{b}$. Then \mathbf{C} is the reproducing kernel for this inner product and the M representers of the M rows of \mathbf{A} are given by $\mathbf{r}_m = \mathbf{C}\mathbf{A}^*\mathbf{e}_m$, $1 \leq m \leq M$. The quadratic form (2.4.3) may be minimized by paraphrasing the spline smoothing procedure, that is, a solution is sought having the form $\mathbf{b} = \overline{\mathbf{b}} + \sum_{m=1}^{M} \beta_m \mathbf{r}_m + \mathbf{g}$ where $\langle \mathbf{r}_m, \mathbf{g} \rangle = 0$, $1 \leq m \leq M$. Some simple algebra leads directly to (2.4.2).

Fiadeiro & Veronis (1984) argued that Wunsch's method is not useful for estimating a velocity field from tracer measurements (see §1.8), since the inverse estimates should suffer from the same defects as severely truncated Fourier series. That is, the subjective appearance of a sum of representers could be misleading, unless corrected by including elements of the null space. The example of a streamfunction with a uniform gradient (a "ramp") was cited: estimates would in general be oscillatory. In reply, Wunsch (1985) proposed that an examination of the resolution matrix $\widehat{\mathbf{V}}\widehat{\mathbf{V}}^*$ (see §1.5) should warn of a potentially misleading inversion. He also showed that choosing a nondiagonal form for the weight matrix \mathbf{W} (see §1.3) led to a reasonable recovery of the "ramp" streamfunction. In effect, the choice of a nondiagonal weight led to the inclusion of elements of the null space associated with a diagonal form. Recognizing a weight as the reciprocal of a covariance, such ideas are natural extensions of the smoothing techniques described in this chapter.

Exercise 2.4.1 Express the resolution matrix $\widehat{\mathbf{V}}\widehat{\mathbf{V}}^*$ in terms of representers. □

3

DATA ASSIMILATION

3.1 Introduction

Most major ocean currents have dynamics which are significantly nonlinear. This precludes the ready development of inverse methods along the lines described in Chapter 1. Accordingly, most attempts to combine oceans models and measurements have followed the practice in operational meteorology: measurements are used to prepare initial conditions for the model, which is then integrated forward in time until further measurements are available. The model is thereupon re-initialized. Such a strategy may be described as sequential. It is clearly the only choice for prediction (that is, genuine forecasting in real time), but the relative simplicity of the approach has led to its adoption for smoothing as well. In the latter situation, measurements are available in some fixed interval $0 \leq t \leq T$, and a best estimate is required at each time t in the interval.

Sequential estimation techniques have become known to meteorologists and oceanographers as data assimilation. There has been extensive development of data-assimilation methods in meteorology, and it is fortunate for oceanographers in particular that the methods are now comprehensively described in the text by Daley (1991). The meteorological problem is especially difficult. First, the synoptic time scale in middle latitudes is only a few days; thus predictions of the synoptic scale must be prepared within a few hours of receipt of the measurements in order to be of any value. This creates great stress on the computing resource. On the other hand, only some dozens of time steps are needed. Second, vortex stretching due to orography,

vortex bending due to shear, and planetary vorticity variations are
so great that quasi-geostrophic models (Pedlosky, 1987, Chapter 3)
are demonstrably inadequate forecast tools. Indeed the flow is far
from geostrophic balance, with Rossby numbers typically of about
unity. The primitive-equation models (Lorenz, 1967) which must be
used in numerical weather prediction have fine vertical resolution, and
consequently can support a gravity-wave structure far richer than is
needed for synoptic interpretation. A major problem in meteorologi-
cal data assimilation is the preparation of initial conditions which do
not generate an unmanageable field of gravity waves.

The task for oceanographers is simpler in certain respects. First,
while there is some demand for genuine forecasting or prediction, there
is a greater emphasis at present on smoothing over a fixed time interval
which corresponds to the duration of a field experiment, for example.
Moreover, the mid-latitude synoptic time scale in the ocean is several
months, and thus several days could be spent in preparing a forecast.
The stress on computing resources is correspondingly much less than
in the meteorological context, and as a consequence sophisticated
methods may be practical. The relative scarcity of oceanographic data
also alleviates the computational burden, but a surplus of oceano-
graphic data would be far more preferable. A second simplification
for oceanographers is that in many circumstances quasi-geostrophic
dynamics describe ocean circulation with an accuracy consistent with
the density and quality both of measurements of thermal and me-
chanical driving and of measurements of the circulation itself. When
primitive-equation ocean models have been used, either the vertical
resolution is minimal, or else the objective is simulation of general
circulation. Consequently, small-scale gravity waves may be crudely
filtered out without loss of important detail. Nevertheless, as ocean
forecasters become more ambitious, they will have to deal with the
difficult subject of initialization of the gravity-wave field, but as always
will be able to draw upon the advances made in meteorology. On the
other hand, the ocean forecaster will always retain the advantage of
not having to forecast on such short time scales.

The purpose of this chapter is to describe a simple data-
assimilation technique, then describe a sophisticated technique, and
finally review applications of each to real oceanographic forecasting.

3.2 Initialization: theory and examples

Quasi-geostrophic ocean models need only values of streamfunctions as initial conditions. Vorticity may be calculated directly from the streamfunction, and may be updated by integrating the vorticity equation in time subject to certain boundary conditions. The updated streamfunction values may be obtained by solving an elliptic or diagnostic problem also subject to certain boundary conditions. The updated streamfunction may then be revised in light of new data, prior to use as a new initial condition for a subsequent integration. Thus initial conditions are obtained, in general, from a first estimate such as model output or even a climatological mean, which is modified in light of the data. Let the prior estimate or first guess for the initial streamfunction be denoted $\phi^{(-)} = \phi^{(-)}(\mathbf{x})$, where \mathbf{x} may be a point in the plane or in space. Let the data be denoted by d_m, $1 \leq m \leq M$, given by

$$d_m = \mathcal{L}_m[\psi] + \epsilon_m, \qquad (1 \leq m \leq M), \qquad (3.2.1)$$

where $\psi = \psi(\mathbf{x})$ is the unknown true streamfunction, \mathcal{L}_m is a known measurement functional and ϵ_m is an unknown measurement error. A revised initial streamfunction $\phi^{(+)}(\mathbf{x})$ may be constructed as a linear combination of the first estimate $\phi^{(-)}$ and the data d_m:

$$\phi^{(+)}(\mathbf{x}) = \phi^{(-)}(\mathbf{x}) + \sum_{m=1}^{M} w_m(\mathbf{x})(d_m - d_m^{(-)}), \qquad (3.2.2)$$

where $w_m(\mathbf{x})$, $1 \leq m \leq M$, are some suitably chosen interpolants, and $d_m^{(-)}$ is the mth measurement of the first estimate:

$$d_m^{(-)} = \mathcal{L}_m[\phi^{(-)}]. \qquad (3.2.3)$$

It remains to choose the smoothing functions $w_m(\mathbf{x})$. Kindle (1986) chose bilinear interpolation, for which high-quality software may be found in scientific subroutine libraries, and for which the theory is well established (Press *et al.*, 1986). However, the choice for $w_m(\mathbf{x})$ favored almost exclusively by oceanographers is that described in Chapter 2, namely, the choice leading to the minimum-variance or best linear unbiased estimator. In this context it is assumed that the prior estimate $\phi^{(-)}$ is an unbiased estimate of the true field:

$$\overline{\phi^{(-)}}(\mathbf{x}) = \overline{\psi}(\mathbf{x}), \qquad (3.2.4)$$

and that the measurement errors have zero mean:

$$\bar{\epsilon}_m = 0, \qquad (1 \leq m \leq M).$$ (3.2.5)

Hence $\overline{\phi}^{(+)}(\mathbf{x})$ is also unbiased:

$$\overline{\phi}^{(+)}(\mathbf{x}) = \overline{\psi}(\mathbf{x}).$$ (3.2.6)

It is also assumed that the errors in the prior estimate are not corre-
lated with the measurement errors:

$$\overline{(\psi(\mathbf{x}) - \phi^{(-)}(\mathbf{x}))\epsilon_m} = 0, \qquad (1 \leq m \leq M).$$ (3.2.7)

Then the error variance

$$\overline{e^+(\mathbf{x})^2} \equiv \overline{(\psi(\mathbf{x}) - \phi^{(+)}(\mathbf{x}))^2}$$ (3.2.8)

is least, given the prior estimate $\phi^{(-)}(\mathbf{x})$ and the data d_m, if

$$\phi^{(+)}(\mathbf{x}_1) = \phi^{(-)}(\mathbf{x}_1) + \mathcal{L}^*_{(\mathbf{x}_2)}\Big[C^{(-)}(\mathbf{x}_1,\mathbf{x}_2)\Big]\Big\{\mathcal{L}_{(\mathbf{x}_3)}\mathcal{L}^*_{(\mathbf{x}_4)}\Big[C^{(-)}(\mathbf{x}_3,\mathbf{x}_4)\Big]$$

$$+\ \overline{\epsilon\epsilon^*}\Big\}^{-1}(\mathbf{d} - \mathbf{d}^{(-)}),$$ (3.2.9)

where \mathcal{L}, \mathbf{d} and ϵ are the vectors of measurement functionals, data
and data errors respectively, while $C^{(-)}$ is the error covariance for the
prior estimate:

$$C^{(-)}(\mathbf{x}_1,\mathbf{x}_2) \equiv \overline{(\psi(\mathbf{x}_1) - \phi^{(-)}(\mathbf{x}_1))(\psi(\mathbf{x}_2) - \phi^{(-)}(\mathbf{x}_2))}.$$ (3.2.10)

The error covariance for $\phi^{(+)}(\mathbf{x})$ is

$$C^{(+)}(\mathbf{x}_1,\mathbf{x}_2) \equiv \overline{(\psi(\mathbf{x}_1) - \phi^{(+)}(\mathbf{x}_1))(\psi(\mathbf{x}_2) - \phi^{(+)}(\mathbf{x}_2))}$$ (3.2.11)

which yields

$$C^{(+)}_{12} = C^{(-)}_{12} - \mathcal{L}^*_3\Big[C^{(-)}_{13}\Big]\Big\{\mathcal{L}_4\mathcal{L}^*_5\Big[C^{(-)}_{45}\Big] + \overline{\epsilon\epsilon^*}\Big\}^{-1}\mathcal{L}_6\Big[C^{(-)}_{62}\Big],$$

(3.2.12)

using an obvious shorthand notation. The unbiased estimate $\phi^{(+)}$ defined by (3.2.9)–(3.2.11) is identical in form to the unbiased estimate ψ_L defined by (2.2.16)–(2.2.17), with $\phi^{(-)}(\mathbf{x})$ replacing $\overline{\psi}(\mathbf{x})$. That is, a prior estimate assumed to be unbiased replaces the true mean field. Further, the prior *error* covariance $C^{(-)}(\mathbf{x}_1, \mathbf{x}_2)$ replaces the true field covariance $C(\mathbf{x}_1, \mathbf{x}_2)$. As mentioned, the choice of interpretation depends upon context. If only a climatological estimate is available for $\overline{\psi}$ then it is natural to identify the estimate with $\overline{\psi}$. Model output containing transient patterns should be identified with $\phi^{(-)}$. Equating a climatological mean with the true mean is as problematic as assuming that a prior estimate is unbiased. A resolution of the problem is seemingly achieved by assuming that the average $\overline{(\)}$ is conditional upon having a given $\phi^{(-)}(\mathbf{x})$ as a prior estimate: see Thiébaux & Pedder (1987), and Daley (1991). However, it would be difficult to estimate the conditional prior error covariance.

Marshall (1985) applied minimum-variance initialization to simulated altimeter data. The problem was compounded by the need to distinguish between the ocean surface and the geopotential surfaces. Malanotte-Rizzoli & Holland (1986, 1988) used (3.2.9) in a simplified form for the assimilation of simulated hydrographic data into a multi-level quasi-geostrophic model. Moore, Cooper & Anderson (1987) used the method of successive corrections (Bergthorsson & Döös, 1955) to assimilate simulated expendable bathythermograph (XBT) measurements collected from ships of opportunity (SOPs) during the Tropical-Oceans Global-Atmosphere Experiment (TOGA). That is, the smoothing formula (3.2.9) was used repeatedly at each initialization, with $\phi^{(+)}$ replacing $\phi^{(-)}$ after the first smoothing. Successively smaller correlation scales were assumed for $C^{(-)}(\mathbf{x}, \mathbf{y})$ in each repetition or "scan." The results were used as (partial) initial conditions for a primitive-equation model of the Indian Ocean.

In a long series of studies, Robinson and co-workers have developed practical data-assimilation methods for use with quasi-geostrophic and primitive-equation models of mesoscale circulation, in several parts of the ocean. The first open-ocean model development may be found in Robinson & Haidvogel (1980), Haidvogel, Robinson & Schulman (1980), and Miller, Robinson & Haidvogel (1983). Applications to the California Current system are described in Mooers & Robinson (1984), Robinson et al. (1984), Smith, Mooers & Robinson (1985), Robinson et al. (1986), and Rienecker, Mooers & Robinson (1987). A remarkable aspect of the 1984 study in particular was

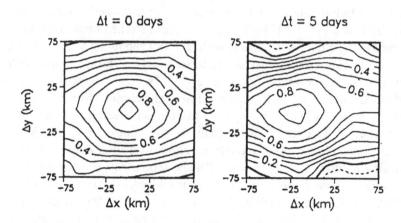

Figure 3.2.1. The space-time covariance functions estimated from surface dynamic height (after Rienecker, *et al.*, 1987).

that forecasts were prepared in real time, the measurements being transferred to computers on shore from ships at sea. Robinson *et al.* (1986) reported several numerical experiments with a multilevel quasi-geostrophic model, and various boundary data. Reasonably accurate forecasts over 14 days were obtained with initial conditions derived by objective analysis of hydrographic data, and with persistent boundary data. That is, the fields appearing in the boundary conditions remained at their initial values during the forecast interval. The forms of these boundary conditions will be discussed in detail in later chapters. Better "forecasts" were obtained with time-continuous boundary data derived by linear interpolation between hydrographic data collected on successive cruises spaced about two weeks apart. These latter "forecasts" are more appropriately described as hindcasts. It may be mentioned that initialization was carried out assuming homogeneous and isotropic statistics, but the data did not show significant anisotropy. Cruise duration was ignored, that is, data collected on a single cruise were assumed to be synoptic. Further experiments with a larger data set are described in Rienecker *et al.* (1987). A region in the California Current system was surveyed hydrographically six times over a two-month period. Some improvement in hindcast skill was gained by allowing for the non-synoptic nature of the data used for initialization. The estimated space-time covariance function for surface dynamic height is shown in Fig. 3.2.1. Some true forecasts were performed, the necessary time-dependent boundary data being

Figure 3.2.2. Nondimensional streamfunction at 50 m resulting from initialization at day 177; (a) statistical forecasts, i.e., statistical extrapolation from day 177 using the time-dependent covariance function shown in Fig.3.2.1; (b) dynamical forecasts using persistent boundary data; (c) dynamical forecasts using statistically forecast boundary data; (d) as for (c) but with assimilation of synoptic data on every other day from day 187 onwards; (e) synoptioc estimates for days 187 and 193: the verifying analysis for (a)-(d) (after Rienecker et al., 1987).

estimated by linear statistical extrapolation using the sample space-time covariance. However, these forecasts were not as accurate as hindcasts using linearly interpolated boundary data (see Fig. 3.2.2). Satellite-derived sea-surface temperature measurements were found at times to be a useful proxy for direct measurements of surface dynamic height. In a more recent study (Rienecker & Miller, 1991), paucity of data continued to be a limiting factor.

Initialization of quasi-geostrophic models of the Gulf Stream is described in Robinson & Leslie (1985), Robinson (1987), and Pinardi & Robinson (1987). The techniques used were essentially those found in the studies of the California Current system, namely, Gauss-Markov smoothing of hydrographic data to provide initial conditions, combined with either persistent, linearly interpolated or statistically extrapolated boundary data.

A quite different approach to initialization was employed by Robinson, Spall & Pinardi (1988). The region of interest was the Gulf Stream, and the model was, again, multilevel quasi-geostrophic. The data were sea-surface temperature (SST), remotely sensed by infrared satellite sensors. However, the initial streamfunction was not obtained by smoothing the data. Instead the description of the field was lumped into a few relatively simple features: a meandering jet (the Gulf Stream), plus a few cold-core or warm-core rings (see Fig. 3.2.3). Simple analytical expressions described the downstream or the swirl velocities in these "feature models." The locations of these features, and the parameters in their flow fields, were determined empirically using the SST data. An example of an initialized streamfunction is shown in Fig. 3.2.4. This so-called feature model method of initialization is intuitively appealing when it is believed that sharp gradients are present in the flow. Obviously a smoothing technique would be particularly ineffective at locating these gradients and estimating their strengths. From a theoretical point of view, the statistics of the circulation and water properties in such a region would be significantly non-Gaussian, to the point almost of being bimodal (slope water versus Sargasso water). Consequently a least-squares estimator such as (3.2.4) would not be the estimator of maximum likelihood; in particular a range of plus or minus one standard deviation would not be a 67% confidence interval. On the other hand, if a numerical model is to be initialized with features possessing strong gradients, then the numerical integration methods and resolution must be chosen accordingly, or else the accuracy of location and the strength of the

Figure 3.2.3. Feature model profiles for Gulf Stream thin jet model: (a) three-dimensional schematic of feature model initialization; (b) along-stream velocity profile as a function of the cross-stream coordinate and depth; (c) horizontal and vertical velocity structure for cold-ring model "a"; R_{max} is the horizontal maximum extension of the ring, and V_{max} is the maximum velocity reached at R_0. Z_{max} is the maximum vertical extension of the ring; (d) same but for cold-ring model "b"; (e) same but for warm-ring model "A"; (f) same but for warm-ring model "B" and "C" (after Robinson, *et al.*, 1988).

gradients will be rapidly lost. Feature models were used by Spall & Robinson (1990) to initialize a regional primitive-equation (PE) model of the Gulf Stream. Large-scale meanders grew less rapidly in the PE

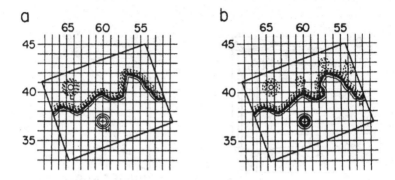

Figure 3.2.4. Initialized geostrophic streamfunction fields at 100 m, for two experiments (a) and (b) (after Robinson, *et al.*, 1988).

model, on monthly time scales, if a quasi-geostrophic model was used for the first three days of simulation.

A novel procedure for locating feature models has been proposed by Mariano (1990). The approach is hodographic: the locations of streamfunction contours are interpolated, as opposed to interpolating streamfunction values.

This very brief outline of the progress of Robinson and co-workers does no justice to the scientific value of their work, such as the diagnostic studies of open-ocean energetics and the synoptic analyses of events including meandering, ring shedding and ring coalescence. The interest here is exclusively in terms of data-assimilation techniques. It should be mentioned that the calculations, especially for the real-time forecasts, required the construction of data-management schemes approaching the complexity of those found in operational numerical weather prediction. An example of an organizational chart may be found in Robinson (1987).

3.3 The Kalman filter

The simple data-assimilation technique described in the previous section requires a prior error covariance $C^{(-)}(\mathbf{x}_1, \mathbf{x}_2)$, associated with the prior streamfunction estimate $\phi^{(-)}(\mathbf{x})$. A posterior error covariance $C^{(+)}(\mathbf{x}_1, \mathbf{x}_2)$ is derived for the posterior estimate $\phi^{(+)}(\mathbf{x})$. The latter estimate is related to $\phi^{(-)}(\mathbf{x})$ and the data \mathbf{d} by the linear smoothing formula (3.2.9). In the applications discussed thus far, $\phi^{(+)}$

has been used as initial data for an integration of a quasi-geostrophic model, forward in time. As new data became available, they were smoothed using the model forecast as $\phi^{(-)}$, and using a prescribed $C^{(-)}(\mathbf{x}_1, \mathbf{x}_2)$. This practice is inconsistent if the error statistics are nonstationary. An evolution equation for the covariance may be derived from the ocean dynamics, with the previous posterior error covariance $C^{(+)}$ being the logical initial value. If the corresponding solution is nonstationary, it offers an alternative to $C^{(-)}$, consistent with the model dynamics. The evolution equation for the covariance is obtained by taking moments of the evolution equation for the state, that is, moments of the ocean model. If the model dynamics are linear, then the covariance equation will be closed, in the sense that it does not involve higher-order moments.

Consider therefore the simple, linear single-layer quasi-geostrophic model

$$\psi_t = L\psi + F + q \,, \tag{3.3.1}$$

where $\psi = \psi(\mathbf{x}, t)$ is the streamfunction depending upon time t and horizontal position $\mathbf{x} = (x, y)$; $F = F(\mathbf{x}, t)$ is an estimate of the external forcing; $q = q(\mathbf{x}, t)$ is the unknown error in the forcing estimate and L is an operator which may be expressed as

$$L = -\mathbf{U} \cdot \mathbf{\nabla} - \beta E^{-1} \frac{\partial}{\partial x} - D_0 + A_H \nabla^2 \,. \tag{3.3.2}$$

In (3.3.2), β is the northward gradient of the Coriolis parameter; \mathbf{U} is a known uniform advecting velocity; D_0 is a non-negative linear or bottom drag coefficient; A_H is a non-negative eddy diffusivity and E is the elliptic operator which relates streamfunction ψ and vorticity ζ:

$$\zeta = E\psi = (\nabla^2 - \rho^{-2})\psi \,, \tag{3.3.3}$$

where ρ is an internal deformation radius. Several alternatives for boundary conditions may be considered, such as

 i) all fields periodic in x and in y;

 ii) ($A_H = 0$) rigid boundary conditions, that is, ψ vanishing on the boundary, or

 iii) ($A_H > 0$) rigid no-slip boundary conditions, with ψ and its normal derivative vanishing on the boundary.

In cases (ii) and (iii) it will be assumed that the normal component of \mathbf{U} vanishes. Nonlinear, open boundary conditions will be considered in subsequent chapters.

Let $\phi(\mathbf{x}, t)$ be the response of the system to the forcing estimate F:

$$\phi_t = L\phi + F\,, \qquad\qquad (3.3.4)$$

subject to the following initial condition at $t = T_k$:

$$\phi(\mathbf{x}, T_k) = \phi^{(k+)}(\mathbf{x})\,. \qquad\qquad (3.3.5)$$

If $\phi^{(k+)}$ is an unbiased estimate of the true streamfunction ψ at $t = T_k$, and if F is an unbiased estimate of the forcing (that is, $\overline{q}(\mathbf{x}, t) = 0$), then ϕ will be an unbiased estimate of ψ for $t \geq T_k$, and the error $\phi(\mathbf{x}, t)' = \psi(\mathbf{x}, t) - \phi(\mathbf{x}, t)$ will satisfy

$$\phi_t' = L\phi' + q\,. \qquad\qquad (3.3.6)$$

The error covariance is

$$C(\mathbf{x}_1, \mathbf{x}_2, t) = \overline{\phi(\mathbf{x}_1, t)'\phi(\mathbf{x}_2, t)'}\,. \qquad\qquad (3.3.7)$$

Taking moments of (3.3.6) yields

$$C_t(\mathbf{x}_1, \mathbf{x}_2, t) = \left(L_{(\mathbf{x}_1)} + L_{(\mathbf{x}_2)}\right) C(\mathbf{x}_1, \mathbf{x}_2, t)$$
$$+ \overline{\phi(\mathbf{x}_1, t)'q(\mathbf{x}_2, t)} + \overline{\phi(\mathbf{x}_2, t)'q(\mathbf{x}_1, t)}\,. \quad (3.3.8)$$

It will now be assumed that, as a time series, $q(\mathbf{x}, t)$ is uncorrelated, stationary white noise:

$$\overline{q(\mathbf{x}_1, t_1)q(\mathbf{x}_2, t_2)} = Q(\mathbf{x}_1, \mathbf{x}_2)\,\delta(t_1 - t_2)\,. \qquad\qquad (3.3.9)$$

It will also be assumed that the forcing error q is not correlated with the initial error:

$$\overline{q(\mathbf{x}_1, t_1)(\psi(\mathbf{x}_2, T_k) - \phi^{(k+)}(\mathbf{x}_2))} = 0\,. \qquad\qquad (3.3.10)$$

Then for the purposes of evaluating the covariances of ϕ and q appearing in (3.3.8), it suffices to approximate the solution of (3.3.6) by

$$\phi(\mathbf{x}, t) \cong \int_{t-0}^{t} q(\mathbf{x}, t_1)\,dt_1\,. \qquad\qquad (3.3.11)$$

Since $\int_0^t \delta(t - t_1) dt_1 = \frac{1}{2}$, (3.3.8) becomes

$$C_t(\mathbf{x}_1, \mathbf{x}_2, t) = (L_{(\mathbf{x}_1)} + L_{(\mathbf{x}_2)}) C(\mathbf{x}_1, \mathbf{x}_2, t) + Q(\mathbf{x}_1, \mathbf{x}_2). \qquad (3.3.12)$$

The appropriate initial condition at $t = T_k$ is

$$\begin{aligned} C(\mathbf{x}_1, \mathbf{x}_2, T_k) &= \overline{(\psi(\mathbf{x}_1, T_k) - \phi^{(k+)}(\mathbf{x}_1))(\psi(\mathbf{x}_2, T_k) - \phi^{(k+)}(\mathbf{x}_2))} \\ &= C^{(k+)}(\mathbf{x}_1, \mathbf{x}_2). \qquad (3.3.13) \end{aligned}$$

The initial streamfunction estimate $\phi^{(k+)}$ with error covariance $C^{(k+)}$ may now be considered. The estimate is assumed to be the result of an initialization procedure such as described in §3.2, that is,

$$\phi^{(k+)}(\mathbf{x}) = \phi^{(k-)}(\mathbf{x}) + \mathbf{K}^{(k)}(\mathbf{x})^* (\mathbf{d}^{(k)} - \mathbf{d}^{(k-)}), \qquad (3.3.14)$$

where

$$\begin{aligned} \mathbf{K}^{(k)}(\mathbf{x}_1) = \{\mathcal{L}_{(\mathbf{x}_3)} \mathcal{L}_{(\mathbf{x}_4)}^* [C^{(k-)}(\mathbf{x}_3, \mathbf{x}_4)] + \overline{\epsilon^k \epsilon^{k*}}\}^{-1} \\ \times \mathcal{L}_{(\mathbf{x}_2)} [C^{(k-)}(\mathbf{x}_1, \mathbf{x}_2)] \qquad (3.3.15) \end{aligned}$$

The measurement functionals \mathcal{L}, data $\mathbf{d}^{(k)}$ and measurement errors ϵ^k at time T_k are related by

$$\mathcal{L}[\psi(\mathbf{x}, T_k)] + \epsilon^k = \mathbf{d}^{(k)}; \qquad (3.3.16)$$

the measurements of the prior streamfunction estimate $\phi^{(k-)}(\mathbf{x})$ are

$$\mathcal{L}[\phi^{(k-)}(\mathbf{x})] = \mathbf{d}^{(k-)}. \qquad (3.3.17)$$

The posterior and prior values of C at time T_k are related by

$$\begin{aligned} C^{(k+)}(\mathbf{x}_1, \mathbf{x}_2) = C^{(k-)}(\mathbf{x}_1, \mathbf{x}_2) \\ - \mathbf{K}^{(k)}(\mathbf{x}_1)^* \mathcal{L}_{(\mathbf{x}_3)} [C^{(k-)}(\mathbf{x}_3, \mathbf{x}_2)]. \qquad (3.3.18) \end{aligned}$$

If no initialization is performed at time $t = T_0$, say, then the initial conditions for (3.3.4) and (3.3.12) are simply

$$\phi(\mathbf{x}, T_0) = \psi^{(0)}(\mathbf{x}), \qquad (3.3.19)$$

$$C(\mathbf{x}_1, \mathbf{x}_2, T_0) = A(\mathbf{x}_1, \mathbf{x}_2), \qquad (3.3.20)$$

where

$$A(\mathbf{x}_1, \mathbf{x}_2) = \overline{a(\mathbf{x}_1)a(\mathbf{x}_2)}, \qquad\qquad (3.3.21)$$

$$a(\mathbf{x}) = \psi(\mathbf{x}, T_0) - \psi^{(0)}(\mathbf{x}). \qquad\qquad (3.3.22)$$

It is assumed that $\psi^{(0)}$ is an unbiased estimate of ψ at $t = T_0$, that is, $\overline{a} = 0$.

In summary, the preceding results yield the following algorithm.

1) $(T_0 \le t < T_1)$ Integrate (3.3.4) and (3.3.12) forward in time, from the initial values (3.3.19) and (3.3.20). Let the forecasts of streamfunction and error covariance at time T_1 serve as the prior estimates $\phi^{(1-)}$ and $C^{(1-)}$ respectively.

2) Obtain posterior estimates $\phi^{(1+)}$ and $C^{(1+)}$ using (3.3.14)–(3.3.18), that is, by smoothing the data $\mathbf{d}^{(1)}$.

3) $(T_1 \le t \le T_2)$ Integrate (3.3.4) and (3.3.12) forward in time, from the initial values $\phi^{(1+)}$ and $C^{(1+)}$ respectively. Let the forecasts at time T_2 serve as the prior estimates $\phi^{(2-)}$ and $C^{(2-)}$.

4) Obtain $\phi^{(2+)}$ and $C^{(2+)}$ as in step (2), and so on.

3.4 Remarks and exercises

1) The algorithm presented in §3.3 is due to Kalman (1960). It is known as the Kalman filter for the model (3.3.1) and measurements (3.2.1). The vector field appearing in (3.3.14) and having the same dimension as \mathcal{L} is known as the Kalman gain. The errors $q(\mathbf{x}, t)$, ϵ^k and $a(\mathbf{x})$ are known as the system noise, data noise and initial noise, respectively.

2) It is usually assumed that q, ϵ^k and a are uncorrelated, thereby assuring that q is not correlated with $\phi^{(k+)'}$. Recall also the stationary white-noise assumption (3.3.9) for the system noise q. The assumption of stationarity may be relaxed, as may the others, but at the expense of greatly complicating an already intricate integration and initialization procedure, and requiring further statistical information.

3) The fields $\phi(\mathbf{x}, t)$, $C(\mathbf{x}, \cdot, t)$ and $C(\cdot, \mathbf{x}, t)$ all satisfy the same boundary conditions as $\psi(\mathbf{x}, t)$ in this example.

Exercise 3.4.1 Show that if A and Q are symmetric in their spatial arguments, then C is symmetric. □

Exercise 3.4.2 Show that if A and Q are positive-definite with respect to their spatial arguments, then C is positive-definite for all $t \geq T_0$. Hint: show that $E_{(\mathbf{x}_1)}E_{(\mathbf{x}_2)}C(\mathbf{x}_1, \mathbf{x}_2, t)$ is positive-definite, where the elliptic operator E is defined by (3.3.3). □

4) The Kalman filter (KF) estimate ϕ for the streamfunction ψ at time $t > T_{k-1}$ depends implicitly upon prior data $\mathbf{d}^{k-1}, \ldots, \mathbf{d}^1$. However, upon the arrival of new data at time T_k, a new estimate $\phi^{(k+)}(\mathbf{x})$ for $\psi(\mathbf{x}, T_k)$ is prepared without revising previous estimates $\phi(\mathbf{x}, t)$ for $t < T_k$ and without explicitly using the prior data. This is a consequence of assuming that the system noise q is white in time. Thus the KF algorithm is a sequential data-assimilation scheme. The terminology "data assimilation" will be reserved henceforth for applications of sequential schemes.

5) Not all measurements may be available at every measurement time. Say datum d_n^k is not available at time T_k, for some n: $1 \leq n \leq M$. This case may be included within (3.3.15) simply by assuming that a worthless datum is available:

$$\overline{\epsilon_n^k \, \epsilon_m^k} = \infty, \qquad (1 \leq m \leq N). \tag{3.4.1}$$

6) In a numerical integration of (3.3.4), the field $\phi(\mathbf{x}, t)$ is replaced by a finite number ν^2 of unknowns at each integration time level. These may be values on a grid with spacing $\propto \nu^{-1}$ or Fourier coefficients cut off at wavenumbers $\propto \nu$, for example. There will be $\frac{1}{2}\nu^2(\nu^2 - 1)$ values in the approximate representation of the error covariance $C(\mathbf{x}_1, \mathbf{x}_2, t)$ at each time level, all of which must be updated. Several copies of this array may be needed in any multilevel integration scheme. Thus severe storage and computational burdens arise for all but the crudest resolution. For this reason, the KF has not come into widespread use in operational meteorology (Lorenc, 1986).

Only the Kalman gain $\mathbf{K}^{(k)}(\mathbf{x})$ appears in the jump condition (3.3.14) for the streamfunction or estimate ϕ. The gain only requires $N\nu^2$ words of storage, which is assuredly much less than $\frac{1}{2}\nu^2(\nu^2 - 1)$. It would, therefore, greatly reduce the computational burden if the error covariance $C(\mathbf{x}_1, \mathbf{x}_2, t)$ were not required. For many purposes,

the error variance $C(\mathbf{x}, \mathbf{x}, t)$ would suffice. However, it is not possible to calculate $\mathbf{K}^{(k)}(\mathbf{x})$ or $C(\mathbf{x}, \mathbf{x}, t)$ without also calculating $C(\mathbf{x}_1, \mathbf{x}_2, t)$ for all \mathbf{x}_1 and \mathbf{x}_2.

Following arguments of Phillips (1986), Cohn & Parrish (1991) reduced the computational burden of a KF severely by truncating the system noise spectrum. Their experiments with a linear atmospheric forecast model and synthetic meteorological data for the region of the United States indicate that the KF yields superior results to conventional forecasts initialized by objective analysis. In the latter, it is recalled, the error covariance is merely prescribed.

7) The KF makes a null estimate for the system noise q between measurement times: $q = 0$, $T_k < t < T_{k+1}$. See (3.3.4).

Exercise 3.4.3 The Kalman-Bucy filter. Suppose that data are available continuously in time:

$$\mathbf{d}(t) = \mathcal{L}[\psi(\mathbf{x}, t)] + \epsilon(t), \qquad (3.4.2)$$

where the data errors $\epsilon(t)$ are white-noise time series:

$$\overline{\epsilon(t_1)\epsilon(t_2)^*} = \mathbf{w}(t_1)^{-1}\delta(t_1 - t_2) \qquad (3.4.3)$$

for some positive-definite symmetric matrix $\mathbf{w}(t)$. Now an unbiased estimate $\phi(\mathbf{x}, t)$ of $\psi(\mathbf{x}, t)$ is sought, satisfying

$$\phi_t(\mathbf{x}, t) = L\phi(\mathbf{x}, t) + F(\mathbf{x}, t) + \mathbf{b}(\mathbf{x}, t)^*(\mathbf{d}(t) - \mathcal{L}_{(\mathbf{x}_1)}[\phi(\mathbf{x}_1, t)]) \quad (3.4.4)$$

for some coefficient vector field $\mathbf{b}(\mathbf{x}, t)$, subject to the initial condition (3.3.19). That is, an unbiased estimate is sought for the system noise $q(t)$ in (3.3.1), using the data $\mathbf{d}(t)$ and measurements of the current state: $\mathcal{L}[\phi]$. Show that the growth rate of the error variance $C(\mathbf{x}, \mathbf{x}, t)$ is least if \mathbf{b} is given by

$$\mathbf{b}(\mathbf{x}_1, t) = \mathbf{K}(\mathbf{x}_1, t) = \mathbf{w}\mathcal{L}_{(\mathbf{x}_2)}[C(\mathbf{x}_1, \mathbf{x}_2, t)], \qquad (3.4.5)$$

where the error covariance $C(\mathbf{x}_1, \mathbf{x}_2, t)$ satisfies

$$\begin{aligned}
C_t(\mathbf{x}_1, \mathbf{x}_2, t) = {} & (L_{(\mathbf{x}_1)} + L_{(\mathbf{x}_2)})\, C(\mathbf{x}_1, \mathbf{x}_2, t) \\
& - \mathcal{L}^*_{(\mathbf{x}_3)}[C(\mathbf{x}_1, \mathbf{x}_3, t)]\mathbf{w}\mathcal{L}_{(\mathbf{x}_4)}[C(\mathbf{x}_4, \mathbf{x}_2, t)] \\
& + Q(\mathbf{x}_1, \mathbf{x}_2),
\end{aligned} \qquad (3.4.6)$$

subject to the initial condition (3.3.20). Recall (3.3.9) and (3.3.21). Assume that

$$\overline{\epsilon(t_1)q(\mathbf{x}, t_2)} = \overline{\epsilon(t_1)a(\mathbf{x})} = \mathbf{0} \qquad (3.4.7)$$

and

$$\overline{q(\mathbf{x}_1, t)a(\mathbf{x}_2)} = 0. \qquad (3.4.8)$$

□

The system (3.4.4)–(3.4.6) for $t \geq T_0$, subject to the initial conditions (3.3.19) and (3.3.20) at $t = T_0$, is due to Kalman & Bucy (1961) and is known as the Kalman-Bucy filter (KBF) for the model (3.3.1) and measurements (3.4.2). If the measurement times for the KF are close, say $T_k = k\Delta T$ for a small time interval ΔT, and if a time-continuous data noise variance is defined to have the values

$$\mathbf{w}(T_k)^{-1} = \Delta T \,\overline{\epsilon^k (\epsilon^k)^*} \qquad (3.4.9)$$

at measurement times, then the KF approximates a KBF, with

$$\mathbf{K}(\mathbf{x}, T_k) \simeq \mathbf{K}^{(k)}(\mathbf{x}). \qquad (3.4.10)$$

The KBF is a sequential scheme. In practice, computation of the quadratic term in the Riccati equation (3.4.6) is extremely time-consuming.

8) The optimal choice for the coefficient $\mathbf{b}(\mathbf{x}, t)$ appearing in (3.4.4) is the Kalman-Bucy gain $\mathbf{K}(\mathbf{x}, t)$ given by (3.4.5). Once again, this choice for \mathbf{b} minimizes the growth rate of the error variance $C(\mathbf{x}, \mathbf{x}, t)$. It is necessary to calculate the error covariance $C(\mathbf{x}_1, \mathbf{x}_2, t)$ in order to determine $\mathbf{K}(\mathbf{x}, t)$, but integration of (3.4.6) has been impractical for all but the crudest numerical approximations. Moreover, the KF and KBF are based on linear models such as (3.3.1), while virtually all realistic ocean models are nonlinear due to advection. A practical alternative to the KBF is simply to guess the smoothing coefficient \mathbf{b} in (3.4.4). The signs of the components of \mathbf{b} should be chosen such that ϕ tends to relax towards the solution for which $\mathcal{L}[\phi] = \mathbf{d}$. This simple scheme, known as "nudging," is sequential and immediately extendable to nonlinear models. An evolution equation for

the error covariance of the "nudged" solution may be derived
for linear models, but integration of the equation and storage of
the covariance are as large tasks as for the KBF. Just as in the
KBF, the nudging error variance cannot be calculated without
first calculating the covariance. Some examples of nudging will
be described in the next chapter.

3.5 Examples of Kalman filters

Kalman filters have been constructed for several prototype ocean
models. Barbieri & Schopf (1982) considered reduced-gravity, equa-
torial Rossby waves (Pedlosky, 1987, Chapter 8) consisting of sin-
gle meridional modes having periodic structure in the zonal direc-
tion. Synthetic data were used, in the form of point measurements
of meridional velocity. Good agreement was found between a sample
covariance for the errors in the filter estimate of the state and the
filter estimate of the error covariance, provided the actual system
noise statistics were consistent with those assumed in the construction
of the filter. The system noise was in fact generated by the use of
an unstable numerical integration scheme. Miller (1986) considered
uniformly advected, barotropic, subtropical Rossby waves (Pedlosky,
1987, Chapter 3) without meridional structure, but having periodic
structure in the zonal direction. Point measurements of either stream-
function or meridional velocity were synthesized. Again, an unstable
numerical integration scheme was used, but explicit system noise
was also introduced in simulations. Experiments were carried out
with open boundary conditions (see Chapter 8). It was found that
without data assimilation, errors grew rapidly. This was attributed
to computational instability, which was in turn regarded as a proxy
for barotropic and baroclinic instability. However, assimilating data
using the Kalman filter kept the error small. Budgell (1987) consid-
ered shallow-water waves in a straight channel of non-uniform cross
section, the waves being uniform across the channel. The system noise
was assumed to be correlated in space; thus its covariance matrix
was effectively rank-deficient. As a consequence, the computational
burden of calculating the filter error covariance matrix was greatly
reduced by the use of a Chandrasekhar-type square-root algorithm.
Indeed, for N state variables (after gridding in space), the largest
matrices which needed to be stored were not of size $N \times N$ but only

$N \times M$, where M is the rank of the $N \times N$ system noise covariance matrix.

Miller & Cane (1989) assimilated real sea-level data into a simple model of equatorial circulation, using the Kalman filter. The model consisted of the linear, reduced-gravity shallow-water equations on the equatorial β-plane, subject to the long-wave approximation (neglect of local meridional accelerations). The meridional structure was represented by a finite sum of meridional modes (Hermite functions); meridional boundaries were assumed at longitudes 120°E and 80°W. The eastern boundary was assumed rigid, but no such condition could be imposed at the western boundary due to the long-wave approximation, which reduced the differential order of the dynamics in the zonal direction. Instead, a condition of vanishing zonal transport was imposed. Without the long-wave approximation, the dynamics allow a very narrow viscous boundary layer at the western boundary. The flow in the layer is approximately non-divergent, so equatorially trapped zonal flow towards a rigid western boundary must carry no transport (Cane & Sarachik, 1981). Miller & Cane included no dissipation in their long-wave dynamics, but their system was able to lose energy by working at the western boundary, which was in effect the outer edge of the viscous boundary layer. Their model was forced by real wind data obtained by subjective analyses of ship observations (Goldenberg & O'Brien, 1981). The error statistics were probably highly inhomogeneous in space, but as a consequence of a lack of data, the system noise was assumed to be homogeneous. A Gaussian shape was adopted by Miller & Cane for the system noise covariance, with principal axes oriented east-west and north-south. The covariance was anisotropic, with a zonal correlation scale of 10° longitude and a meridional scale of 2° latitude. The amplitude was consistent with an rms wind error of about 2 m s^{-1}, which was believed to be appropriate for the data set. Miller & Cane took sea-level data from island tide gauges (Wyrtki *et al.*, 1988) at various islands in the Pacific: see Fig. 3.5.1. The signals of interest were anomalies in the monthly mean sea level, relative to the climatological annual cycle. These surface anomalies should be manifestations of thermocline displacements of opposite sign (and much greater amplitude) associated with El Niño-Southern Oscillation events (Philander, 1990). The data are believed to have rms errors of about 3 cm, and were assumed to be uncorrelated both in time and between islands. The locations of the latter are shown in Fig. 3.5.1. The island data are shown in Fig. 3.5.2, along

Figure 3.5.1. The tropical Pacific, showing locations of tide gauges from which data are drawn for assimilation (after Miller & Cane, 1989).

Figure 3.5.2. Comparison of observations (*) of monthly mean sea-level anomaly, model forecast (thin line) and Kalman filter forecast (thick line), in centimeters (after Miller & Cane, 1989).

with Miller & Cane's model forecast and also their Kalman filter forecast. The filter clearly tracked the data accurately at these islands, including the 1982 El Niño event in particular. Such a result is to be expected. The rms anomaly at Christmas Island, for example, is 10 cm, so the relative error variance in the data is $(3/10)^2 = 0.09$. The rms wind speed is about 5 m s^{-1}, so the relative error variance in the wind speed is $(2/5)^2 = 0.16$. Moreover, the model is forced by pseudo-stresses which are quadratic in the wind speed, so the relative variance of the system noise is 0.32 or about three times the relative variance of the measurement noise. The model and filter forecasts were also compared at four other islands where data were available, but not included in the assimilation scheme: see Fig. 3.5.3. In effect, these data were assigned infinite error variance. The filter tracked

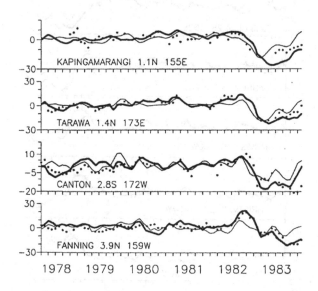

Figure 3.5.3. Comparison of observations (*) of monthly mean sea-level anomaly, model forecast (thin line) and Kalman filter forecast at islands not participating in data assimilation, in centimeters (after Miller & Cane, 1989).

the 1982 El Niño event best at Fanning Island, owing to data at nearby Christmas Island. This influence was made clear in maps of the Kalman gain for each contributing island: see Fig. 3.5.4.

The application of Kalman filtering to sea-level observations derived from satellites (see Chapter 4) is attracting growing interest (Miller, 1989; Fu, Vazquez & Perigaud, 1991).

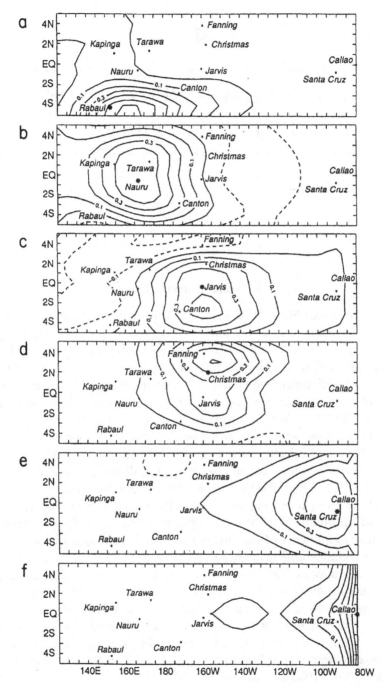

Figure 3.5.4. Contour maps of influence of data in centimeters at tide gauge stations marked (*). Dashed lines represent negative values (after Miller & Cane, 1989).

4

THE SPATIAL STRUCTURE OF
THE KALMAN FILTER

4.1 Spatial structure

The Kalman filter (KF) was originally developed for simple dynamical systems, such as the few ordinary differential equations representing the motion of a projectile. Ocean models are represented by partial differential equations, which are equivalent to infinite systems of ordinary differential equations. These are known as "distributed parameter systems" in the engineering literature (e.g., Aziz, Wingate & Balas, 1977). There is a question of convergence, equivalent to determining whether the KF estimate of the state (here, the ocean circulation) is physically realizable. Unsatisfactory estimates are obtained even at modest spatial resolution if realizability conditions are not met. Typically, the estimates are strongly influenced by the data only in the immediate neighborhood of the measurement site. In that case, the approximate estimation procedure has served little purpose. The range of influence of the data is correctly determined by the scales of the dynamics and also those of the system noise covariance. A small range may be the consequence of realistic choices for the scales, in which case even exact KF estimation would serve little purpose. The objective of this section is to analyze the relationship between the range of influence of the data (that is, the spatial scales of the Kalman gain) and the scales of the dynamics, the system noise, and, to a lesser extent, the initial noise. The discussion follows Bennett & Budgell (1987, 1989).

To expedite the analysis of scales, the single-layer quasi-geostrophic model (3.3.1) will be used in conjunction with the choice (i) for boundary conditions, namely, periodicity of all fields in the x- and y-directions. Specifically, let the period be 2π in each direction, so the domain is $-\pi < x, y < \pi$. Thus (3.3.1) should be regarded as being in dimensionless form. The scale analysis is made explicit if the fields are replaced by their Fourier series expansions:

$$\psi(x, y, t) = \sum_{n=-\infty}^{\infty} \sum_{m=-\infty}^{\infty} \psi_{n,m}(t) e^{-i(nx+my)} , \qquad (4.1.1a)$$

which is more compactly expressed by

$$\psi(\mathbf{x}, t) = \sum_{\mathbf{n}} \psi_{\mathbf{n}}(t) e^{-i\mathbf{n}\cdot\mathbf{x}} . \qquad (4.1.1b)$$

The Fourier coefficients $\psi_{\mathbf{n}}$ are related to the field ψ by

$$\psi_{\mathbf{n}}(t) = \psi_{n,m}(t) = (2\pi)^{-2} \int_{\pi}^{\pi} dx \int_{-\pi}^{\pi} dy \, \psi(x, y, t) e^{i(nx+my)} \qquad (4.1.2a)$$

$$= (2\pi)^{-2} \int d\mathbf{x} \, \psi(\mathbf{x}, t) e^{i\mathbf{n}\cdot\mathbf{x}} . \qquad (4.1.2b)$$

Since $\psi(\mathbf{x}, t)$ is real, there is a conjugate symmetry amongst its Fourier coefficients:

$$\psi_{-\mathbf{n}}(t) = \psi_{\mathbf{n}}(t)^* , \qquad (4.1.3)$$

where the superscript $*$ denotes the complex conjugate. (Hereafter, if the superscripted quantity is a real vector or matrix, then the asterisk denotes the transposed quantity; if the superscripted quantity is a complex vector or matrix, then the asterisk denotes the transposed conjugate quantity.) In terms of Fourier coefficients, the evolution equation (3.3.1) for the streamfunction becomes

$$\frac{d}{dt}\psi_{\mathbf{n}} = L_{\mathbf{n}}\psi_{\mathbf{n}} + F_{\mathbf{n}} + q_{\mathbf{n}} , \qquad (4.1.4)$$

where the operator L has been replaced by the complex number $L_{\mathbf{n}}$:

$$L_{\mathbf{n}} = i\left\{\mathbf{U}\cdot\mathbf{n} - \beta n\left(\nu^2 + \rho^{-2}\right)^{-1}\right\} - D_0 - A_H \nu^2 ,$$

$$(\nu^2 = |\mathbf{n}|^2 = n^2 + m^2) , \qquad (4.1.5)$$

while $F_\mathbf{n}$, $q_\mathbf{n}$ are the Fourier coefficients for the estimated forcing F and unknown forcing error or system noise q. The data are vectors, generated by a vector $\boldsymbol{\mathcal{L}}$ of linear, spatial measurement functionals acting at discrete data times T_k, $(k = 1, 2, 3, \ldots)$. Thus measurement may be expressed as

$$\boldsymbol{\mathcal{L}}_{(\mathbf{x})}\left[\psi(\mathbf{x}, T_k)\right] = \sum_\mathbf{n} \boldsymbol{\mathcal{L}}_\mathbf{n} \psi_\mathbf{n}(T_k), \qquad (4.1.6)$$

$$(k = 1, 2, 3, \ldots),$$

where $\boldsymbol{\mathcal{L}}_\mathbf{n}$ is the complex vector

$$\boldsymbol{\mathcal{L}}_\mathbf{n} = \boldsymbol{\mathcal{L}}_{(\mathbf{x})}[e^{i\mathbf{n}\cdot\mathbf{x}}]. \qquad (4.1.7)$$

Functions of two positions have Fourier expansions of the form

$$Q(\mathbf{x}_1, \mathbf{x}_2) = \sum_{\mathbf{n}_1} \sum_{\mathbf{n}_2} Q_{\mathbf{n}_1, \mathbf{n}_2} e^{i(\mathbf{n}_1 \cdot \mathbf{x}_1 - \mathbf{n}_2 \cdot \mathbf{x}_2)}, \qquad (4.1.8)$$

where

$$Q_{\mathbf{n}_1, \mathbf{n}_2} = (2\pi)^{-4} \int d\mathbf{x}_1 \int d\mathbf{x}_2 \, Q(\mathbf{x}_1, \mathbf{x}_2) e^{-i(\mathbf{n}_1 \cdot \mathbf{x}_1 - \mathbf{n}_2 \cdot \mathbf{x}_2)}. \qquad (4.1.9)$$

Since $Q(\mathbf{x}, \mathbf{y})$ is real and symmetric in \mathbf{x} and \mathbf{y}, its Fourier coefficients have the following symmetries:

$$Q_{\mathbf{n}_1, \mathbf{n}_2} = Q^*_{-\mathbf{n}_1, -\mathbf{n}_2} = Q^*_{\mathbf{n}_2, \mathbf{n}_1}. \qquad (4.1.10)$$

The Fourier representation of the KF algorithm becomes

i) initial conditions at $t = T_0$:

$$\phi_\mathbf{n}(T_0) = \phi_\mathbf{n}^{(0)}, \qquad (4.1.11)$$

$$C_{\mathbf{n}_1, \mathbf{n}_2}(T_0) = A_{\mathbf{n}_1, \mathbf{n}_2}; \qquad (4.1.12)$$

ii) evolution equations $(T_{k-1} < T < T_k,\ 1 \le k)$:

$$\frac{d}{dt}\phi_\mathbf{n} = L_\mathbf{n}\phi_\mathbf{n} + F_\mathbf{n}, \qquad (4.1.13)$$

$$\frac{d}{dt}C_{\mathbf{n}_1, \mathbf{n}_2} = (L_{\mathbf{n}_1} + L^*_{\mathbf{n}_2})C_{\mathbf{n}_1, \mathbf{n}_2} + Q_{\mathbf{n}_1, \mathbf{n}_2}; \qquad (4.1.14)$$

iii) jump conditions at time $t = T_k$ $(1 \le k)$:

$$\phi_{\mathbf{n}}^{(k+)} = \phi_{\mathbf{n}}^{(k-)} + \mathbf{K}_{\mathbf{n}}^{(k)*}(\mathbf{d}^{(k)} - \sum_{\mathbf{n_2}} \mathcal{L}_{\mathbf{n_2}} \phi_{\mathbf{n_2}}^{(k-)}), \quad (4.1.15)$$

$$C_{\mathbf{n_1},\mathbf{n_2}}^{(k+)} = C_{\mathbf{n_1},\mathbf{n_2}}^{(k-)} - \mathbf{K}_{\mathbf{n_1}}^{(k)*} \sum_{\mathbf{n_3}} \mathcal{L}_{\mathbf{n_3}} C_{\mathbf{n_3},\mathbf{n_2}}^{(k-)}, \qquad (4.1.16)$$

where

$$\mathbf{K}_{\mathbf{n_1}}^{(k)} = \left\{ \sum_{\mathbf{n_2}} \sum_{\mathbf{n_3}} \mathcal{L}_{\mathbf{n_2}} \mathcal{L}_{\mathbf{n_3}}^* C_{\mathbf{n_2},\mathbf{n_3}}^{(k-)} + \overline{\epsilon^k(\epsilon^k)^*} \right\}^{-1} \sum_{\mathbf{n_4}} \mathcal{L}_{\mathbf{n_4}}^* C_{\mathbf{n_4},\mathbf{n_1}}^{(k-)}.$$
$$(4.1.17)$$

The behavior of $\phi_{\mathbf{n}}(t)$ at high wavenumber ν is required, since this determines the smoothness of $\phi(\mathbf{x},t)$. For $T_0 \le t < T_1$, analytical solutions are easily found for $\phi_{\mathbf{n}}$ in terms of $\phi_{\mathbf{n}}^{(0)}$ and $F_{\mathbf{n}}$. It is assumed that the initial and forcing fields are well behaved (rapidly decaying) for large ν. Since $L_{\mathbf{n}}$ has the asymptotic behavior

$$L_{\mathbf{n}} \sim i\mathbf{U} \cdot \mathbf{n} - D_0 - \nu^2 A_H, \qquad \text{as } \nu \to \infty, \qquad (4.1.18)$$

the propagator $\exp(L_{\mathbf{n}}t)$ will also be well behaved, as will the solution $\phi_{\mathbf{n}}$. Indeed, dependence upon $\phi_{\mathbf{n}}^{(0)}$ will be lost as t increases, and even more so as data are assimilated at times $t = T_k$ provided the data are good. It suffices, therefore, to restrict attention to the asymptotic behavior of $\phi_{\mathbf{n}}(t)$ for high wavenumber ν and large time t. The behavior is determined by the KF gain $\mathbf{K}_{\mathbf{n}}^{(k)}$ appearing in the jump condition (4.1.15), which in turn is related to the error cross-spectrum $C_{\mathbf{n_1},\mathbf{n_2}}$ by (4.1.17).

Assume that the system noise has a red cross-spectrum, that is, $Q_{\mathbf{n_1},\mathbf{n_2}} \to 0$ as $\nu_1, \nu_2 \to \infty$. Assume that the measurement coefficients $\mathcal{L}_{\mathbf{n}}$ are bounded for large ν. It will then be consistent to assume that $C_{\mathbf{n_1},\mathbf{n_2}}$ and $\mathbf{K}_{\mathbf{n}}^{(k)}$ are also red, in which case the product term in the jump condition (4.1.16) is relatively small at high wavenumber, and so $C_{\mathbf{n_1},\mathbf{n_2}}$ is approximately continuous in time at high wavenumber. Provided $A_H > 0$ or $D_0 > 0$, it follows from (4.1.14) that the error cross-spectrum has the limit

$$C_{\mathbf{n_1},\mathbf{n_2}}(t) \sim -(L_{\mathbf{n_1}} + L_{\mathbf{n_2}}^*)Q_{\mathbf{n_1},\mathbf{n_2}} \qquad (4.1.19)$$

as ν_1, ν_2 and $t \to \infty$. Hence $C_{\mathbf{n}_1,\mathbf{n}_2}$ is at least as red as $Q_{\mathbf{n}_1,\mathbf{n}_2}$, and by (4.1.17), $\mathbf{K}_{\mathbf{n}}^{(k)}$ is as red as $C_{\mathbf{n}_1,\mathbf{n}_2}$. Note that the matrix inverse in (4.1.17) is independent of wavenumber.

4.2 Measurement at a single position

For simplicity, suppose there is only one measurement and that it is simply the value of the streamfunction at the origin. Then \mathcal{L} is just a scalar functional \mathcal{L}, and $\mathcal{L}[\psi(\mathbf{x}, t)] = \psi(\mathbf{0}, t)$. The Fourier coefficients $\mathcal{L}_{\mathbf{n}}$ are the scalars $\mathcal{L}_{\mathbf{n}}$, all equal to unity. The gain is the vector $K_{\mathbf{n}}^{(k)}$, given by

$$K_{\mathbf{n}}^{(k)} = (\mu^{(k)})^{-1} \sum_{\mathbf{n}_2} C_{\mathbf{n},\mathbf{n}_2}^{(k-)}, \qquad (4.2.1)$$

where

$$\mu^{(k)} = \left\{ C^{(k-)}(\mathbf{0}, \mathbf{0}) + \overline{\epsilon^2} \right\}, \qquad (4.2.2)$$

$\overline{\epsilon^2}$ being, for all k, the measurement error variance at $t = T_k$. Next assume that the system noise q is statistically homogeneous:

$$Q(\mathbf{x}_1, \mathbf{x}_2) = Q(\mathbf{x}_1 - \mathbf{x}_2), \qquad (4.2.3)$$

so

$$Q_{\mathbf{n}_1,\mathbf{n}_2} = Q_{\mathbf{n}_1} \delta_{\mathbf{n}_1,\mathbf{n}_2}, \qquad (4.2.4)$$

where $Q_{\mathbf{n}}$ is the Fourier coefficient for $Q(\mathbf{z})$. In the high-wavenumber, large time limit, the error cross-spectrum becomes

$$C_{\mathbf{n}_1,\mathbf{n}_2}(t) \sim C_{\mathbf{n}_1,\mathbf{n}_2}(\infty) = -(2\mathrm{Re}\ L_{\mathbf{n}_1})^{-1} Q_{\mathbf{n}_1} \delta_{\mathbf{n}_1,\mathbf{n}_2}, \qquad (4.2.5)$$

and the gain becomes

$$K_{\mathbf{n}}^{(k)} \sim K_{\mathbf{n}}^{(\infty)} = -(2\mu^{(\infty)}\ \mathrm{Re}\ L_{\mathbf{n}})^{-1} Q_{\mathbf{n}}. \qquad (4.2.6)$$

Three limiting cases will now be examined.

a) Non-diffusive limit

If $A_H = 0$ (no diffusion), $\mathbf{U} = \mathbf{0}$ (no advection), but $D_0 > 0$ (bottom drag present), (4.2.6) yields

$$K_{\mathbf{n}}^{(k)} \sim K_{\mathbf{n}}^{(\infty)} = \left(2\mu^{(\infty)} D_0\right)^{-1} Q_{\mathbf{n}} \qquad (4.2.7)$$

as $\nu,\ k \to \infty$. Summing the Fourier series leads to

$$K^{(k)}(\mathbf{x}) \sim K^{(\infty)}(\mathbf{x}) = \left(2\mu^{(\infty)} D_0\right)^{-1} Q(\mathbf{x}) \qquad (4.2.8)$$

as $|\mathbf{x}| \to 0$ and $k \to \infty$. That is, the behavior of the gain in the neighborhood of the measurement site is given by the covariance of the system noise. Consider a system noise spectrum with the algebraic asymptotic behavior

$$Q_{\mathbf{n}} \sim Q_0 \nu^{-b}, \qquad (4.2.9)$$

as $\nu \to \infty$. The corresponding spatial structures of the gain, streamfunction estimate ϕ and vorticity estimate $\nabla^2 \phi$ are given in Table 4.2.1 for various $b > 0$.

Table 4.2.1										
$(A_H = 0,\ D_0 > 0,\ Q_n \sim \nu^{-b})$ $K,\ \phi \sim \text{constant} + \text{constant} \times h(\mathbf{x})$ $\nabla^2 \phi \sim \text{constant} + \text{constant} \times j(\mathbf{x})$										
b	$h(\mathbf{x})$	$j(\mathbf{x})$								
0	$\delta(\mathbf{x})$	$\nabla^2 \delta(\mathbf{x})$								
2	$ln\,	\mathbf{x}	$	$	\mathbf{x}	^{-1}$				
4	$	\mathbf{x}	^2\,ln\,	\mathbf{x}	$	$ln\,	\mathbf{x}	$		
6	$	\mathbf{x}	^4\,ln\,	\mathbf{x}	$	$	\mathbf{x}	^2\,ln\,	\mathbf{x}	$

For example, if a spatially continuous map of the streamfunction is required, then b must exceed 2; if a continuous map of vorticity is required, then b must exceed 4. The above analysis is invalid if the system noise spectrum is white ($b = 0$); in that case the product term in (4.1.16) cannot be neglected as $\nu \to \infty$. Bennett & Budgell (1987) computed K and $\nabla^2 K$ for $Q_{\mathbf{n}} = (\nu^2 + \rho^{-2})^{-b/2}$; $\rho = 1$; $b = 0, 4, 6$ and wavenumber truncations $\nu \le 8\sqrt{2}$, $16\sqrt{2}$ and $32\sqrt{2}$. Both K and

$$K(x,0,\infty) \qquad \nabla^2 K(x,0,\infty)$$

Figure 4.2.1. (a), (b) Two-dimensional equilibrium streamfunction and vorticity Kalman gains for $\rho = 1$, $\beta = 1$, $\Delta T = 5$, $A_H = 0$, $Q_{\mathbf{n}} \equiv Q_0$ and $\nu = 8\sqrt{2}$; (c), (d) as for (a), (b) except $\nu = 16\sqrt{2}$; (e), (f) as for (a), (b) except $\nu = 32\sqrt{2}$ (after Bennett & Budgell, 1987).

$\nabla^2 K$ exhibit singular behavior at $\mathbf{x} = 0$ as ν increases, for $b = 0$ (Fig. 4.2.1). Only $\nabla^2 K$ is singular as ν increases, for $b = 4$ (Fig. 4.2.2), while both K and $\nabla^2 K$ converge to smooth fields as ν increases, for $b = 6$ (Fig. 4.2.3). The gain K is broader as b increases (redder spectra); thus the (good) data at the origin have a broader influence. The analytical estimates, in Table 4.2.1, of asymptotic behavior near $\mathbf{x} = 0$ are verified with remarkable accuracy, even at such low resolution.

Figure 4.2.2. As for Fig. 4.2.1 except $Q_{\mathbf{n}} = Q_0(\nu^2 + \rho^{-2})^{-2}$.

Time series for the corresponding total error variance

$$C(t) = (2\pi)^{-2} \int d\mathbf{x}\, C(\mathbf{x}, \mathbf{x}, t) = \sum_{\mathbf{n}} C_{\mathbf{n},\mathbf{n}}(t) \qquad (4.2.10)$$

are shown in Fig. 4.2.4. The Kalman filter is evidently more effective as b increases.

Curtain & Pritchard (1978) investigated the spatial structure of the Kalman-Bucy filter (KBF), requiring only that the system noise estimate be square-integrable. Accordingly they imposed square-

$K(x,0,\infty)$ $\nabla^2 K(x,0,\infty)$

Figure 4.2.3. As for Fig. 4.2.1 except $Q_{\mathbf{n}} = Q_0(\nu^2 + \rho^{-2})^{-3}$.

summability conditions on the noise cross-spectrum $Q_{\mathbf{n},\mathbf{m}}$; these conditions do not ensure that the filter estimate of the state (here, the streamfunction or the vorticity) is a smooth field.

b) Diffusive limit

If $A_H > 0$, the asymptotic estimate (4.2.6) for the gain at high wavenumber yields

$$K_{\mathbf{n}}^{(k)} \sim K_{\mathbf{n}}^{(\infty)} = (2\mu^{(\infty)} A_H \nu^2)^{-1} Q_{\mathbf{n}}. \qquad (4.2.11)$$

The behavior of $\phi_{\mathbf{n}}^{(k+)}$ should also be $O(\nu^{-2} Q_{\mathbf{n}})$, and $\phi_{\mathbf{n}}(t)$ should

Figure 4.2.4a. Time-series total streamfunction variance, and time series of total error variance for a one-dimensional Kalman filter with wavenumber spectral index $b = 0$ and $b = 2$. Other parameters: $\rho = 1$, $\beta = 1$, $\Delta T = 5$, $A_H = 0$ and w $= 1$ (after Bennett & Budgell, 1989).

become much redder for $T_k < t$ due to the effects of diffusion. The behavior at the data times is of the greatest concern. If $Q_\mathbf{n} \sim \nu^{-2}$, then the anticipated behavior is

$$\phi^{(k+)}(\mathbf{x}, t) \sim \text{constant} + \text{constant} \times |\mathbf{x}|^2 \, ln \, |\mathbf{x}|$$

as $|\mathbf{x}| \to 0$ and $t \to \infty$. However, if measurement is rapid or continuous, numerical solutions indicate a quite different behavior. A boundary-layer structure appears near the data point. Some results

Figure 4.2.4b. As for Fig. 4.2.4a, for $b = 4$ and $b = 6$ (after Bennett & Budgell, 1989).

from a one-dimensional filter calculation are shown in Figs. 4.2.5 to 4.2.7. For $b = 0$, $K(x, 0, \infty)$ exhibits a steep but finite cusp at $x = 0$; for $b = 2$, K_{xx} has a cusp while for $b = 4$ both K and K_{xx} appear to have smooth limits as resolution is increased. An explanation may be found by inspecting the steady-state nonlinear dynamics of $C_{n_1,n_2}^{(k-)}$. Indeed, if the sampling interval ΔT is small, then (4.1.14) may be approximated by

$$C_{n_1,n_2}^{(-)} = C_{n_1,n_2}^{(+)} + \Delta T (L_{n_1} + L_{n_2}^*) C_{n_1,n_2}^{(-)} + \Delta T Q_{n_1,n_2} . \qquad (4.2.12)$$

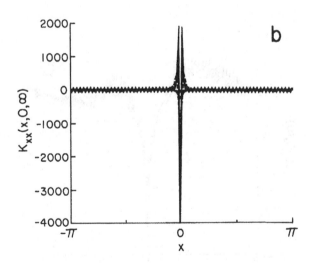

Figure 4.2.5. (a) One-dimensional equilibrium streamfunction Kalman gain $K(x, 0, \infty)$ and (b) vorticity gain $K_{xx}(x, 0, \infty)$ for ν = 8, 16, 32, 64 and 128. Other parameters: $\rho = 1$, $\beta = 1$, $A_H = 7.5 \times 10^{-3}$, $Q_n \equiv Q_0$ (after Bennett & Budgell, 1987).

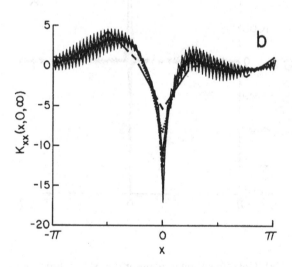

Figure 4.2.6. As for Fig. 4.2.5 except $Q_n = Q_0(n^2 + \rho^{-2})^{-1}$.

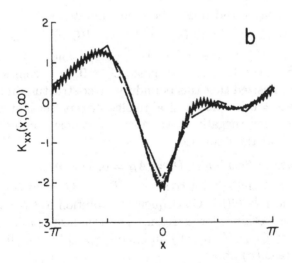

Figure 4.2.7. As for Fig. 4.2.5 except $Q_n = Q_0(n^2 + \rho^{-2})^{-2}$.

The superfluous superscripts k, $k - 1$ have been discarded in this analysis of the steady-state case. Combining (4.1.16) and (4.2.12) yields

$$(L_{\mathbf{n_1}} + L_{\mathbf{n_2}})C^{(-)}_{\mathbf{n_1},\mathbf{n_2}} - (\mu \Delta T)^{-1} \sum_{\mathbf{n_3}} C^{(-)}_{\mathbf{n_1},\mathbf{n_3}} \sum_{\mathbf{n_4}} C^{(-)}_{\mathbf{n_4},\mathbf{n_2}}$$

$$+ Q_{\mathbf{n_1},\mathbf{n_2}} = 0, \qquad (4.2.13)$$

where $\mu = C^{(-)}(\mathbf{0},\mathbf{0}) + \overline{\epsilon^2}$. Approximating $L_{\mathbf{n}}$ by $-A_H \nu^2$, the coordinate representation of (4.2.13) is

$$A_H(\nabla^2_{(\mathbf{x_1})} + \nabla^2_{(\mathbf{x_2})})C^{(-)}(\mathbf{x_1},\mathbf{x_2}) - (\mu \Delta T)^{-1}C^{(-)}(\mathbf{x_1},\mathbf{0})C^{(-)}(\mathbf{0},\mathbf{x_2})$$

$$+ Q(\mathbf{x_1},\mathbf{x_2}) = 0, \quad (4.2.14)$$

for $\mathbf{x_1}, \mathbf{x_2} \simeq \mathbf{0}$. The boundary-layer structure is revealed by neglecting the inhomogeneous term Q in (4.2.14) and by seeking separable solutions of the form $C^{(-)}(\mathbf{x_1},\mathbf{x_2}) = B(\mathbf{x_1})B(\mathbf{x_2})$ where $B(\mathbf{x}) = \exp[-\alpha(|x| + |y|)]$. It follows that

$$\alpha^2 = (4A_H \mu \Delta T)^{-1}C^{(-)}(\mathbf{0},\mathbf{0}). \qquad (4.2.15)$$

The boundary-layer width α^{-1} becomes infinite as $\Delta T \to \infty$. The width approaches the limit $\left(4A_H \overline{\epsilon^2} \Delta T / C^{(-)}(\mathbf{0},\mathbf{0})\right)^{1/2}$ as $\Delta T \to 0$, since $\overline{\epsilon^2} = O(\Delta T)^{-1}$ (see Exercise 3.4.3), and has the limit $(4A_H \Delta T)^{1/2}$ as the KF error variance $C^{(-)}(\mathbf{0},\mathbf{0})$ becomes large. It should be emphasized that this boundary-layer structure in the KF at the data site $\mathbf{x} = \mathbf{0}$ has no physical reality. It represents steady diffusion of information (negative covariance) balancing nonlinear growth of information at the data site.

c) *Adiabatic flow limit:* $D_0 = A_H = 0$, $\mathbf{U} \neq \mathbf{0}$

In this case, the system coefficient $L_{\mathbf{n}} \sim i\mathbf{U} \cdot \mathbf{n}$ as $\nu \to \infty$, but $L_{\mathbf{n}}$ is not uniformly $O(\nu)$. Consequently, evolution in time only leads to a spatial smoothing downstream. The roughest spatial structure will be cross-stream $(\mathbf{U} \cdot \mathbf{n} = 0)$ and is described by one-dimensional analogies of case (a) above.

4.3 Altimetry

In the next decade, altimeters mounted on earth-orbiting satellites will be a major source of essential oceanographic data with global

coverage. Detailed descriptions of these sophisticated microwave rangefinders may be found in Stewart (1985) and Chelton, Freilich & Haines (1996). It is an elementary remark that the position of the satellite must be known before the height of the ocean surface can be inferred. In practice a critical aspect of data processing is the estimation of the celestial geopotential surface (the geoid) on which the satellite orbits, and also the marine geopotentials which define hydrostatic balance. Setting aside such fundamental issues of accuracy for the moment, let it be assumed that altimeters provide measurements of the height of the ocean surface along tracks under the orbit of the satellite. These data could be readily assimilated into a primitive-equation ocean circulation model which includes a free surface, simply by "nudging" the evolution equation which expresses the kinematic free-surface condition (see Chapter 3 of Pedlosky, 1987; also Remark *8*, §3.4). If the primitive-equation model, as is usually the case, has a "rigid lid," then altimetric data may be interpreted as values for the pressure under the lid, and may be used to nudge the momentum equations at least in the model's shallowest level. Quasi-geostrophic models may be nudged with altimeter data if the latter are interpreted as upper-level values of the streamfunction. Holland & Malanotte-Rizzoli (1989), for example, have carried out nudging experiments with a multi-level, nonlinear quasi-geostrophic model and simulated altimeter data in order to investigate the merits of various orbit patterns.

The choice of a nudging coefficient is always subjective, but the KBF may be used for guidance since the latter is the optimal sequential assimilation scheme for time-continuous data. Accordingly, the KBF will now be developed for an idealized quasi-geostrophic model and an idealized model for the altimetric data. It will be shown that seemingly significant effects, omitted in the idealization, are in fact inessential to the analysis. Consider, then, the simple linear, single-level quasi-geostrophic model (3.3.1), (3.3.2). The idealized data consist of streamfunction values at all times along a single track (Fig. 4.3.1) oriented from south to north:

$$\alpha(y,t) = \psi(X,y,t) + \epsilon(y,t), \qquad (4.3.1)$$

where X is the fixed longitude of the track, y is the latitude along the track, and α and ϵ are, respectively, the altimeter data and error. It is implied by (4.3.1) that the satellite has effectively infinite speed,

Figure 4.3.1. Single, idealized altimeter track along "longitude X" in a double-periodic domain.

data being available simultaneously along the track. Certainly the speed vastly exceeds that of advection or propagation in the model. There is a significant delay between data on different tracks, but it suffices here to consider a single track, provided that the system noise covariance scales are smaller than the track spacing.

The finite set of measurement functionals \mathcal{L} must be replaced by a continuous set. The data error is assumed to have a vanishing mean and to be white in time:

$$\overline{\epsilon(y_1, t_1)\, \epsilon(y_2, t_2)} = \sigma(y_1, y_2)\, \delta(t_1 - t_2)\,. \qquad (4.3.2)$$

The KBF error covariance satisfies

$$C_t(\mathbf{x}_1, \mathbf{x}_2, t) = (L_{(\mathbf{x}_1)} + L_{(\mathbf{x}_2)})C(\mathbf{x}_1, \mathbf{x}_2, t) - (2\pi)^{-2} \int dy_3 \int dy_4$$
$$\times\, C(\mathbf{x}_1, X, y_3, t)\, w(y_3, y_4)\, C(X, y_4, \mathbf{x}_2, t) + Q(\mathbf{x}_1, \mathbf{x}_2)\,, \quad (4.3.3)$$

where w is the inverse of σ:

$$(2\pi)^{-1} \int dy_3\, w(y_1, y_3)\sigma(y_3, y_2) = \delta(y_1 - y_2)\,. \qquad (4.3.4)$$

The KBF estimate of the streamfunction satisfies

$$\phi_t(\mathbf{x}, t) = L\phi(\mathbf{x}, t) + F(\mathbf{x}, t)$$
$$+ (2\pi)^{-1}\int dy_2 K(\mathbf{x}, y_2, t)\{\alpha(y_2, t) - \phi(X, y_2, t)\}, \quad (4.3.5)$$

where the KBF gain is given by

$$K(\mathbf{x}, y_2, t) = (2\pi)^{-1}\int dy_3 C(\mathbf{x}, X, y_3, t)w(y_3, y_2). \quad (4.3.6)$$

The last term in (4.3.5) is the KBF estimate of the system noise $q(\mathbf{x}, t)$; compare (4.3.5) with (3.3.1). In general the data influence the noise estimate off the track, and the influence spreads to other latitudes. The positive gain (4.3.6) is the optimal choice for a nudging coefficient; any other would lead to more rapid growth of the error variance $C(\mathbf{x}, \mathbf{x}, t)$. The KBF will now be examined in an extreme case.

First assume that the data errors are not correlated as functions of latitude, that is,

$$\sigma(y_1, y_2) = \sigma_0\, \delta(y_1 - y_2), \quad (4.3.7)$$

so $w(y_1, y_2) = \sigma_0^{-1}\delta(y_1 - y_2)$, and thus

$$K(\mathbf{x}, y_2, t) = (2\pi\sigma_0)^{-1}C(\mathbf{x}, X, y_2, t). \quad (4.3.8)$$

There are two limits to be considered:

a) *Non-diffusive limit*

If the system noise q is homogeneous and has the spectrum

$$Q_\mathbf{n} \sim Q_0\nu^{-2}, \quad\quad\quad \text{as } \nu \to \infty, \quad (4.3.9)$$

then the analysis of §4.2 shows that, for $A_H = 0$,

$$C(\mathbf{x}_1, \mathbf{x}_2, t) \sim (a_1 D_0)^{-1}Q_0\, ln|\mathbf{x}_1 - \mathbf{x}_2| \quad (4.3.10)$$

where a_1 and subsequent a_2, a_3, \ldots are numerical constants. The vorticity equation for the KBF becomes, approximately,

$$
\begin{aligned}
E\phi_t(\mathbf{x}, t) = {} & EL\phi(\mathbf{x}, t) + EF(\mathbf{x}, t) \\
& - (a_2 \sigma_0 D_0)^{-1} Q_0 \delta(x - X)\{\alpha(y, t) - \phi(X, y, t)\}, \quad (4.3.11)
\end{aligned}
$$

for $x \cong X$ and $t \to \infty$. This is nudging with direct insertion in the vorticity equation. The negative nudging coefficient is consistent with the appearance of the negative operator E in (4.3.11). Recall that E has only negative eigenvalues. In a finite-difference approximation to (4.3.11), the nudging term on the right-hand side becomes $-(a_2 \sigma_0 D_0 \Delta x)^{-1} Q_0 \delta_{n,n_X}\{\alpha_m - \phi_{n_X,m}\}$ at time t, where n and m are longitudinal and latitudinal grid indices respectively, and n_X is the index for the track.

The nudging coefficient in (4.3.11) is proportional to $(\sigma_0 D_0)^{-1} Q_0$; that is, it is proportional to the ratio of the streamfunction error variance $Q_0/(2D_0)$ to the data error variance σ_0. Thus, if the ratio is relatively large (bad dynamics or good data), the solution of (4.3.11) relaxes quickly to the data. In the other extreme, data have little influence on the solution.

b) Diffusive limit

If the system noise q is white in space:

$$
Q_{\mathbf{n}} \sim Q_0, \qquad\qquad \text{as } \nu \to \infty, \qquad\qquad (4.3.12a)
$$

or, equivalently,

$$
Q(\mathbf{x}_1, \mathbf{x}_2) = Q(\mathbf{x}_1 - \mathbf{x}_2) \sim Q_0 \, \delta(\mathbf{x}_1 - \mathbf{x}_2), \qquad \text{as } \mathbf{x}_1 \to \mathbf{x}_2, \quad (4.3.12b)
$$

then the linear analysis of §4.2 suggests that for $A_H > 0$, (4.3.10) and (4.3.11) hold with $(a_2 \sigma_0 D_0)$ replaced by $(a_3 \sigma_0 A_H)$. (Note that the dimensions of Q_0 in (4.3.12) are not those of the Q_0 appearing in (4.3.9).) However, as mentioned above, numerical experiments have shown that a nonlinear analysis is necessary. Such an analysis is simpler if applied to the KBF equation (4.3.3).

First, the linear steady-state solution is

$$
C(\mathbf{x}_1, \mathbf{x}_2) = -(L_{(\mathbf{x}_1)} + L_{(\mathbf{x}_2)})^{-1} Q(\mathbf{x}_1, \mathbf{x}_2). \qquad\qquad (4.3.13)
$$

For $A_H > 0$ it has the property that

$$E_{(\mathbf{x}_1)}C(\mathbf{x}_1, \mathbf{x}_2) \sim -a_4 A_H^{-1} Q_0 \, \delta(\mathbf{x}_1 - \mathbf{x}_2) \qquad (4.3.14)$$

as $(\mathbf{x}_1 - \mathbf{x}_2) \to 0$, when Q is given by (4.3.12). Together with (4.3.8), this property implies direct insertion of data in the vorticity equation for the KBF, as in (4.3.11). Second, there is a "free" $(Q = 0)$ steady-state solution of (4.3.3) of the form $C(\mathbf{x}_1, \mathbf{x}_2) = B(\mathbf{x}_1)B(\mathbf{x}_2)$, where $B(\mathbf{x})$ satisfies

$$LB = \theta^2 B \qquad (4.3.15)$$

for some separation constant θ. Guided by (4.3.13) and (4.3.15), a steady-state solution of (4.3.3) is sought in the form

$$C(\mathbf{x}_1, \mathbf{x}_2) = G(\mathbf{x}_1)G(\mathbf{x}_2) - (L_{(\mathbf{x}_1)} + L_{(\mathbf{x}_2)})^{-1}Q(\mathbf{x}_1, \mathbf{x}_2) \qquad (4.3.16)$$

for some function $G(\mathbf{x})$. Substituting (4.3.16) into (4.3.3), making the white data noise assumption (4.3.7), allowing the simple cancellation of the Q's, neglecting the remaining inhomogeneous term due to its relative spatial smoothness, and separating dependences on \mathbf{x}_1 and \mathbf{x}_2 yields a linear equation for $G(\mathbf{x})$:

$$LG(\mathbf{x}) - \theta^2 G(\mathbf{x}) + a_5(\sigma_0 A_H)^{-1} Q_0 \, E^{-1} \delta(x - X) G(X, y) \cong 0. \qquad (4.3.17)$$

The last term in (4.3.17) is a contribution from the quadratic term in (4.3.3). Recalling that L is dominated by the diffusion operator

$$L \sim A_H \nabla^2, \qquad (4.3.18)$$

it is evident that the essential fine structure of G in the x-direction (normal to the satellite track) satisfies

$$G_{xxxx} \propto \delta(x - X), \qquad (4.3.19)$$

which implies that G is smoother than the linear solution (4.3.13): compare (4.3.14) and (4.3.19). It is concluded that the linear result (4.3.13) is, to leading order, the asymptotic description of the behavior of the vorticity gain near the track when the system noise q is white in space ($Q \propto \delta$).

In summary, direct insertion of streamfunction data into the vorticity equation is consistent with Kalman filtering under the assumptions that, first, the data noise is δ-correlated along the track and, second, $Q_n \sim Q_0 \nu^{-2}$ (for $A_H = 0$) or $Q_n \sim Q_0$ (for $A_H > 0$). It has also been assumed as usual that the initial noise, data noise and system noise are uncorrelated while the latter two noise types are δ-correlated in time. The spatial white-noise model (4.3.7) for the data errors is the most obviously suspect, since significant geoid errors are included and these are correlated over planetary scales. Indeed in a regional model it might be more sensible to assume perfect correlation of the data errors along the track. Errors in the corrections for tropospheric moisture would be correlated on the atmospheric synoptic scale (several thousand kilometers). Streamfunction system noise spectra of $O(\nu^0)$ or $O(\nu^{-2})$, as $\nu \to \infty$, correspond to wind stress error spectra of $O(\nu^2)$ or $O(\nu^0)$. On the other hand, even a perfectly known wind stress field would have to be smoothed before being used to force an ocean circulation model, since the latter would inevitably have limited physical and numerical resolution. Consequently the difference between true and estimated forcing fields would also be smooth. In other words, a data-assimilation scheme such as trackwise nudging, which is based on an assumption of significant system noise at high wavenumbers, is ill-conceived.

Given plausible wind statistics and data statistics, it would be feasible to compute a time-asymptotic Kalman gain for a linearized model on a coarse grid. Such a gain could reasonably be used as a nudging "function" for the original linear model, on its finer grid.

The preceding analyses have been based on the simple, linear single-layer quasi-geostrophic model. However, the results extend immediately to layered and stratified quasi-geostrophic models, since horizontal fine structure is not influenced by quasi-geostrophic potential vorticity. That is, the operator $E = \nabla^2 - \rho^{-2}$ may be approximated by ∇^2; fine-structure scales are smaller than the internal deformation radius ρ. On the other hand, the system noise fields at different levels may be correlated, in which case the Kalman gain for surface data will not vanish below the surface layer. That is, altimeter data should be inserted in subsurface layers as well. Computation of the Kalman gain would be a major task and would be sensitive to estimates of the vertical correlation scales of the system noise.

Variants of the preceding ideas may be found in Mellor & Ezer (1991), who assimilated altimetric data into a primitive-equation (PE)

model of the Gulf Stream region. Covariances of the free-surface displacement with subsurface fields were estimated from model results. This is loosely equivalent to solving the PE version of the error covariance equation (4.3.3), linearized about the mean circulation, in the case of worthless data ($w \equiv 0$). The system noise (for which Q is the covariance) arises from the model variability on the shorter spatial and temporal scales, owing principally to barotropic and baroclinic instability.

Exercise 4.3.1 Assume that vorticity data are available along a track. Find conditions under which trackwise nudging with the vorticity data is asymptotically equivalent to a Kalman filter. □

5

GENERALIZED INVERSES OF
DYNAMICAL MODELS

5.1 Introduction

Conventional ocean modeling consists of solving the model equations as accurately as possible, and then comparing the results with observations. While encouraging levels of qualitative agreement have been obtained, as a rule there is significant quantitative disagreement owing to many sources of error: model formulation, model inputs, computation and the data themselves. Computational errors aside, the errors made both in formulating the model and in specifying its inputs usually exceed the errors in the data. Thus it is unsatisfactory to have a model solution which is uninfluenced by the data. In the spirit of the inverse methods in Chapter 1, the approach which is developed here finds the ocean circulation providing the best fit simultaneously to the model equations and to the data. The best fit is defined in a weighted least-squares sense, with weights reflecting prior estimates of the various standard errors. Once unknown errors are explicitly included in the model equations and the data, the problem of finding the circulation is underdetermined, and so the least-squares fit may be regarded as a generalized inverse of the combined dynamics and observing system.

Finding the generalized inverse may also be regarded as a smoothing problem. The smoothing norm involves the differential operators for the model equations as well as derivatives of, or covariances for, the errors or residuals in the equations. Indeed, it is shown that

the generalized inverse is equivalent to Gauss-Markov smoothing in space and time, based on the space-time covariance of model solutions forced by random fields having prescribed covariances. However, the development of the generalized inverse will reverse the order of Chapter 2: the representer solution is constructed first. The representers yield expressions for the error covariances of estimates of the circulation, the model forcing, the initial and boundary values and the data. The reduction in the penalty functional defining the best fit is also expressed using representers. The method is shown to be closely related to the Backus-Gilbert techniques for solid-earth geophysics.

Two other methods for finding the generalized inverse are discussed: control-theoretic approaches based on the Kalman-Bucy filter, and gradient descent.

The linear quasi-geostrophic model of §3.3 is the working example throughout this chapter.

5.2 The field, the data, and the model

For simplicity, consider smoothing a scalar-valued field of space and time, namely a streamfunction $\psi = \psi(\mathbf{x}, t)$. As before, let the M-dimensional data vector \mathbf{d} consist of the values of $\boldsymbol{\mathcal{L}}[\psi]$, where $\boldsymbol{\mathcal{L}}$ is a vector of M measurement functionals acting on the streamfunction, plus an error vector $\boldsymbol{\epsilon}$:

$$\mathbf{d} = \boldsymbol{\mathcal{L}}[\psi(\mathbf{x}, t)] + \boldsymbol{\epsilon} \, . \tag{5.2.1}$$

Where necessary, the notation $\boldsymbol{\mathcal{L}}$ will be extended to $\boldsymbol{\mathcal{L}}_{(\mathbf{x}, t)}$ to indicate the appropriate field arguments. The data are assumed to have been collected over a time interval $0 \le t \le T$; they may include values of the streamfunction or velocity at discrete times t_k, or weighted averages over the entire interval such as Fourier coefficients. Similarly, the data may be values at isolated points \mathbf{X}_j, or weighted averages over paths or spatial subregions. The essential assumption is that there is only a finite number M of all such observations.

Once again, the model is a linear, single-level, reduced-gravity quasi-geostrophic model expressed in the form

$$\frac{\partial \psi}{\partial t} = L\psi + F + q \, , \tag{5.2.2}$$

where $F = F(\mathbf{x}, t)$ is a prior estimate of the forcing, and $q = q(\mathbf{x}, t)$ is the unknown error either in the dynamics or in the forcing variously referred to as the dynamical misfit, dynamical residual, or stream-function system noise. The linear operator L is

$$L = -\mathbf{U} \cdot \nabla - \beta E^{-1} \frac{\partial}{\partial x} - D_0 + A_H \nabla^2 , \qquad (5.2.3)$$

where \mathbf{U} is a known uniform advecting velocity, β is the northward gradient of the Coriolis parameter, D_0 is a bottom drag coefficient, and A_H an eddy diffusivity, while E is the elliptic operator

$$E = \nabla^2 - \rho^{-2} , \qquad (5.2.4)$$

ρ being the internal deformation radius. An appropriate initial condition for the model is

$$\psi(\mathbf{x}, 0) = \Psi(\mathbf{x}) + a(\mathbf{x}) \qquad (5.2.5)$$

where $\Psi(\mathbf{x})$ is a prior estimate of the initial streamfunction and $a(\mathbf{x})$ is an unknown initial error. It remains to prescribe boundary conditions. In the interests of simplicity it will be assumed as before that all fields, including the advecting velocity \mathbf{U}, are periodic in x and y with a period of 2π in each direction. Thus the spatial domain is the square $-\pi \leq x, y \leq \pi$, and the model (5.2.2)–(5.2.4) is in dimensionless form.

5.3 The generalized inverse: Euler-Lagrange equations

The model solution corresponding to vanishing errors $q(\mathbf{x}, t)$ and $a(\mathbf{x})$ does not, as a rule, agree with the data. It is natural therefore to define a generalized inverse of the model and data as a streamfunction providing the best fit to the model and data. The best fit may, for example, be defined in a weighted least-squares sense. The smoothness of the inverse may be ensured by suitable choice of weights. To begin, define a quadratic penalty functional by

$$\mathcal{J}[\psi] = (2\pi)^{-4} \int_0^T dt_1 \int_0^T dt_2 \int d\mathbf{x}_1 \int d\mathbf{x}_2 \, q(\mathbf{x}_1, t_1) W(\mathbf{x}_1, t_1, \mathbf{x}_2, t_2) q(\mathbf{x}_2, t_2)$$

$$+ (2\pi)^{-4} \int d\mathbf{x}_1 \int d\mathbf{x}_2 \, a(\mathbf{x}_1) V(\mathbf{x}_1, \mathbf{x}_2) a(\mathbf{x}_2) + \boldsymbol{\epsilon}^* \mathbf{w} \boldsymbol{\epsilon} , \qquad (5.3.1)$$

where ϵ, q and a are related to ψ by (5.2.1), (5.2.2) and (5.2.5), respectively. The spatial integrals in (5.3.1) are double integrals over the square domain, and factors of $(2\pi)^{-2}$ are included for the convenience of spatial Fourier series expansions. The weights W, V and \mathbf{w} must be positive-definite and symmetric. In the most general choice of quadratic penalty functional, there would be products of q with a and ϵ, and a with ϵ. However, such off-diagonal terms greatly clutter the inversion and are neglected in most applications, corresponding to the assumption of uncorrelated errors. Indeed, there is little prospect of providing credible estimates for such covariances, although it is plausible that they are small.

The calculus of variations (see §5.11 and Courant & Hilbert, 1953) may be used to show that if $\hat{\psi}(\mathbf{x}, t)$ is a local extremum of the penalty functional $\mathcal{J}[\psi]$ defined by (5.3.1), then $\hat{\psi}$ satisfies

$$\frac{\partial}{\partial t}\hat{\psi}(\mathbf{x}, t) = L\hat{\psi}(\mathbf{x}, t) + F(\mathbf{x}, t)$$

$$+ (2\pi)^{-2}\int_0^T dt_2 \int d\mathbf{x}_2\, Q(\mathbf{x}, t, \mathbf{x}_2, t_2)\, \lambda(\mathbf{x}_2, t_2) \quad (5.3.2)$$

subject to

$$\hat{\psi}(\mathbf{x}, 0) = \Psi(\mathbf{x}) + (2\pi)^{-2}\int d\mathbf{x}_2\, A(\mathbf{x}, \mathbf{x}_2)\, \lambda(\mathbf{x}_2, 0)\,, \quad (5.3.3)$$

where the adjoint variable λ satisfies the adjoint equation

$$-\frac{\partial}{\partial t}\lambda(\mathbf{x}, t) = L^\dagger \lambda(\mathbf{x}, t)$$

$$+ (2\pi)^2 \mathcal{L}^*_{(\mathbf{x}_2, t_2)}\left[\delta(\mathbf{x} - \mathbf{x}_2)\delta(t - t_2)\right] \mathbf{w}(\mathbf{d} - \mathcal{L}[\hat{\psi}]) \quad (5.3.4)$$

subject to

$$\lambda(\mathbf{x}, T) = 0\,. \quad (5.3.5)$$

In (5.3.2) and (5.3.3), Q and A are the inverses of W and V respectively:

$$(2\pi)^{-4}\int_0^T dt_2 \int d\mathbf{x}_2\, Q(\mathbf{x}_1, t_1, \mathbf{x}_2, t_2)\, W(\mathbf{x}_2, t_2, \mathbf{x}_3, t_3)$$

$$= \delta(\mathbf{x}_1 - \mathbf{x}_3)\delta(t_1 - t_3)\,, \quad (5.3.6)$$

$$(2\pi)^{-4} \int dx_2\, A(x_1, x_2) V(x_2, x_3) = \delta(x_1 - x_3) . \qquad (5.3.7)$$

In (5.3.4), L^\dagger is the operator adjoint to L, differing from L only in that the signs of all first derivatives are reversed. Note that when deriving (5.3.2)–(5.3.5), it is necessary to substitute for ϵ, q and a in (5.3.1) using the definitions (5.2.1), (5.2.2) and (5.2.5), before taking the first variation of \mathcal{J} with respect to ψ.

The system (5.3.2)–(5.3.5) will be referred to here as the Euler-Lagrange equations; strictly speaking, (5.3.4) is *the* Euler-Lagrange equation. Inspection of (5.3.2) and (5.3.3) shows that the generalized inverse estimates of the prior errors q and a are, respectively,

$$\hat{q}(x, t) = (2\pi)^{-2} \int_0^T dt_2 \int dx_2\, Q(x, t, x_2, t_2)\, \lambda(x_2, t_2) \qquad (5.3.8)$$

and

$$\hat{a}(x) = (2\pi)^{-2} \int dx_2\, A(x, x_2)\, \lambda(x_2, 0) . \qquad (5.3.9)$$

For the homogeneous problem $F(x, t) \equiv 0$, $\Psi(x) \equiv 0$ and $d = 0$, it may be shown that solutions $\hat{\psi}$ and λ of (5.3.2)–(5.3.5) satisfy the identity

$$(2\pi)^{-4} \int_0^T dt_1 \int_0^T dt_2 \int dx_1 \int dx_2\, \lambda(x_1, t)\, Q(x_1, t_1, x_2, t_2)\, \lambda(x_2, t_2)$$

$$+ (2\pi)^{-4} \int dx_1 \int dx_2\, \lambda(x_1, 0)\, A(x_1, x_2)\, \lambda(x_2, 0)$$

$$+ \mathcal{L}^*[\hat{\psi}]\, w \mathcal{L}[\hat{\psi}] = 0 . \qquad (5.3.10)$$

Consequently, $\lambda(x, t) \equiv 0$, $\hat{\psi}(x, 0) \equiv 0$ and $\mathcal{L}[\hat{\psi}] = 0$. Again for the homogeneous case, it may be shown that

$$\frac{d}{dt} \int dx\, \hat{\psi}(x, t)^2 \le 0 . \qquad (5.3.11)$$

Hence the homogeneous problem has only the trivial solution: $\hat{\psi}(x, t) \equiv 0$. It may be concluded that the inhomogeneous problem has a unique solution, and so \mathcal{J} has a unique local extremum. It may

be inferred from the quadratic nature of \mathcal{J} that the local extremum must be the global minimum.

Exercise 5.3.1 Using the inverse relationships (5.3.6) and (5.3.7), eliminate the adjoint variable λ from the Euler-Lagrange equations (5.3.2)–(5.3.5), arriving at a second-order two-point boundary-value problem for $\hat{\psi}$ in the time interval $0 \leq t \leq T$. Show that the second-order operator involves the forward operator $\frac{\partial}{\partial t} - L$, the backward operator $\frac{-\partial}{\partial t} - L^{\dagger}$, the weights W and V, and the measurement functionals \mathcal{L}. □

There is a special case of interest in which the prior estimate $\Psi(\mathbf{x})$ for the initial streamfunction is given no weight in the penalty functional: $V(\mathbf{x}_1, \mathbf{x}_2) \equiv 0$. Then (5.3.5) is replaced by

$$\lambda(\mathbf{x}, 0) = 0. \qquad (5.3.12)$$

The identity (5.3.10) still holds for the homogeneous problem, except that the second term (the integral over initial values of λ) is no longer present. Consequently, $\mathcal{L}[\hat{\psi}] = \mathbf{0}$ and $\lambda(\mathbf{x}, t) \equiv 0$, but $\hat{\psi}(\mathbf{x}, 0)$ is indeterminate. It follows that for an inhomogeneous problem the streamfunction estimate $\hat{\psi}$ interpolates the data ($\mathcal{L}[\hat{\psi}] = \mathbf{d}$) and satisfies the dynamics exactly ($\hat{q}(\mathbf{x}, t) \equiv 0$), but is not uniquely determined (Bennett & Miller, 1991).

Exercise 5.3.2
(*This exercise is strongly recommended; a solution may be found in §5.11. It provides a simple setting for the exploration of material introduced in this section, and in subsequent sections and chapters.*)
Consider simple wave propagation in the horizontal:

$$u_t + u_x = F + f, \qquad (5.3.13)$$

where $u(x, t)$ is the true wave amplitude, $F(x, t)$ is a prior estimate of the forcing, and $f(x, t)$ is the error in that estimate. The region of interest is the entire x-axis: $-\infty < x < \infty$, and u vanishes as $x \to \pm\infty$. The initial condition is

$$u(x, 0) = I(x) + i(x), \qquad (5.3.14)$$

where I is the prior initial estimate and i is the prior initial error. Assume that pointwise data are available:

$$d_m = u(X_m, T_m) + \epsilon_m , \qquad (1 \le m \le M), \qquad (5.3.15)$$

where d_m is a datum and ϵ_m is a measurement error $(1 \le m \le M)$. Formulate (5.3.13)–(5.3.15) as a generalized inverse problem, by defining a suitable quadratic penalty functional. Derive the Euler-Lagrange equations for the wave amplitude $\hat{u}(x, t)$ which minimizes the functional.

 □

5.4 The representer solution

The Euler-Lagrange equations (5.3.2)–(5.3.5) comprise a two-point boundary-value problem in the time interval $0 \le t \le T$. It may be reduced to $2M + 1$ initial-value problems, by expressing the solution of $\hat{\psi}$ in the form

$$\hat{\psi}(\mathbf{x}, t) = \psi_F(\mathbf{x}, t) + \hat{\mathbf{b}}^* \, \mathbf{r}(\mathbf{x}, t) , \qquad (5.4.1)$$

where ψ_F is the prior estimate of the streamfunction ψ, \mathbf{r} is a vector of representer fields and $\hat{\mathbf{b}}$ is a vector of coefficients. The prior estimate ψ_F satisfies

$$\frac{\partial}{\partial t}\psi_F = L\psi_F + F , \qquad (5.4.2)$$

subject to

$$\psi_F(\mathbf{x}, 0) = \Psi(\mathbf{x}) . \qquad (5.4.3)$$

The representer vector \mathbf{r} satisfies

$$\frac{\partial}{\partial t}\mathbf{r}(\mathbf{x}, t) = L\mathbf{r}(\mathbf{x}, t)$$

$$+ (2\pi)^{-2} \int_0^T dt_2 \int d\mathbf{x}_2 \, Q(\mathbf{x}, t, \mathbf{x}_2, t_2) \, \boldsymbol{\alpha}(\mathbf{x}_2, t_2) , \qquad (5.4.4)$$

subject to

$$\mathbf{r}(\mathbf{x}, 0) = (2\pi)^{-2} \int d\mathbf{x}_2 \, A(\mathbf{x}, \mathbf{x}_2) \, \boldsymbol{\alpha}(\mathbf{x}_2, 0) . \qquad (5.4.5)$$

The adjoint representer vector $\boldsymbol{\alpha}$ satisfies

$$-\frac{\partial}{\partial t}\boldsymbol{\alpha}(\mathbf{x},t) = L^{\dagger}\boldsymbol{\alpha}(\mathbf{x},t) + (2\pi)^2\,\boldsymbol{\mathcal{L}}_{(\mathbf{x}_2,t_2)}\left[\delta(\mathbf{x}-\mathbf{x}_2)\,\delta(t-t_2)\right], \quad (5.4.6)$$

subject to

$$\boldsymbol{\alpha}(\mathbf{x},T) = 0. \qquad (5.4.7)$$

An integration of (5.4.6) and (5.4.7) backward in time from $t = T$ yields $\boldsymbol{\alpha}$; a subsequent forward integration of (5.4.4) and (5.4.5) from $t = 0$ yields \mathbf{r}.

Substitution of the representer expression (5.4.1) into the Euler-Lagrange equations (5.3.2)–(5.3.5) yields a linear problem for the vector $\hat{\mathbf{b}}$ of representer coefficients:

$$\mathbf{P}\hat{\mathbf{b}} = \mathbf{h}, \qquad (5.4.8)$$

where

$$\mathbf{P} = \mathbf{R} + \mathbf{w}^{-1}, \qquad (5.4.9)$$
$$\mathbf{R} = \boldsymbol{\mathcal{L}}[\mathbf{r}^*], \qquad (5.4.10)$$

and \mathbf{h} is the vector of prior misfits:

$$\mathbf{h} = \mathbf{d} - \boldsymbol{\mathcal{L}}[\psi_F]. \qquad (5.4.11)$$

The nature of the $M \times M$ representer matrix \mathbf{R} may be determined by considering the representer $\Gamma(\mathbf{x},t,\mathbf{x}',t')$, and its adjoint $\gamma(\mathbf{x},t,\mathbf{x}',t')$, for the evaluation functional \mathcal{L}' at the point (\mathbf{x}',t'):

$$\mathcal{L}'\left[\psi(\mathbf{x},t)\right] = \psi(\mathbf{x}',t'). \qquad (5.4.12)$$

Since

$$\mathcal{L}'_{(\mathbf{x}_2,t_2)}\left[\delta(\mathbf{x}-\mathbf{x}_2)\,\delta(t-t_2)\right] = \delta(\mathbf{x}-\mathbf{x}')\,\delta(t-t'), \qquad (5.4.13)$$

it follows as a textbook exercise that γ is the influence function for the forward problem. For example, the prior estimate ψ_F may be expressed as

$$\psi_F(\mathbf{x},t) = (2\pi)^{-2}\int_0^t dt' \int d\mathbf{x}'\,\gamma(\mathbf{x}',t',\mathbf{x},t)\,F(\mathbf{x}',t')$$

$$+ (2\pi)^{-2}\int d\mathbf{x}'\,\gamma(\mathbf{x}',0,\mathbf{x},t)\,\Psi(\mathbf{x}'). \qquad (5.4.14)$$

Moreover, Γ itself may be expressed as

$$
\Gamma(\mathbf{x}_1, t_1, \mathbf{x}_2, t_2) = (2\pi)^{-4} \int_0^{t_1} dt_3 \int d\mathbf{x}_3 \int_0^{t_2} dt_4 \int d\mathbf{x}_4 \, \gamma(\mathbf{x}_3, t_3, \mathbf{x}_1, t_1)
$$
$$
\times Q(\mathbf{x}_3, t_3, \mathbf{x}_4, t_4) \, \gamma(\mathbf{x}_4, t_4, \mathbf{x}_2, t_2)
$$
$$
+ (2\pi)^{-4} \int d\mathbf{x}_3 \int d\mathbf{x}_4 \, \gamma(\mathbf{x}_3, 0, \mathbf{x}_1, t_1)
$$
$$
\times A(\mathbf{x}_3, \mathbf{x}_4) \, \gamma(\mathbf{x}_4, 0, \mathbf{x}_2, t_2) \,. \qquad (5.4.15)
$$

Inspection of (5.4.15) reveals that Γ is symmetric; it may be shown that Γ is also positive-definite. It is easily seen that

$$
\mathbf{r}(\mathbf{x}, t) = \boldsymbol{\mathcal{L}}_{(\mathbf{x}', t')} \left[\Gamma(\mathbf{x}, t, \mathbf{x}', t') \right] \qquad (5.4.16)
$$

and so

$$
\mathbf{R} = \boldsymbol{\mathcal{L}}_{(\mathbf{x}, t)} \boldsymbol{\mathcal{L}}^*_{(\mathbf{x}', t')} \left[\Gamma(\mathbf{x}, t, \mathbf{x}', t') \right] \,. \qquad (5.4.17)
$$

It may be concluded from (5.4.17) that the representer matrix \mathbf{R} is symmetric and positive-definite. Since the data weight \mathbf{w} has these properties, it follows that \mathbf{P} also has them. Hence there is a unique solution for $\hat{\mathbf{b}}$, which must yield the unique solution $\hat{\psi}$ of the Euler-Lagrange equations. The conditioning of \mathbf{P} is discussed in Chapter 6.

Exercise 5.4.1 Find equations for Γ and γ, analogous to those for \mathbf{r} and $\boldsymbol{\alpha}$. Prove that γ is the influence function for the forward model, and derive the expression (5.4.15) for Γ. Prove that Γ is positive-definite, by exploiting the uniqueness of solutions of the forward model. □

In summary, the representer expression for the minimum of $\mathcal{J}[\psi]$ is

$$
\hat{\psi}(\mathbf{x}, t) = \psi_F(\mathbf{x}, t) + \mathbf{r}(\mathbf{x}, t)^* \left(\mathbf{R} + \mathbf{w}^{-1} \right)^{-1} \left(\mathbf{d} - \boldsymbol{\mathcal{L}}[\psi_F] \right), \qquad (5.4.18)
$$

and the corresponding estimate for the forcing error q is

$$
\hat{q}(\mathbf{x}, t) = \mathbf{s}(\mathbf{x}, t)^* \left(\mathbf{R} + \mathbf{w}^{-1} \right)^{-1} \left(\mathbf{d} - \boldsymbol{\mathcal{L}}[\psi_F] \right), \qquad (5.4.19)
$$

where \mathbf{s} is a vector of model residuals associated with the representers \mathbf{r}:

$$
\mathbf{s} = \left(\frac{\partial}{\partial t} - L \right) \mathbf{r} \,. \qquad (5.4.20)
$$

In terms of Γ, the streamfunction estimate is

$$\hat{\psi}(\mathbf{x}, t) = \psi_F(\mathbf{x}, t) + \boldsymbol{\mathcal{L}}^*_{(\mathbf{x}_1, t_1)} \Big[\Gamma(\mathbf{x}, t, \mathbf{x}_1, t_1) \Big]$$

$$\times \Big\{ \boldsymbol{\mathcal{L}}_{(\mathbf{x}_2, t_2)} \boldsymbol{\mathcal{L}}^*_{(\mathbf{x}_3, t_3)} \Big[\Gamma(\mathbf{x}_2, t_2, \mathbf{x}_3, t_3) \Big] + \mathbf{w}^{-1} \Big\}^{-1} (\mathbf{d} - \boldsymbol{\mathcal{L}}[\psi_F]); \quad (5.4.21)$$

an analogous expression holds for the estimated forcing error \hat{q}.

It is not necessary to evaluate the representer expression (5.4.1) in order to find $\hat{\psi}(\mathbf{x}, t)$. Using the result

$$\mathbf{w}\big(\mathbf{d} - \boldsymbol{\mathcal{L}}[\hat{\psi}]\big) = \hat{\mathbf{b}}, \qquad (5.4.22)$$

the right-hand side of (5.3.4) may be explicitly evaluated, and so λ may be found by a backward integration. Then $\hat{\psi}$ follows from a forward integration, for a total of $2M + 3$ integrations (one for ψ_F, $2M$ for the representer vector \mathbf{r} and hence the representer matrix \mathbf{R}, and two more for $\hat{\psi}$). The last two integrations obviate the need to store the representer vector $\mathbf{r}(\mathbf{x}, t)$ for all \mathbf{x} and t. By overwriting variables during computation, only one scalar field over (\mathbf{x}, t) need be stored at any stage in the construction of $\hat{\psi}$. Since the calculations of the M representers are entirely independent of each other, the algorithm is readily adapted to multi-processor computers: see Chapter 10.

5.5 Geometrical formulation

The representer expression (5.4.1) for the solution $\hat{\psi}$ of the Euler-Lagrange equations contains M disposable parameters $\hat{b}_1, \ldots, \hat{b}_M$. The choice (5.4.8) leads to the minimum of the quadratic penalty functional $\mathcal{J}[\psi]$. Not every streamfunction may be expressed in the form (5.4.1). For example, suppose that $\boldsymbol{\mathcal{L}}$ is comprised entirely of evaluation functionals such that $\mathcal{L}_M\big[\psi(\mathbf{x}, t)\big] = \psi(\mathbf{X}_m, T_m)$, $1 \leq m \leq M$. Then the field

$$\psi(\mathbf{x}, t) = \psi_F(\mathbf{x}, t) + \prod_{m=1}^{M} |\mathbf{x} - \mathbf{X}_m| \, (t - T_m) \qquad (5.5.1)$$

cannot be so expressed. A class of fields has been discarded by seeking the minimum of $\mathcal{J}[\psi]$ as a solution of the Euler-Lagrange

equations: the following geometrical formulation characterizes the discarded fields.

To begin, define an inner product by

$$\langle \theta, \phi \rangle = (2\pi)^{-4} \int_0^T dt_1 \int_0^T dt_2 \int dx_1 \int dx_2 \Big\{ \frac{\partial}{\partial t_1} \theta(x_1, t_1)$$

$$- L\theta(x_1, t_1) \Big\} W(x_1, t_1, x_2, t_2) \Big\{ \frac{\partial}{\partial t_2} \phi(x_2, t_2) - L\phi(x_2, t_2) \Big\}$$

$$+ (2\pi)^{-4} \int dx_1 \int dx_2 \, \theta(x_1, 0) V(x_1, x_2) \phi(x_2, 0) \qquad (5.5.2)$$

for any pair of functions $\theta = \theta(x, t)$ and $\phi = \phi(x, t)$. It may be readily shown that the representer $\Gamma(x, t, x', t')$ for the evaluation functional at (x', t') has the property

$$\Big\langle \Gamma(x, t, x', t'), \psi(x, t) \Big\rangle = \psi(x', t') \qquad (5.5.3)$$

for any field $\psi = \psi(x, t)$. That is, Γ is the reproducing kernel (rk) for the inner product. Recalling that the representers r may be constructed by measuring Γ:

$$r(x, t) = \mathcal{L}_{(x', t')} \big[\Gamma(x, t, x', t') \big], \qquad (5.5.4)$$

it follows that, for any field ψ,

$$\Big\langle r(x, t), \psi(x, t) \Big\rangle = \mathcal{L}[\psi]. \qquad (5.5.5)$$

That is, r represents the measurement functionals \mathcal{L} in the sense of the inner product. Hence the penalty functional \mathcal{J} may be expressed in terms of the inner product and representers:

$$\mathcal{J}[\psi] = \Big\langle \psi - \psi_F, \psi - \psi_F \Big\rangle + \Big(d - \langle \psi, r \rangle \Big)^* w \Big(d - \langle \psi, r \rangle \Big). \qquad (5.5.6)$$

Now any field $\psi = \psi(x, t)$ whatsoever may be expressed as a linear combination of ψ_F, the representers r, and a field g orthogonal to the representers:

$$\psi(x, t) = \psi_F(x, t) + r(x, t)^* b + g(x, t), \qquad (5.5.7)$$

where
$$\langle \mathbf{r}, g \rangle = \mathbf{0} \,. \tag{5.5.8}$$

The field g may be regarded as *unobservable*. (If ψ were given, then the coefficients \mathbf{b} would be determined by (5.5.8).) Substituting (5.5.7) into (5.5.6) yields

$$\mathcal{J}[\psi] = \mathbf{b}^{*}\mathbf{R}\mathbf{b} + \langle g, g \rangle + (\mathbf{R}\mathbf{b} - \mathbf{h})^{*}\mathbf{w}(\mathbf{R}\mathbf{b} - \mathbf{h}) \,, \tag{5.5.9}$$

where \mathbf{R} is the positive-definite, symmetric representer matrix

$$\mathbf{R} = \langle \mathbf{r}, \mathbf{r}^{*} \rangle = \mathcal{L}[\mathbf{r}^{*}] \,, \tag{5.5.10}$$

and \mathbf{h} is again the residual for measurements of the prior streamfunction estimate ψ_F:
$$\mathbf{h} = \mathbf{d} - \mathcal{L}[\psi_F] \,. \tag{5.5.11}$$

Thus far all that has been achieved is the compact expression (5.5.9) for \mathcal{J}. The disposable parameters are the coefficient vector \mathbf{b} and the field g. If ψ minimizes \mathcal{J}, then clearly $\langle g, g \rangle = 0$; as a consequence of the uniqueness of solutions of the forward problem,

$$g(\mathbf{x}, t) \equiv 0 \,. \tag{5.5.12}$$

The unobservable field g must be discarded, reducing \mathcal{J} from the infinite-dimensional quadratic functional (5.3.1) to the finite-dimensional quadratic form

$$\mathcal{J}[\mathbf{b}] = \mathbf{b}^{*}\mathbf{R}\mathbf{b} + (\mathbf{R}\mathbf{b} - \mathbf{h})^{*}\mathbf{w}(\mathbf{R}\mathbf{b} - \mathbf{h}) \,. \tag{5.5.13}$$

Minimization of \mathcal{J} may be completed after rearranging (5.5.13) as

$$\mathcal{J}[\mathbf{b}] = (\mathbf{b} - \hat{\mathbf{b}})^{*}\mathbf{S}(\mathbf{b} - \hat{\mathbf{b}}) + \hat{\mathcal{J}} \,, \tag{5.5.14}$$

where
$$\mathbf{S} = \mathbf{R} + \mathbf{R}\mathbf{w}\mathbf{R} \,, \tag{5.5.15}$$
$$\mathbf{S}\hat{\mathbf{b}} = \mathbf{R}\mathbf{w}\mathbf{h} \,, \tag{5.5.16}$$
and

$$\hat{\mathcal{J}} = \mathbf{h}^*\mathbf{w}\mathbf{h} - \hat{\mathbf{b}}\mathbf{S}\hat{\mathbf{b}}\,. \qquad (5.5.17)$$

Inspection of (5.5.14) shows that \mathbf{S} is the Hessian matrix associated with the finite-dimensional penalty function $\mathcal{J}[\mathbf{b}]$; that is, \mathbf{S} is the matrix of second partial derivatives of \mathcal{J} with respect to \mathbf{b}. The optimal choice for \mathbf{b} is

$$\mathbf{b} = \hat{\mathbf{b}} = \mathbf{P}^{-1}\mathbf{h}\,, \qquad (5.5.18)$$

where again $\mathbf{P} = \mathbf{R} + \mathbf{w}^{-1}$, and the minimum value of \mathcal{J} is

$$\hat{\mathcal{J}} = \mathcal{J}[\hat{\psi}] = \mathbf{h}^*\mathbf{P}^{-1}\mathbf{h}\,. \qquad (5.5.19)$$

Note that the prior estimate of \mathcal{J} is

$$\mathcal{J}_F = \mathcal{J}[\psi_F] = \mathbf{h}^*\mathbf{w}\mathbf{h}\,. \qquad (5.5.20)$$

In summary, $\mathcal{J}[\psi]$ is minimized by

$$\hat{\psi}(\mathbf{x}, t) = \psi_F(\mathbf{x}, t) + \mathbf{r}(\mathbf{x}, t)^* \, \mathbf{P}^{-1}\mathbf{h}\,, \qquad (5.5.21)$$

in agreement with (5.4.18): the geometrical construction of $\hat{\psi}$ is equivalent to solving the Euler-Lagrange equations for $\hat{\psi}$. Any other approach to minimizing \mathcal{J} must search through the infinite-dimensional space of unobservable fields, even though the null field (5.5.12) is the only correct choice.

5.6 Error estimates for the generalized inverse

In order to develop error estimates for the generalized inverse $\hat{\psi}$, it will be necessary to make assumptions about the errors ϵ, q and a in the prior estimates \mathbf{d}, F and Ψ for the data, forcing and initial conditions respectively. These assumptions are:

$$\bar{\epsilon} = \mathbf{0}\,, \qquad \bar{q}(\mathbf{x}, t) = \bar{a}(\mathbf{x}) = 0\,, \qquad (5.6.1)$$

$$\overline{\epsilon q(\mathbf{x}, t)} = \overline{\epsilon a(\mathbf{x})} = \mathbf{0}\,, \qquad \overline{q(\mathbf{x}_1, t)a(\mathbf{x}_2)} = 0\,, \qquad (5.6.2)$$

and

$$\overline{\epsilon\epsilon^*} = \mathbf{w}^{-1}, \qquad \overline{q(\mathbf{x}_1, t_1)q(\mathbf{x}_2, t_2)} = Q(\mathbf{x}_1, t_1, \mathbf{x}_2, t_2),$$
$$\overline{a(\mathbf{x}_1)a(\mathbf{x}_2)} = A(\mathbf{x}_1, \mathbf{x}_2), \tag{5.6.3}$$

where \mathbf{w}^{-1}, Q and A are the inverses of \mathbf{w}, W and V. The means in (5.6.1)–(5.6.3) may be formally regarded as moments of distributions conditional upon the prior estimates \mathbf{d}, F and Ψ. Hence

$$\overline{\psi} = \psi_F, \quad \overline{\mathbf{d}} = \mathcal{L}[\overline{\psi}_F], \quad \overline{\mathbf{h}} = \mathbf{0}, \tag{5.6.4}$$

where ψ, satisfying (5.2.2) and (5.2.5), is the true streamfunction. The error covariance for $\hat{\psi}$ is defined by

$$C_{\hat{\psi}\hat{\psi}}(\mathbf{x}_1, t_1, \mathbf{x}_2, t_2) \equiv$$
$$\overline{\left(\psi(\mathbf{x}_1, t_1) - \hat{\psi}(\mathbf{x}_1, t_1)\right)\left(\psi(\mathbf{x}_2, t_2) - \hat{\psi}(\mathbf{x}_2, t_2)\right)}. \tag{5.6.5}$$

Substituting for $\hat{\psi}$ using the representer expression (5.4.1) yields

$$C_{\hat{\psi}\hat{\psi}} = C_{\psi\psi} + e^{(2)} + e^{(3)} \tag{5.6.6}$$

where

$$C_{\psi\psi}(\mathbf{x}_1, t_1, \mathbf{x}_2, t_2) \equiv \overline{(\psi(\mathbf{x}_1, t_1) - \overline{\psi}(\mathbf{x}_1, t_1))(\psi(\mathbf{x}_2, t_2) - \overline{\psi}(\mathbf{x}_2, t_2))}, \tag{5.6.7}$$

$$e^{(2)}(\mathbf{x}_1, t_1, \mathbf{x}_2, t_2) = -2\overline{\{\psi(\mathbf{x}_1, t_1)' \, \hat{\mathbf{b}}^*\}} \, \mathbf{r}(\mathbf{x}_2, t_2), \tag{5.6.8}$$

and

$$e^{(3)}(\mathbf{x}_1, t_1, \mathbf{x}_2, t_2) = \mathbf{r}(\mathbf{x}_1, t_1)^* \overline{\hat{\mathbf{b}}\hat{\mathbf{b}}^*} \, \mathbf{r}(\mathbf{x}_2, t_2). \tag{5.6.9}$$

Note that $\overline{\mathbf{h}} = \mathbf{0}$ implies $\overline{\hat{\mathbf{b}}} = \mathbf{0}$ (the representers are not random variables here). The three contributions to $C_{\hat{\psi}\hat{\psi}}$ will now be obtained in turn.

The streamfunction fluctuation $\psi' = \psi - \overline{\psi}$ may be expressed in terms of q and a using the influence function γ:

$$\psi(\mathbf{x}, t)' = (2\pi)^{-2} \int_0^t dt_1 \int d\mathbf{x}_1 \, \gamma(\mathbf{x}_1, t_1, \mathbf{x}, t) \, q(\mathbf{x}_1, t_1)$$

$$+ (2\pi)^{-2} \int d\mathbf{x}_1 \, \gamma(\mathbf{x}_1, 0, \mathbf{x}, t) \, a(\mathbf{x}_1). \tag{5.6.10}$$

It follows from (5.6.2), (5.6.3), (5.6.7) and (5.4.15) that

$$C_{\psi\psi}(\mathbf{x}_1, t_1, \mathbf{x}_2, t_2) \equiv \overline{\psi(\mathbf{x}_1, t_1)'\, \psi(\mathbf{x}_2, t_2)'} = \Gamma(\mathbf{x}_1, t_1, \mathbf{x}_2, t_2). \quad (5.6.11)$$

So the covariance of the true streamfunction ψ is the rk for the inner product $\langle\ ,\ \rangle$. Reduction of $e^{(2)}$ begins with (5.4.8). Then (5.2.1), (5.4.15), (5.5.4) and (5.6.3) yield

$$e^{(2)}(\mathbf{x}_1, t_1, \mathbf{x}_2, t_2) = -2\mathbf{r}(\mathbf{x}_1, t_1)^*\mathbf{P}^{-1}\mathbf{r}(\mathbf{x}_2, t_2). \quad (5.6.12)$$

A similar manipulation yields

$$e^{(3)}(\mathbf{x}_1, t_1, \mathbf{x}_2, t_2) = \mathbf{r}(\mathbf{x}_1, t_1)^*\mathbf{P}^{-1}\mathbf{r}(\mathbf{x}_2, t_2). \quad (5.6.13)$$

Combining (5.6.11)–(5.6.13) yields the streamfunction error covariance $C_{\hat\psi\hat\psi}$ in terms of the rk and the representers:

$$C_{\hat\psi\hat\psi}(\mathbf{x}_1, t_1, \mathbf{x}_2, t_2) = \Gamma(\mathbf{x}_1, t_1, \mathbf{x}_2, t_2)$$
$$- \mathbf{r}(\mathbf{x}_1, t_1)^*\mathbf{P}^{-1}\mathbf{r}(\mathbf{x}_2, t_2). \quad (5.6.14)$$

Now let \hat{q} be the generalized inverse estimate of the forcing error q, given by (5.3.8), that is,

$$\frac{\partial}{\partial t}\hat{\psi} = L\hat{\psi} + F + \hat{q}. \quad (5.6.15)$$

The error covariance for \hat{q} is

$$C_{\hat{q}\hat{q}}(\mathbf{x}_1, t_1, \mathbf{x}_2, t_2) \equiv \overline{\big(q(\mathbf{x}_1, t_1) - \hat{q}(\mathbf{x}_1, t_1)\big)\big(q(\mathbf{x}_2, t_2) - \hat{q}(\mathbf{x}_2, t_2)\big)}. \quad (5.6.16)$$

An explicit expression for $C_{\hat{q}\hat{q}}$ may be found by applying the model operator to (5.6.14):

$$C_{\hat{q}\hat{q}}(\mathbf{x}_1, t_1, \mathbf{x}_2, t_2) = Q(\mathbf{x}_1, t_1, \mathbf{x}_2, t_2) - \mathbf{s}(\mathbf{x}_1, t_1)^*\mathbf{P}^{-1}\mathbf{s}(\mathbf{x}_2, t_2), \quad (5.6.17)$$

where, again, s is the residual for r:

$$\frac{\partial}{\partial t}\mathbf{r} = L\mathbf{r} + \mathbf{s}. \quad (5.6.18)$$

The error covariance for the generalized inverse estimate \hat{a} of the initial error a is

$$C_{\hat{a}\hat{a}}(\mathbf{x}_1, \mathbf{x}_2) \equiv \overline{\big(a(\mathbf{x}_1) - \hat{a}(\mathbf{x}_1)\big)\big(a(\mathbf{x}_2) - \hat{a}(\mathbf{x}_2)\big)}\,. \qquad (5.6.19)$$

An explicit expression for $C_{\hat{a}\hat{a}}$ may be found by evaluating (5.6.14) at $t_1 = t_2 = 0$, and by making use of (5.6.11):

$$C_{\hat{a}\hat{a}}(\mathbf{x}_1, \mathbf{x}_2) = A(\mathbf{x}_1, \mathbf{x}_2) - \mathbf{r}(\mathbf{x}_1, 0)^*\mathbf{P}^{-1}\mathbf{r}(\mathbf{x}_2, 0)\,. \qquad (5.6.20)$$

Finally, applying the measurement functionals \mathcal{L} to (5.6.14) yields the error covariance of the inverse estimate $\hat{\epsilon}$ $\big(\equiv \mathbf{d} - \mathcal{L}[\hat{\psi}]\big)$ for the measurement error ϵ:

$$\mathbf{C}_{\hat{\epsilon}\hat{\epsilon}} \equiv \overline{(\epsilon - \hat{\epsilon})(\epsilon - \hat{\epsilon})^*} = \mathbf{R} - \mathbf{R}\mathbf{P}^{-1}\mathbf{R}$$
$$= \mathbf{w}^{-1} - \mathbf{w}^{-1}\mathbf{P}^{-1}\mathbf{w}^{-1}\,. \qquad (5.6.21)$$

The first terms Γ, Q, A and \mathbf{w}^{-1}, in (5.6.14), (5.6.17), (5.6.20) and (5.6.21), respectively, are the covariances of the unknowns ψ', q, a and ϵ. Recall that $\psi' \equiv \psi - \bar{\psi} = \psi - \psi_F$. Thus, Γ, Q, A and \mathbf{w}^{-1} are the *prior* error covariances, that is, the error covariances of the prior estimates ψ_F, F, Ψ and \mathbf{d} for the true streamfunction ψ, the true forcing $F + q$, the true initial value $\Psi + a$ and true measurements $\mathcal{L}[\psi]$. The second terms in (5.6.14), (5.6.17), (5.6.20) and (5.6.21) are the covariance reductions due to inversion, i.e., the *explained* covariances. The differences $C_{\hat{\psi}\hat{\psi}}$, $C_{\hat{q}\hat{q}}$, $C_{\hat{a}\hat{a}}$ and $\mathbf{C}_{\hat{\epsilon}\hat{\epsilon}}$ between prior and explained covariances are the *posterior* error covariances.

This completes the task of finding posterior error covariances in terms of the rk Γ, the representers \mathbf{r} and their residuals \mathbf{s}, and the representer matrix \mathbf{R}. These in turn depend upon the model operator $\partial/\partial t - L$, the measurement functionals \mathcal{L}, plus the inverses Q, A and \mathbf{w}^{-1} of the weights W, V and \mathbf{w} respectively. In particular, the actual prior estimates F, Ψ and \mathbf{d} do not appear, thus the posterior error covariances may be determined before performing an inversion.

A procedure for calculating \mathbf{r}, \mathbf{s} and \mathbf{R} has been described in §5.4. Thus $C_{\hat{q}\hat{q}}$ and $C_{\hat{\epsilon}\hat{\epsilon}}$ are determined. However $C_{\hat{\psi}\hat{\psi}}$ depends upon Γ, the computation of which is a formidable task even for moderate resolution in space and time. There is a tremendous amount of information in (5.6.14); even just the error variance, given by

$$C_{\hat{\psi}\hat{\psi}}(\mathbf{x}, t, \mathbf{x}, t) = \Gamma(\mathbf{x}, t, \mathbf{x}, t) - \mathbf{r}(\mathbf{x}, t)^*\mathbf{P}^{-1}\mathbf{r}(\mathbf{x}, t)\,, \qquad (5.6.22)$$

would suffice. However, it is not possible to derive an evolution equation for $\Gamma(\mathbf{x}, t, \mathbf{x}, t)$ in closed form, that is, without reference to two-point functions.

Estimates of the variance (5.6.22) could be made by statistical simulation, using pseudo-random fields $q = q(\mathbf{x}, t)$ and $a = a(\mathbf{x})$ consistent with (5.6.1)–(5.6.3). The representers could be estimated at the same time, since (5.6.11) implies that

$$\mathbf{r}(\mathbf{x}, t) = \overline{\psi'(\mathbf{x}, t) \, \boldsymbol{\mathcal{L}}[\psi']} \tag{5.6.23}$$

and

$$\mathbf{R} = \overline{\boldsymbol{\mathcal{L}}[\psi'] \, \boldsymbol{\mathcal{L}}[\psi']^*} \,. \tag{5.6.24}$$

That is, the representers are the streamfunction–data covariances. Note that the model must be integrated once per sample. The standard errors in the sample of N estimates are $O(N^{-\frac{1}{2}})$; if such an error is sufficiently small and if $N << 2M + 3$, then statistical simulation would be more efficient than the deterministic procedure for computing the representers described in §5.3. The latter procedure, it is emphasized, does not yield the rk Γ.

Once again, the identity relation (5.6.11) between the rk Γ and the streamfunction covariance $C_{\hat{\psi}\hat{\psi}}$ shows that the generalized inverse (5.4.18) for the data, model and initial conditions is identical to the space-time Gauss-Markov smoother for the data using the covariance of the measurement errors and the covariance of solutions of the model and initial conditions.

The representer analysis may also be used to express the reduction in the penalty functional due to minimization. The prior value is, again,

$$\mathcal{J}_F = \mathcal{J}[\psi_F] = \mathbf{h}^* \mathbf{w} \mathbf{h} \,, \tag{5.6.25}$$

where \mathbf{h} is the prior data misfit, while the minimal or posterior value is, again,

$$\hat{\mathcal{J}} = \mathcal{J}[\hat{\psi}] = \mathbf{h}^* \, \mathbf{P}^{-1} \mathbf{h} \,. \tag{5.6.26}$$

Now the prior data misfit \mathbf{h} may be expressed as

$$\mathbf{h} = \mathbf{d} - \boldsymbol{\mathcal{L}}[\psi_F] = \boldsymbol{\mathcal{L}}[\psi'] + \boldsymbol{\epsilon} \,, \tag{5.6.27}$$

thus its covariance is

$$\overline{\mathbf{h}\mathbf{h}^*} = \overline{\boldsymbol{\mathcal{L}}[\psi'] \, \boldsymbol{\mathcal{L}}[\psi']^*} + \overline{\boldsymbol{\epsilon}\boldsymbol{\epsilon}^*} = \mathbf{R} + \mathbf{w}^{-1} = \mathbf{P} \,, \tag{5.6.28}$$

and so the mean value of $\mathcal{J}[\psi_F]$ is

$$\overline{\mathcal{J}_F} = \text{Tr}(\mathbf{w}^{\frac{1}{2}}\overline{\mathbf{h}\mathbf{h}^*}\mathbf{w}^{\frac{1}{2}}) \tag{5.6.29}$$

$$= \text{Tr}(\mathbf{w}^{\frac{1}{2}}\mathbf{R}\mathbf{w}^{\frac{1}{2}}) + M, \tag{5.6.30}$$

where $\text{Tr}(\mathbf{A})$ denotes the trace of matrix \mathbf{A}. In particular, $\text{Tr}(\mathbf{I}) = M$ when \mathbf{I} is the $M \times M$ unit matrix. The mean value of $\widehat{\mathcal{J}}$ is

$$\overline{\widehat{\mathcal{J}}} = \text{Tr}(\mathbf{I}) = M. \tag{5.6.31}$$

If the weight matrix is just a multiple of the unit matrix: $\mathbf{w} = w\mathbf{I}$ (uncorrelated measurement errors with common variance w^{-1}), then the mean reduction in the penalty functional is simply

$$\overline{\mathcal{J}_F} - \overline{\widehat{\mathcal{J}}} = \text{Tr}(\mathbf{w}^{\frac{1}{2}}\mathbf{R}\mathbf{w}^{\frac{1}{2}}) = w \sum_{m=1}^{M} \lambda_m, \tag{5.6.32}$$

where $\lambda_1, \dots, \lambda_M$ are the real, positive eigenvalues of the representer matrix \mathbf{R}. In practice, the mean reduction may exceed $\overline{\widehat{\mathcal{J}}}$ by orders of magnitude.

Exercise 5.6.1 Show that the error covariance for the inverse estimates of the representer coefficients is the inverse of the Hessian matrix \mathbf{S}:

$$\overline{(\mathbf{b} - \hat{\mathbf{b}})(\mathbf{b} - \hat{\mathbf{b}})^*} = \mathbf{S}^{-1}. \tag{5.6.33}$$

□

Exercise 5.6.2 Use (5.6.33) to construct (5.6.21), which is the error covariance for the inverse estimate of the streamfunction measurements. □

Exercise 5.6.3 Can (5.6.14), (5.6.17) and (5.6.20), that is, the error covariances for the inverse estimates of the streamfunction, forcing and initial values, respectively, be constructed from (5.6.33)? □

5.7 The sweep algorithm

Representers reduce the Euler-Lagrange equations (5.3.2)–(5.3.5) from a two-point boundary value problem in time to a finite number

of initial-value problems and a problem in linear algebra. The sweep algorithm of Gelfand & Fomin (1963) also reduces the Euler-Lagrange equations to a number of initial-value problems.

Consider the special case

$$Q(\mathbf{x}_1, t_1, \mathbf{x}_2, t_2) = Q(\mathbf{x}_1, \mathbf{x}_2)\,\delta(t_1 - t_2)\,. \qquad (5.7.1)$$

That is, the system noise $q(\mathbf{x}, t)$ is white in time: see (5.6.3). Then (5.3.2) simplifies to

$$\frac{\partial}{\partial t}\hat{\psi}(\mathbf{x}, t) = L\hat{\psi}(\mathbf{x}, t) + F(\mathbf{x}, t)$$

$$+ (2\pi)^{-2}\int d\mathbf{x}_2\, Q(\mathbf{x}, \mathbf{x}_2)\, \lambda(\mathbf{x}_2, t)\,. \qquad (5.7.2)$$

Moreover, assume that the measurements are taken at discrete times $t = t_k$, $(0 \le t_k \le T, 1 \le k \le M_T)$, and that there are M_X spatial measurement functionals defined at each time t_k, so that the data may be expressed as

$$\boldsymbol{\mathcal{L}}_{(\mathbf{x})}[\psi(\mathbf{x}, t_k)] + \boldsymbol{\epsilon}^k = \mathbf{d}^k\,, \qquad (1 \le k \le M_T)\,, \qquad (5.7.3)$$

where the measurement vector $\boldsymbol{\mathcal{L}}$, error vector $\boldsymbol{\epsilon}^k$ and data vector \mathbf{d}^k are all of length M_X. Finally, assume that data errors at different times are uncorrelated, but have the same covariance matrix:

$$\overline{\boldsymbol{\epsilon}^k(\boldsymbol{\epsilon}^j)^*} = \mathbf{w}^{-1}\delta_{kj}\,, \qquad (5.7.4)$$

where \mathbf{w} is now an $M_X \times M_X$ positive-definite symmetric matrix. With all these assumptions, (5.3.4) becomes

$$-\frac{\partial}{\partial t}\lambda(\mathbf{x}, t) = L^{\dagger}\lambda(\mathbf{x}, t)$$

$$+ (2\pi)^2\boldsymbol{\mathcal{L}}^*_{(\mathbf{x}_2)}\left[\delta(\mathbf{x} - \mathbf{x}_2)\right]\sum_{k=1}^{M_T}\delta(t - t_k)\mathbf{w}\left(\mathbf{d}^k - \boldsymbol{\mathcal{L}}_{(\mathbf{x}_3)}[\hat{\psi}(\mathbf{x}_3, t_k)]\right).$$

$$(5.7.5)$$

It is evident from (5.7.5) that λ has simple discontinuities in time. That is,

$$-\frac{\partial}{\partial t}\lambda(\mathbf{x}, t) = L^{\dagger}\lambda(\mathbf{x}, t)\,, \qquad (t_k < t < t_{k+1})\,, \qquad (5.7.6)$$

and

$$-\lambda(\mathbf{x}, t_{k-}) + \lambda(\mathbf{x}, t_{k+})$$
$$= (2\pi)^2 \boldsymbol{\mathcal{L}}^*_{(\mathbf{x}_2)} \left[\delta(\mathbf{x} - \mathbf{x}_2)\right] \mathbf{w}\left(\mathbf{d}^k - \boldsymbol{\mathcal{L}}_{(\mathbf{x}_3)} \left[\hat{\psi}(\mathbf{x}_3, t_k)\right]\right),$$
$$(1 \le k \le M_T). \qquad (5.7.7)$$

In (5.7.7), $\lambda(\mathbf{x}, t_{k\pm}) = \lim_{\epsilon \to 0} \lambda(\mathbf{x}, t_k \pm \epsilon)$. Note that while λ is in general discontinuous, $\hat{\psi}$ is continuous but has a discontinuous time derivative at each data time.

Since the linear, two-point boundary-value problem (5.7.2), (5.3.3), (5.7.6), (5.7.7), (5.3.5) has a unique solution, no generality is lost if a solution is sought in a particular form, such as a linear relationship between $\hat{\psi}$ and λ:

$$\hat{\psi}(\mathbf{x}, t) = (2\pi)^{-2} \int d\mathbf{x}_2 \, P(\mathbf{x}, \mathbf{x}_2, t) \lambda(\mathbf{x}_2, t) + \phi(\mathbf{x}, t) \qquad (5.7.8)$$

for some as yet undetermined "slope" $P = P(\mathbf{x}_1, \mathbf{x}_2, t)$ and "intercept" $\phi = \phi(\mathbf{x}, t)$. Substituting (5.7.8) into (5.7.2), using (5.7.6), and equating integrands proportional to λ yields

$$\frac{\partial}{\partial t} P(\mathbf{x}_1, \mathbf{x}_2, t) = (L_{(\mathbf{x}_1)} + L_{(\mathbf{x}_2)}) P(\mathbf{x}_1, \mathbf{x}_2, t) + Q(\mathbf{x}_1, \mathbf{x}_2), \quad (5.7.9)$$

$$\frac{\partial}{\partial t} \phi(\mathbf{x}, t) = L\phi(\mathbf{x}, t) + F(\mathbf{x}, t), \qquad (t_k < t < t_{k+1}). \quad (5.7.10)$$

Substituting (5.7.7) into (5.7.8) and invoking the continuity of $\hat{\psi}$ yields

$$P(\mathbf{x}_1, \mathbf{x}_2, t_{k+}) - P(\mathbf{x}_1, \mathbf{x}_2, t_{k-}) = -\boldsymbol{\mathcal{L}}^*_{(\mathbf{x}_3)} \left[P(\mathbf{x}_1, \mathbf{x}_3, t_{k-})\right] \Big\{ \boldsymbol{\mathcal{L}}_{(\mathbf{x}_4)} \boldsymbol{\mathcal{L}}^*_{(\mathbf{x}_5)}$$
$$\times \left[P(\mathbf{x}_4, \mathbf{x}_5, t_{k-})\right] + \mathbf{w}^{-1} \Big\}^{-1}$$
$$\times \boldsymbol{\mathcal{L}}_{(\mathbf{x}_6)} \left[P(\mathbf{x}_6, \mathbf{x}_2, t_{k-})\right], \qquad (5.7.11)$$

$$\phi(\mathbf{x}, t_{k+}) - \phi(\mathbf{x}, t_{k-}) = -\mathbf{K}^{(k)}(\mathbf{x})^* \left(\mathbf{d}^k - \boldsymbol{\mathcal{L}}_{(\mathbf{x}_2)} [\phi(\mathbf{x}_2, t_{k-})]\right),$$
$$(5.7.12)$$

where

$$\mathbf{K}^{(k)}(\mathbf{x}) = \mathbf{w} \boldsymbol{\mathcal{L}}_{(\mathbf{x}_2)} \left[P(\mathbf{x}, \mathbf{x}_2, t_{k+})\right]. \qquad (5.7.13)$$

Substituting (5.7.8) into (5.3.3) yields

$$P(\mathbf{x}_1, \mathbf{x}_2, 0) = A(\mathbf{x}_1, \mathbf{x}_2), \qquad (5.7.14)$$

$$\phi(\mathbf{x}, 0) = \Psi(\mathbf{x}). \qquad (5.7.15)$$

The system (5.7.9)–(5.7.15) may be recognized as the Kalman filter for the model (5.2.2), initial condition (5.2.5) and data (5.7.3), with ϕ being the KF estimate of the streamfunction and P the KF error covariance. This reduction is due to Meditch (1970); the algorithm (5.7.8) for sweeping two-point boundary-value problems into initial-value problems is due to Gelfand & Fomin (1963). To the extent that the first term in (5.7.8) may be neglected, the KF streamfunction ϕ provides a suboptimal approximation to the generalized inverse estimate $\hat{\psi}$. As a consequence of (5.3.5), $\phi = \hat{\psi}$ at $t = T$, but ϕ does not minimize the penalty functional \mathcal{J}. However, ϕ does minimize the jump (5.7.11) in the error covariance for sequential estimates: see Chapter 3.

Having integrated (5.7.9)–(5.7.15) for ϕ and P, the adjoint variable λ may be found by integration of (5.7.6) backward from (5.3.5). The jump condition (5.7.7) for λ at the data times becomes explicit (Bryson & Ho, 1975) if the sweep substitution is made for $\hat{\psi}$:

$$\lambda(\mathbf{x}, t_{k-}) - \lambda(\mathbf{x}, t_{k+}) = -(2\pi)^2 \mathcal{L}^*_{(\mathbf{x}_2)} \left[\delta(\mathbf{x} - \mathbf{x}_2)\right]$$
$$\times \left\{ \mathbf{w} \left(\mathbf{d}^k - \mathcal{L}_{(\mathbf{x}_3)} \left[\phi(\mathbf{x}_3, t_{k+})\right] \right) \right.$$
$$\left. - (2\pi)^{-2} \int d\mathbf{x}_4 \mathbf{K}^{(k)}(\mathbf{x}_4) \lambda(\mathbf{x}_4, t_{k+}) \right\}. \quad (5.7.16)$$

Note that only the Kalman gain $\mathbf{K}^{(k)}$ need be stored for this purpose: the covariance P is not required.

Once λ has been found, $\hat{\psi}$ may be most efficiently determined by integrating (5.7.2) forward in time from (5.3.3): see Fig. 5.7.1. This alternative to explicit evaluation of $\hat{\psi}$ using the sweep algorithm (5.7.8) is attractive since it avoids having to store P for all times. Thus the largest structures which must be stored are $\lambda(\mathbf{x}, t)$, which may be overwritten with $\hat{\psi}(\mathbf{x}, t)$, plus the KF estimate $\phi(\mathbf{x}, t_{k+})$ and gain $\mathbf{K}^{(k)}(\mathbf{x})$ at data times. Note that a symmetric, positive-definite $M_X \times M_X$ matrix must be inverted at each data time: see (5.7.11).

Figure 5.7.1. Directions of integration in time, for the filter and smoother algorithms described in §5.7.

There is a second alternative to explicit evaluation of the sweep algorithm. The latter may be inverted in favor of λ, and then substituted into (5.7.2), yielding

$$\frac{\partial}{\partial t}\hat{\psi}(\mathbf{x},t) = L\hat{\psi}(\mathbf{x},t) + F(\mathbf{x},t)$$
$$+ (2\pi)^{-2}\int d\mathbf{x}_2\, Q(\mathbf{x},\mathbf{x}_2)(2\pi)^{-2}\int d\mathbf{x}_3 O(\mathbf{x}_2,\mathbf{x}_3,t)$$
$$\times \left\{\hat{\psi}(\mathbf{x}_3,t) - \phi(\mathbf{x}_3,t)\right\}, \tag{5.7.17}$$

where O is the inverse of P:

$$(2\pi)^{-4}\int d\mathbf{x}_2\, O(\mathbf{x}_1,\mathbf{x}_2,t)\, P(\mathbf{x}_2,\mathbf{x}_3,t) = \delta(\mathbf{x}_1 - \mathbf{x}_3). \tag{5.7.18}$$

The appropriate final condition for (5.7.17) is

$$\hat{\psi}(\mathbf{x},T) = \phi(\mathbf{x},T). \tag{5.7.19}$$

Backward integration of (5.7.17) from (5.7.19) is well conditioned, since a high-wavenumber, large-time asymptotic analysis shows that (5.7.17) reduces to

$$\frac{\partial\hat{\psi}}{\partial t} \sim -2L(\hat{\psi} - \phi). \tag{5.7.20}$$

Inversion of P is in general very demanding of computing resources, especially as it must be performed for every time step in the backward integration of (5.7.17). By implication, P must be stored at every

time step during forward integration of (5.7.9). On the other hand, an evolution equation may be derived for O from (5.7.9):

$$\frac{\partial}{\partial t}O(\mathbf{x}_1,\mathbf{x}_2,t) = -(L_{(\mathbf{x}_1)} + L_{(\mathbf{x}_2)})O(\mathbf{x}_1,\mathbf{x}_2,t)$$

$$- (2\pi)^{-4}\int d\mathbf{x}_3 \int d\mathbf{x}_4\, O(\mathbf{x}_1,\mathbf{x}_3,t)$$

$$\times Q(\mathbf{x}_3,\mathbf{x}_4,t)\, O(\mathbf{x}_4,\mathbf{x}_2,t)\,. \tag{5.7.21}$$

The final condition is provided by

$$(2\pi)^{-4}\int d\mathbf{x}_3\, O(\mathbf{x}_1,\mathbf{x}_3,T)\, P(\mathbf{x}_3,\mathbf{x}_2,T) = \delta(\mathbf{x}_1 - \mathbf{x}_2)\,. \tag{5.7.22}$$

That is, after forward integration for P, backward integration yields O. It is only necessary to store and invert P at $t = T$.

Finally, it is possible to derive an evolution equation for $C_{\hat{\psi}\hat{\psi}}$, the error covariance for $\hat{\psi}$, starting from (5.7.17). The result is

$$\frac{\partial}{\partial t}C_{\hat{\psi}\hat{\psi}}(\mathbf{x}_1,\mathbf{x}_2,t) = (L_{(\mathbf{x}_1)} + L_{(\mathbf{x}_2)})C_{\hat{\psi}\hat{\psi}}(\mathbf{x}_1,\mathbf{x}_2,t)$$

$$+ (2\pi)^{-4}\int d\mathbf{x}_3 \int d\mathbf{x}_4\, Q(\mathbf{x}_1,\mathbf{x}_3,t)\, O(\mathbf{x}_3,\mathbf{x}_4,t)\, C_{\hat{\psi}\hat{\psi}}(\mathbf{x}_4,\mathbf{x}_2,t)$$

$$+ (2\pi)^{-4}\int d\mathbf{x}_3 \int d\mathbf{x}_4\, Q(\mathbf{x}_2,\mathbf{x}_3,t)\, O(\mathbf{x}_3,\mathbf{x}_4,t)\, C_{\hat{\psi}\hat{\psi}}(\mathbf{x}_4,\mathbf{x}_1,t)$$

$$- Q(\mathbf{x}_1,\mathbf{x}_2,t)\,. \tag{5.7.23}$$

A suitable final condition for (5.7.23) is

$$C_{\hat{\psi}\hat{\psi}}(\mathbf{x}_1,\mathbf{x}_2,T) = P(\mathbf{x}_1,\mathbf{x}_2,T)\,. \tag{5.7.24}$$

Again, asymptotic analysis may be used to show that backward integration of (5.7.23) is well conditioned: each term of the form $QOC_{\hat{\psi}\hat{\psi}}$ in (5.7.23) reduces to $-(L_1 + L_2)C_{\hat{\psi}\hat{\psi}}^{12}$, in an obvious shorthand notation. However, evaluation of the two $QOC_{\hat{\psi}\hat{\psi}}$ terms in (5.7.23) is extremely expensive in computer time. Note that the variance $C_{\hat{\psi}\hat{\psi}}(\mathbf{x}_1,\mathbf{x}_1,t)$ cannot be determined without a knowledge of the covariance $C_{\hat{\psi}\hat{\psi}}(\mathbf{x}_1,\mathbf{x}_2,t)$. It is remarkable that the measurement processes do not appear explicitly in (5.7.23), although they implicitly

influence O, the inverse of P. Also O is, like P, discontinuous in time, but $C_{\hat{\psi}\hat{\psi}}$ only has a discontinuous time derivative.

The procedure (5.7.17–19), (5.7.21–24) for obtaining $\hat{\psi}$ and $C_{\hat{\psi}\hat{\psi}}$, given ϕ and P, is the Rauch-Tung-Streibel (RTS) algorithm. The generalized inverse $\hat{\psi}$ is also known as the RTS smoother or fixed-interval Kalman smoother (Gelb, 1974).

Time series of smoother error variance $C_{\hat{\psi}\hat{\psi}}(x, x, t)$ for a one-dimensional model (Bennett & Budgell, 1989) are compared with filter error variance $P(x, x, t)$ in Fig. 4.2.4. They agree at time $t = T$, in accordance with (5.7.24); at all other times the smoother should be more accurate than the filter. Gaspar & Wunsch (1989) applied the RTS smoothing algorithm to a linear barotropic Rossby wave model (Pedlosky, 1987, Chapter 3) and GEOSAT altimeter data. The model was expressed as a propagator for five free-wave modes, with additive error:

$$\mathbf{q}(t_k + \Delta t) = \mathbf{A}(t_k)\,\mathbf{q}(t_k) + \mathbf{w}(t_k)\,, \qquad (5.7.25)$$

where $\mathbf{q}(t_k)$ is the vector of wave amplitudes at time t_k, $\mathbf{A}(t_k)$ is the propagator matrix or "state-transition matrix" and $\mathbf{w}(t_k)$ is the "process error." The smoother explained about 6%–15% of the data variance, according to the choice of process error statistics and other parameters. The results bear a qualitative resemblance to Fig. 4.2.4.

5.8 The method of Backus and Gilbert

The construction of a generalized inverse for a linear ocean model combined with linear measurements bears a strong resemblance to the methods used by Backus & Gilbert (1970) and Backus (1971) for inferring models of the earth's composition from seismic and other data. However, the resemblance is incomplete as there are important conceptual differences.

For simplicity alone, suppose that the first estimate F for the model forcing is in error, while the initial streamfunction estimate Ψ in (5.2.5) is correct; so

$$a(\mathbf{x}) \equiv 0\,, \qquad A(\mathbf{x}_1, \mathbf{x}_2) \equiv 0\,. \qquad (5.8.1)$$

Then the true streamfunction ψ, satisfying (5.2.2) and (5.2.5), is given

by

$$\psi(\mathbf{x}, t) = \psi_F(\mathbf{x}, t) + (2\pi)^{-2} \int_0^t dt_1 \int d\mathbf{x}_1\, \gamma(\mathbf{x}, t, \mathbf{x}_1, t_1)\, q(\mathbf{x}_1, t_1),\quad (5.8.2)$$

where the first-guess streamfunction ψ_F satisfies (5.4.2) and (5.4.3). The measured values of the true streamfunction are therefore

$$\mathcal{L}[\psi] = \mathcal{L}[\psi_F] + \{\alpha, q\},\qquad (5.8.3)$$

where

$$\alpha(\mathbf{x}, t) = \mathcal{L}_{(\mathbf{x}_1, t_1)}\left[\gamma(\mathbf{x}_1, t_1, \mathbf{x}, t)\right],\qquad (5.8.4)$$

and the inner product $\{\ ,\ \}$ is defined by

$$\{\theta, \phi\} = (2\pi)^{-2} \int_0^T dt \int d\mathbf{x}\, \theta(\mathbf{x}, t)\phi(\mathbf{x}, t).\qquad (5.8.5)$$

Consequently, *error-free* data \mathbf{d} are related to q by

$$\{\alpha, q\} = \mathbf{h} = \mathbf{d} - \mathcal{L}[\psi_F],\qquad (5.8.6)$$

and thus any linear decomposition of q in terms of α must be of the form

$$q(\mathbf{x}, t) = \alpha(\mathbf{x}, t)^* \{\alpha, \alpha^*\}^{-1}\mathbf{h} + \theta(\mathbf{x}, t).\qquad (5.8.7)$$

In (5.8.7), θ is orthogonal to α:

$$\{\alpha, \theta\} = \mathbf{0},\qquad (5.8.8)$$

but is otherwise unknown. At this point, no assumptions have been made about q, other than the tacit assumption that the inner product (5.8.6) is finite. However, the model and data have already been used, so θ is unconstrained aside from the orthogonality condition (5.8.8). In the solid-earth geophysical problems considered by Backus & Gilbert the role of q is taken by the earth's composition, which has presumably been steady since seismic profiling began and is likely to remain so

for the foreseeable future. It makes no sense, therefore, to regard the composition of the earth as a random variable; at most there may be plausible bounds for q such as

$$\| q \| = \{q, q\}^{\frac{1}{2}} \leq B\,. \tag{5.8.9}$$

Combining (5.8.7)–(5.8.9) yields

$$\| \theta \|^2 \leq B^2 - \mathbf{h}^* \{\boldsymbol{\alpha}, \boldsymbol{\alpha}^*\}^{-1} \mathbf{h}\,; \tag{5.8.10}$$

any field $\hat{\theta}$ satisfying (5.8.8) and (5.8.10) qualifies as an estimate of θ.

Recalling that $\boldsymbol{\alpha}$ is related to the fundamental solution γ via the measurement functionals \mathcal{L}, see (5.8.4), it is evident that $\boldsymbol{\alpha}(\mathbf{x}, t)$ may have discontinuities at some order or even be singular. Accordingly, a decomposition of q in terms of $\boldsymbol{\alpha}$, such as (5.8.7), may be unacceptable on physical grounds. Backus suggests that there may be a linear operator Q, termed a "quelling," such that $Q\boldsymbol{\alpha}$ is acceptably well behaved: in the context of fields dependent on space and time, Q may be defined by a smoothing kernel $Q(\mathbf{x}_1, t_1, \mathbf{x}_2, t_2)$:

$$(Q\boldsymbol{\alpha})(\mathbf{x}_1, t_1) = (2\pi)^{-2} \int_0^T dt_2 \int d\mathbf{x}_2\, Q(\mathbf{x}_1, t_1, \mathbf{x}_2, t_2) \boldsymbol{\alpha}(\mathbf{x}_2, t_2)\,. \tag{5.8.11}$$

Then a linear decomposition of q in terms of $Q\boldsymbol{\alpha}$ must be of the form

$$q(\mathbf{x}, t) = Q\boldsymbol{\alpha}(\mathbf{x}, t)^* \{\boldsymbol{\alpha}, Q\boldsymbol{\alpha}^*\}^{-1} \mathbf{h} + \phi(\mathbf{x}, t)\,, \tag{5.8.12}$$

where ϕ is orthogonal to $\boldsymbol{\alpha}$:

$$\{\boldsymbol{\alpha}, \phi\} = \mathbf{0}\,, \tag{5.8.13}$$

but is otherwise unknown. Again, a bound on q such as (5.8.9) would constrain ϕ. If Q is identified with the inverse of the weighting function W in the inner product $\langle\ ,\ \rangle$, see (5.5.2), then it is readily shown that

$$Q\boldsymbol{\alpha} = \mathbf{s} \tag{5.8.14}$$

and

$$\{\alpha, Q\alpha^*\} = \mathbf{R}, \qquad (5.8.15)$$

where s is the vector of residuals associated with the representers r, see (5.4.20), and \mathbf{R} is the representer matrix for the case considered here: $A \equiv 0$ (error-free initial conditions). The choice $\phi = 0$ in the decomposition (5.8.12) yields

$$\hat{q}(\mathbf{x}, t) = \mathbf{s}(\mathbf{x}, t)^* \mathbf{R}^{-1}(\mathbf{d} - \boldsymbol{\mathcal{L}}[\psi_F]), \qquad (5.8.16)$$

which is formally equivalent to the generalized inverse estimate for q (5.4.19), given error-free data $(\mathbf{w}^{-1} = \mathbf{0})$.

The preceding discussion is easily extended to include estimation of initial streamfunction errors $a(\mathbf{x})$. The smoothing kernel $A(\mathbf{x}_1, \mathbf{x}_2)$ may be introduced as a quelling for $\alpha(\mathbf{x}, 0)$. Thus the mathematical machinery for the inverse theory developed in §5.4 is formally indistinguishable from the Backus-Gilbert method, at least for error-free data. In §5.4, the components of q and a orthogonal to the quelled fields $Q\alpha(\mathbf{x}, t)$ and $A\alpha(\mathbf{x}, 0)$, respectively, are set to zero in order to minimize the penalty functional \mathcal{J}. Again, Backus (1971) proposed in the solid-earth context that fields such as q and a should be subject only to bounds. Moreover, for the purposes of obtaining just a streamfunction estimate without error estimates, Q and A need only be regarded as smoothing kernels. If model residuals and initial errors have finite norms, then physical realizability of the residuals and initial errors can be ensured, even though these norms involve the reciprocal roughening kernels W and V, respectively, as in (5.5.2).

It is necessary to introduce covariances for residuals, initial errors and data, with Q, A and \mathbf{w}^{-1} being the respective natural choices, only if error statistics are required for the inverse estimates: see §5.6. Such prior covariances for unknown fields may have no meaning in solid-earth geophysics, but ocean forcing and initial fields may well be regarded as stationary random processes, for which statistical parameters can be estimated *a priori* as time averages of archived data.

The Backus-Gilbert method can include random measurement errors, albeit awkwardly (Backus, 1970a,b). Suppose the objective is to estimate for q some spatial or temporal moment m defined by the averaging kernel $\mu(\mathbf{x}, t)$. That is, an estimate is required for $m = \{\mu, q\}$. Now μ may be decomposed in terms of α (or perhaps the

quelled basis $Q\alpha$, if necessary):

$$\mu = \{\alpha, \mu\}^*\alpha + (\mu - \{\alpha, \mu\}^*\alpha) \qquad (5.8.17)$$

$$= \mu^{\|} + \mu^{\perp} \qquad (5.8.18)$$

in an obvious notation. Further, any field ϕ in the space spanned by the basis α may be expressed as

$$\mu^{\|} + \phi = \mathbf{c}^*\alpha \qquad (5.8.19)$$

for some coefficients \mathbf{c}. Hence the moment of interest is

$$m = \{\mu, q\} = \mathbf{c}^*\{\alpha, q\} + \{\mu^{\perp} - \phi, q\}. \qquad (5.8.20)$$

If the data \mathbf{d} have errors ϵ, as in (5.2.1), then $\{\alpha, q\} = \mathbf{h} - \epsilon$. Estimating m as $\hat{m} = \mathbf{c}^*\mathbf{h}$ yields an error $-\mathbf{c}^*\epsilon + \{\mu^{\perp} - \phi, q\}$ with variance

$$\sigma^2 = \mathbf{c}^*\mathbf{w}^{-1}\mathbf{c} + \{\mu^{\perp} - \phi, q\}^2, \qquad (5.8.21)$$

where \mathbf{w}^{-1} is the covariance of ϵ. Combining the bound (5.8.9) for q with the definition (5.8.19) for ϕ and taking the worst case, namely $q = \mu^{\perp} - \phi$, yields

$$\sigma^2 \le \mathbf{c}^*\mathbf{w}^{-1}\mathbf{c} + \left[\,\|\mu^{\perp}\|^2 + \|\mu^{\|} - \mathbf{c}^*\alpha\|^2\,\right]B^2. \qquad (5.8.22)$$

The preceding bound for σ^2 is an inhomogeneous quadratic polynomial in \mathbf{c}; to paraphrase Backus (1970a): "evidently the observer should choose those constants to minimize that polynomial. Only when $\mathbf{w}^{-1} = 0$ will he choose the constants so the $\mathbf{c}^*\alpha = \mu^{\|}$." Such a choice would yield

$$\hat{m} = \{\mu, \alpha\}^*\{\alpha, \alpha^*\}^{-1}\mathbf{h}; \qquad (5.8.23)$$

that is, for error-free data, the estimate \hat{m} of the moment m is just the moment of the estimate

$$\hat{q} = \alpha(\mathbf{x}, t)^*\{\alpha, \alpha^*\}^{-1}\mathbf{h}. \qquad (5.8.24)$$

Note that a null estimate has been made for the part of q orthogonal to α.

Exercise 5.8.1 Suppose that the streamfunction satisfying the ocean model (5.2.2), (5.2.5) is subject to linear inequality constraints at certain times, and data are available as in (5.2.1). Show that the inequality constraints merely extend the finite-dimensional inverse problem (5.4.8) for the minimal representer coefficients $\hat{\mathbf{b}}$ to a problem such as considered in §1.7. □

5.9 Gradient descent

The generalized inverse of an ocean model and a finite set of data has been defined as the field which provides the best fit simultaneously to the dynamics and to the data. The fit has been defined in the weighted least-squares sense of the quadratic penalty functional \mathcal{J}: see (5.3.1). The best fit differs from the first-guess field by a sum of the finite set of representer fields associated with the measurement functionals: see (5.4.1). This result depends upon the linearity of both the dynamics and the measurement functionals. The error estimates in §5.6 show that the representer expansion is in one-to-one correspondence with Gauss-Markov smoothing, the representers being field-measurement covariances. Thus the generalized inverse (the minimum of \mathcal{J}) could be calculated using covariances generated by statistical simulation. The generalized inverse is also the Kalman smoother for the model and data, and so may be calculated using the sweep algorithm described in §5.7.

There is a fourth method for finding the minimum of the penalty functional \mathcal{J}. This last method is conceptually the simplest of all: a search is made in the direction of the gradient of \mathcal{J} until the magnitude of the gradient is acceptably small. The functional \mathcal{J} maps the space-time streamfunction field ψ into the scalar $\mathcal{J}[\psi]$; thus the gradient of \mathcal{J} must be defined in a function space. The definition is really quite simple, being based on inner products (see, for example, Talagrand & Courtier, 1987), but the concept is unlikely to be familiar to ocean modelers and so will be avoided here. Instead, the sought-after field ψ will be lumped into a vector belonging to a space of large but finite dimension. Lumping is achieved by considering only the values of ψ on a regular space-time grid, and by making finite-difference approximations to the derivatives in the equations for the

ocean model. Such a presentation has been adopted by Thacker & Long (1988), for example. However, Thacker & Long's formulation of the model, as a system of finite-difference equations of general form, will not be used here. Instead, the finite-difference equations will have the same form as the ocean model. On the other hand, in the interests of clarity alone, the role of the ocean model will be taken by an *ordinary* differential equation resembling the partial differential equation (5.2.2).

Consider the function of time $\psi = \psi(t)$, which for convenience will be referred to as a "streamfunction." It obeys the ordinary differential equation

$$\frac{d\psi}{dt} = L\psi + F + q \,, \tag{5.9.1}$$

where $F = F(t)$ and $q = q(t)$ are, respectively, a prior estimate for the forcing and the error in that estimate. The coefficient L is a real, negative constant. An initial condition for (5.9.1) is

$$\psi(0) = \Psi + a \,, \tag{5.9.2}$$

where Ψ and a are, respectively, a prior estimate of the initial value and the error in that estimate. A simple forward difference approximation to (5.9.1), (5.9.2) is:

$$\frac{\psi_{n+1} - \psi_n}{\Delta t} = L\psi_n + F_n + q_n \,, \tag{5.9.3}$$

or, equivalently,

$$\psi_{n+1} = (1 + \Delta t L)\psi_n + \Delta t(F_n + q_n) \tag{5.9.4}$$

and

$$\psi_0 = \Psi + a \,, \tag{5.9.5}$$

where

$$\psi_n = \psi(t_n) = \psi(n\Delta t) \,, \tag{5.9.6}$$

for example, Δt being a suitably small time interval. Assume that "data" are available at every time step *within* the interval $0 \le t \le T = N\Delta t$:

$$d_n = \psi_n + \epsilon_n \,, \qquad (1 \le n \le N - 1) \,, \tag{5.9.7}$$

where d_n is a streamfunction datum, and ϵ_n is a measurement error. Allowing data at times $t_0 = 0$ and $t_N = T$ would merely complicate the following algebra.

A generalized inverse of (5.9.3)–(5.9.7) may be defined as the minimum of a quadratic penalty function analogous to (5.3.1):

$$\mathcal{J}_N[\psi] = \Delta t W |\mathbf{q}|^2 + V a^2 + \Delta t w |\epsilon|^2 , \qquad (5.9.8)$$

where ψ is the column vector $(\psi_0, \ldots, \psi_N)^*$, while \mathbf{q} is $(q_0, \ldots, q_{N-1})^*$ and ϵ is $(\epsilon_1, \ldots, \epsilon_{N-1})^*$. The weights W, V and w are real, non-negative constants; \mathbf{q}, a and ϵ are related to ψ by (5.9.3), (5.9.5) and (5.9.7), respectively. The expression (5.9.8) is compact, but it is necessary to substitute for \mathbf{q}, a and ϵ in order to calculate the following gradient of \mathcal{J} with respect to $\hat{\psi}$. A necessary condition for the quadratic form \mathcal{J}_N to have a minimum at $\psi = \hat{\psi}$ is that

$$\nabla \mathcal{J}_N[\hat{\psi}] = \mathbf{0} , \qquad (5.9.9)$$

where the nth component of $\nabla \mathcal{J}_N$ in \Re^{N+1} is

$$(\nabla \mathcal{J}_N)_n = \frac{\partial \mathcal{J}_N}{\partial \psi_n} , \qquad (0 \le n \le N) . \qquad (5.9.10)$$

Suppose $\psi^{(1)}$ is an estimate of the best-fit streamfunction $\hat{\psi}$. Then a better estimate is

$$\psi^{(2)} = \psi^{(1)} - \theta \nabla \mathcal{J}_N[\psi^{(1)}] \qquad (5.9.11)$$

for some small positive value θ. Further improvements may be obtained by continuing down the gradient until $|\nabla \mathcal{J}_N|^2$ is acceptably small.

Minimization by stepping down the gradient could hardly be simpler conceptually, but it entails much computation if each component of $\nabla \mathcal{J}_N$ is approximated by a partial difference:

$$\frac{\partial \mathcal{J}_N}{\partial \psi_n} \cong \big\{ \mathcal{J}_N[\psi_0, \ldots, \psi_n + \Delta \psi_n, \ldots, \psi_N]$$
$$- \mathcal{J}_N[\psi_0, \ldots, \psi_n, \ldots, \psi_N] \big\}/\Delta \psi_n , \qquad (0 \le n \le N) . \qquad (5.9.12)$$

Since there are $O(N)$ operations involved in evaluating (5.9.12), for each component, a total of $O(N^2)$ operations are required per step. However, an order-of-magnitude reduction in the computation may be achieved by algebraic evaluation of $\nabla \mathcal{J}_N$:

$$\frac{\partial \mathcal{J}_N}{\partial \psi_0} = 2\Delta t W \left(\frac{\psi_1 - \psi_0}{\Delta t} - L\psi_0 - F_0 \right) \left(-\frac{1}{\Delta t} - L \right) + 2V(\psi_0 - \Psi),$$

$$(5.9.13a)$$

$$\frac{\partial \mathcal{J}_N}{\partial \psi_n} = -2W \left(\frac{\psi_{n+1} - \psi_n}{\Delta t} - L\psi_n - F_n \right) \left(1 + \Delta t L \right)$$
$$+ 2W \left(\frac{\psi_n - \psi_{n-1}}{\Delta t} - L\psi_{n-1} - F_{n-1} \right)$$
$$- 2\Delta t w(d_n - \psi_n), \qquad (1 \leq n \leq N - 1), \qquad (5.9.13b)$$

and

$$\frac{\partial \mathcal{J}_N}{\partial \psi_N} = 2W \left(\frac{\psi_N - \psi_{N-1}}{\Delta t} - L\psi_{N-1} - F_{N-1} \right). \qquad (5.9.13c)$$

Defining $\boldsymbol{\lambda} = (\lambda_0, \ldots, \lambda_{N-1})^*$ by

$$\lambda_n = W \left(\frac{\psi_{n+1} - \psi_n}{\Delta t} - L\psi_n - F_n \right), \qquad (0 \leq n \leq N - 1), \quad (5.9.14)$$

reduces (5.9.13) to

$$\frac{\partial \mathcal{J}_N}{\partial \psi_0} = -2\lambda_0 (1 + L\Delta t) + 2V(\psi_0 - \Psi), \qquad (5.9.15a)$$

$$\frac{\partial \mathcal{J}_N}{\partial \psi_n} = 2\Delta t \left[\frac{\lambda_{n-1} - \lambda_n}{\Delta t} - L\lambda_n - w(d_n - \psi_n) \right], \quad (5.9.15b)$$

$$(1 \leq n \leq N - 1),$$

and

$$\frac{\partial \mathcal{J}_N}{\partial \psi_N} = 2\lambda_{N-1}. \qquad (5.9.15c)$$

Thus, setting the right-hand sides of (5.9.15) to zero yields backward-difference approximations to Euler-Lagrange equations analogous to

(5.3.2)–(5.3.5). The difference equations are *exactly* satisfied by extrema of \mathcal{J}_N. Evaluation of $\boldsymbol{\nabla}\mathcal{J}_N[\boldsymbol{\psi}]$ using (5.9.14) and (5.9.15) requires only $O(N)$ operations, per step. Applying this approach to finding the gradient of (5.3.1) in lieu of direct evaluation would reduce the operation count from $O(N^2 \times N_x^2 \times N_y^2)$ to $O(N \times N_x \times N_y)$, where N_x and N_y are measures of spatial resolution in the two space coordinates.

There is a special case of the preceding algorithm which has attracted much attention in the oceanographic literature (see, for example, Thacker & Long, 1988; Tziperman & Thacker, 1989; Schröter, 1989; Sheinbaum & Anderson, 1990a,b). The first-guess forcing estimate F is assumed to be perfect, that is, the error q is assumed to vanish identically. This case may be recovered here by examining \mathcal{J}_N in the limit of arbitrarily large dynamical weight: $W \to \infty$. Then λ_n as defined by (5.9.14), that is,

$$\boldsymbol{\lambda} = W\mathbf{q}, \qquad (5.9.16)$$

becomes the Lagrange multiplier for the *strong constraint* $\mathbf{q} = \mathbf{0}$ in the minimization of

$$\mathcal{K}_N[\boldsymbol{\psi}] = Va^2 + \Delta t w|\boldsymbol{\epsilon}|^2. \qquad (5.9.17)$$

Again, $\boldsymbol{\psi}$ is related to \mathbf{q}, a and $\boldsymbol{\epsilon}$ by (5.9.3)–(5.9.7). In this case \mathcal{J}_N becomes

$$\mathcal{J}_N[\boldsymbol{\psi},\boldsymbol{\lambda}] = \Delta t \boldsymbol{\lambda}^*\mathbf{q} + \mathcal{K}_N[\boldsymbol{\psi}]. \qquad (5.9.18)$$

The gradient of \mathcal{J}_N with respect to $\boldsymbol{\psi}$ is still given by (5.9.15); the gradient with respect to $\boldsymbol{\lambda}$ is simply $\mathbf{q} = \mathbf{0}$, that is, the dynamics hold exactly. In general, both $\boldsymbol{\lambda}$ and $\boldsymbol{\psi}$ must be known in order to evaluate the gradient. However, it is advantageous to exploit the following necessary conditions for an extremum of \mathcal{J}_N, namely

$$\frac{\partial \mathcal{J}_N}{\partial \psi_N} = 2\lambda_{N-1} = 0, \qquad (5.9.19a)$$

$$\frac{\partial \mathcal{J}_N}{\partial \psi_n} = 2\Delta t\left[\frac{\lambda_{n-1} - \lambda_n}{\Delta t} - L\lambda_n - w(d_n - \psi_n)\right]$$

$$= 0, \qquad (1 \le n \le N-1), \qquad (5.9.19b)$$

and

$$\frac{\partial \mathcal{J}_N}{\partial \lambda_n} = \psi_{n+1} - (1 + L\Delta t)\psi_n - \Delta t F_n$$

$$= 0, \qquad (0 \le n \le N - 1).\tag{5.9.20}$$

The only non-zero component of $\nabla \mathcal{J}_N$ is now

$$\frac{\partial \mathcal{J}_N}{\partial \psi_0} = -2\lambda_0(1 + L\Delta t) + 2V\psi_0.\tag{5.9.21}$$

An elegant procedure may be used to evaluate this component, given an initial value ψ_0 for the streamfunction. First, (5.9.20) is used to find $\psi_1, \psi_2, \ldots, \psi_N$ by "forward integration." Next, (5.9.19) is used to find $\lambda_{N-1}, \lambda_{N-2}, \ldots, \lambda_0$ by "backward integration." Finally, (5.9.21) is evaluated using the given value of ψ_0 and the computed value of λ_0. If the result is not acceptably small, the given value of ψ_0 may be perturbed in the obvious way; the algorithm is then used to find the perturbed value of λ_0 and hence a reduced value of $\partial \mathcal{J}_N / \partial \psi_0$. As mentioned, each such step requires two integrations, one of the model (5.9.20), and one of the adjoint model (5.9.19), involving $O(N)$ operations. When this algorithm is applied to the quasi-geostrophic model with $N_x \times N_y$ spatial resolution, the gradient of \mathcal{J} with respect to the initial streamfunction $\psi(\mathbf{x}, 0)$ has $N_x \times N_y$ components, but these may all be evaluated using just two model integrations. The operation count is $O(N \times N_x \times N_y)$ per step.

5.10 Remarks

1) The preceding discussion of gradient descent methods has two main points. If the dynamical weight is finite: $W < \infty$, then it is necessary to calculate the gradient of the penalty functional \mathcal{J} with respect to the entire state $\psi(\mathbf{x}, t)$. In discrete form the gradient has $N \times N_x \times N_y$ components, which may be efficiently evaluated by *substitution* into the model and then into the adjoint model equation. The total operation count is $O(N \times N_x \times N_y)$ per step. If on the other hand the dynamical weight is infinite: $W \to \infty$, then it is only necessary to evaluate the gradient of \mathcal{J} with respect to the initial streamfunction $\psi(\mathbf{x}, 0)$. In discrete form the gradient has $N_x \times N_y$ components, which may be evaluated by *integration* of both the model and the adjoint model

equations. The total operation count is $O(N \times N_x \times N_y)$ per step.

It is particularly noteworthy in the special case $W \to \infty$ that the two-point boundary-value problem for the extremum of \mathcal{J} has been replaced with two initial-value problems, by the device of not setting $\partial \mathcal{J}/\partial \psi_0$ to zero. However, many such pairs of initial-value problems must be integrated while descending the gradient.

2) Gradient descent is degenerate if the initial weight vanishes ($V \equiv 0$). Such a choice implies that the prior estimate of the initial streamfunction is worthless, and that arbitrarily large initial values may be chosen without direct penalty. Indeed it is clear that provided the initial resolution is sufficiently fine that there are more initial values than data values, then it is possible to choose $\psi(\mathbf{x}, 0)$ such that $\psi(\mathbf{x}, t)$ *interpolates* the data. Consequently $\lambda(\mathbf{x}, t) \equiv 0$, and a null estimate is made for the dynamical error $q(\mathbf{x}, t)$ whether or not the dynamical weight W is infinite. See the remarks following Exercise 5.3.1. Thus, while gradient descent yields a streamfunction which interpolates data, the streamfunction is not unique. It depends upon the start of the descent path, even if the dynamics are linear, since any initial streamfunction correction which is unobservable (all measured values vanish) may be added without penalty. In practice the streamfunction will be ill-conditioned: the gradient descent will stop before precise interpolation has been achieved, so the result will include finely structured fields which are almost unobservable (Bennett & Miller, 1991). It is important to note that the number of independent data may be far fewer than the total number of data, hence the optimal choice of initial conditions may be effectively underdetermined even though the choice is formally overdetermined. A unique inverse $\hat{\psi}$ obtains if the initial weight V is positive, and the inverse is better conditioned if V generates a smoothing norm, that is, if \mathcal{J} penalizes fine spatial structure in the initial streamfunction (Courtier & Talagrand, 1987). Excessive smoothing may eventually degrade conditioning, however, as discussed in Chapter 6.

3) The operation count for the representer expansion of $\hat{\psi}$ is $O(M \times N \times N_x \times N_y)$, where M is the number of measurements. This is comparable to a gradient descent of M steps. The operation count for the sweep algorithm is dominated by the calculation of

the filter error covariance matrix. The latter requires $O(M^2 \times N \times N_x^2 \times N_y^2)$. The factor of M^2 arises from the need to evaluate the quadratic term in the evolution equation for the covariance.

4) Gradient descent produces a monotonically decreasing sequence of values for the penalty functional. Since the values of the function must be non-negative (this will be the case in practice if the step sizes are sufficiently small), it follows that the sequence has a limit. The corresponding sequence of streamfunction estimates $\{\psi^{(n)}\}_{n=1,\infty}$ will be well-behaved functions of position and time provided \mathcal{J} is defined by a smoothing norm, as in (5.3.1).

5) When (5.9.3), (5.9.5) and (5.9.6) are substituted into (5.9.8), \mathcal{J}_N is explicitly an inhomogeneous quadratic form in ψ. The form is positive-definite, and hence may be used to define a multivariate normal distribution function for ψ. Consequently $\hat{\psi}$, which minimizes \mathcal{J}_N, is the maximum-likelihood estimate of ψ. Moreover, the covariance of the error $\psi - \hat{\psi}$ is the matrix inverse of the Hessian form \mathcal{H} as defined by

$$\mathcal{H}_{jk} = \frac{\partial^2 \mathcal{J}_H}{\partial \psi_j \partial \psi_k}, \qquad (1 \le j, \ k \le N). \qquad (5.10.1)$$

In practice \mathcal{H} has been evaluated by perturbing the gradient vector $\nabla \mathcal{J}_N$ (Thacker, 1989; Tziperman & Thacker, 1989; Schröter, 1989). For large models \mathcal{H} is a very large array; its calculation and inversion become prohibitive tasks. Note that calculation and inversion of the entire Hessian is necessary even to find just the variance of the error $\psi - \hat{\psi}$.

Much of the computation required to find \mathcal{H} must be redundant, since (5.10.1) includes variations with respect to the unobservable projections of ψ. The representer expression (5.6.17) for the model error covariance only involves the $M \times M$ matrix $\mathbf{R} + \mathbf{w}^{-1}$, which is in turn closely related to \mathbf{S}, the Hessian of \mathcal{J} with respect to the representer coefficients \mathbf{b} (see (5.5.14), (5.5.15)). That is, (5.6.17) only involves the Hessian for the observables. However, the expression (5.6.14) for the streamfunction error covariance also involves the reproducing kernel Γ, which is also the streamfunction covariance. It therefore includes the unobservables. In practice Γ may be estimated *a priori* by statistical simulation: see (5.6.11).

6) Preconditioning methods (Tarantola, 1987; Tziperman & Thacker, 1989) could solve the problem of reducing the rank of the Hessian \mathcal{H} during gradient descent, but the choice of preconditioning matrices is not yet clear.

5.11 The calculus of variations: a tutorial

The one-dimensional wave model described in Exercise 5.3.2 may be formulated as a simple inverse problem, by seeking the field $\hat{u}(x,t)$ which minimizes the quadratic functional

$$\mathcal{J}[u] = W_f \int_{-\infty}^{\infty} dx \int_0^T dt (u_t + u_x - F)^2 + W_i \int_{-\infty}^{\infty} dx (u_0 - I)^2$$

$$+ W_\epsilon \sum_{m=1}^{M} (d_m - u_m)^2 , \qquad (5.11.1)$$

where W_f, W_i and W_ϵ are constant, positive weights, while $u_0(x) \equiv u(x,0)$ and $u_m \equiv u(X_m, T_m)$, $1 \le m \le M$. If $\hat{u}(x,t)$ is an extremum of \mathcal{J}, then for any small perturbation $v(x,t)$,

$$\mathcal{J}[\hat{u} + v] - \mathcal{J}[\hat{u}] = O(v^2) . \qquad (5.11.2)$$

That is,

$$W_f \int_{-\infty}^{\infty} dx \int_0^T dt (\hat{u}_t + \hat{u}_x - F)(v_t + v_x) + W_i \int_{-\infty}^{\infty} dx (\hat{u}_0 - I) v_0$$

$$- W_\epsilon \sum_{m=1}^{M} (d_m - \hat{u}_m) v_m = 0 , \qquad (5.11.3)$$

where $\hat{u}_0(x) \equiv \hat{u}(x,0)$, $v_0(x) \equiv v(x,0)$ and $v_m \equiv v(X_m, T_m)$, $1 \le m \le M$. Integrating by parts in (5.11.3) yields

$$\int_{-\infty}^{\infty} dx [\lambda v]_{t=0}^{t=T} - \int_{-\infty}^{\infty} dx \int_0^T dt (\lambda_t + \lambda_x) v + W_i \int_{-\infty}^{\infty} dx (\hat{u}_0 - I) v_0$$

$$- W_\epsilon \sum_{m=1}^{M} (d_m - \hat{u}_m) \int_{-\infty}^{\infty} dx \int_0^T dt\, v(x,t)\delta(x - X_m)\delta(t - T_m) = 0 ,$$

$$(5.11.4)$$

where the variable $\lambda(x,t)$ is defined by

$$\hat{u}_t(x,t) + \hat{u}_x(x,t) \equiv F(x,t) + W_f^{-1}\lambda(x,t). \tag{5.11.5}$$

Note that the value v_m of the perturbation v at (X_m, T_m) has been expressed in terms of integrals of $v(x,t)$ with Dirac delta-functions as kernels. Since v is arbitrary, it is permissible to vary $v(x,t)$, $v(x,0)$ and $v(x,T)$ independently, leading to the following conditions for an extremum of $\mathcal{J}[u]$:

$$-\lambda_t(x,t) - \lambda_x(x,t) = W_\epsilon \sum_{m=1}^{M}(d_m - \hat{u}_m)\delta(x - X_m)\delta(t - T_m), \tag{5.11.6}$$

$$-\lambda(x,0) + W_i\big(\hat{u}(x,0) - I(x)\big) = 0 \tag{5.11.7}$$

and

$$\lambda(x,T) = 0. \tag{5.11.8}$$

The conditions (5.11.5)–(5.11.8) are the Euler-Lagrange equations for extrema of \mathcal{J}; strictly speaking, it is (5.11.6) which is *the* Euler-Lagrange equation. The extreme residuals for the model and initial condition are, respectively,

$$\hat{f}(x,t) = W_f^{-1}\lambda(x,t) \tag{5.11.9}$$

and

$$\hat{i}(x) = W_i^{-1}\lambda(x,0). \tag{5.11.10}$$

The model is introduced into (5.11.1) as a weak constraint, that is, only a weighted best fit to the model is sought. It may be introduced as a strong constraint, which is satisfied exactly, by appending the strong constraint to the rest of the penalty functional using a linear Lagrange multiplier $\sigma = \sigma(x,t)$:

$$\mathcal{K}[u,\sigma] = 2\int_{-\infty}^{\infty} dx \int_{0}^{T} dt\, \sigma(u_t + u_x - F) + W_i \int_{-\infty}^{\infty} dx(u_0 - I)^2$$
$$+ W_\epsilon \sum_{m=1}^{M}(d_m - u_m)^2. \tag{5.11.11}$$

Then \mathcal{K} is stationary at $u = \hat{u}$ and $\sigma = \hat{\sigma}$, with respect to independent perturbations of u and σ, provided that

$$\hat{u}_t(x,t) + \hat{u}_x(x,t) = F(x,t)\,, \tag{5.11.12}$$

$$-\hat{\sigma}_t(x,t) - \hat{\sigma}_x(x,t) = W_\epsilon \sum_{m=1}^{M}(d_m - \hat{u}_m)\delta(x - X_m)\delta(t - T_m)\,,$$

$$\tag{5.11.13}$$

$$-\hat{\sigma}(\mathbf{x},0) + W_i\big(\hat{u}(x,0) - I(x)\big) = 0\,, \tag{5.11.14}$$

and

$$\hat{\sigma}(x,T) = 0\,. \tag{5.11.15}$$

The Euler-Lagrange equations (5.11.12)–(5.11.15) for the strong-constraint inverse problem may also be recovered from the weak-constraint equations (5.11.5)–(5.11.8) in the limit as $W_f \rightarrow \infty$: the model is given infinite weight relative to the initial values and data. For a detailed discussion of the theory of the first variation as outlined above, see Courant & Hilbert (1953). A crucial issue is the sense in which $v(x,t)$, $v(x,0)$ and $v(x,T)$ may be varied independently, if the choice for v is restricted to smooth functions of x and t.

6

ANTENNA ANALYSIS

6.1 Introduction

A generalized inverse estimate of ocean circulation is a field which nearly solves an ocean circulation model and which nearly fits a finite set of data. It is intuitively obvious that if several of the data were collected at almost the same place or at almost the same time, yet had widely differing values, then fitting the data closely would be very difficult. Any field which did fit would tend to fluctuate wildly in an extended neighborhood of the observing sites. However, if the data were assumed to possess significantly large errors, then a relatively well behaved field might be able to come consistently close to the data: see Fig. 6.1.1. Nevertheless it would be concluded that there is redundancy in the data.

The concepts "almost the same place or time," "significantly large error" and "well behaved" are defined relative to the scales and amplitudes of the circulation, as characterized by the stream-function, for example. In a generalized inverse estimate the scales and amplitudes are those of the initial errors, the dynamical errors, and the internal scales of the dynamics. The initial and dynamical information is contained in covariances such as A and Q in §5.6, and in the dynamical operator L appearing in §5.2. The measurements are completely described by the functionals \mathcal{L}, and the measurement error covariance matrix such as \mathbf{w}^{-1}, also defined in §5.6.

It is shown in this chapter that for linear dynamics, the redundancies in the observing system or "antenna" may be determined from these parameters and operators. The independent and the redundant

Figure 6.1.1. Closely spaced observations may be smoothed if the data errors are large enough.

observations are not necessarily associated with individual functionals \mathcal{L}_i or "antenna elements," but in general are linear combinations of the \mathcal{L}_i and will be referred to as "principal antenna elements." The corresponding linear combinations of representers and their residuals will be named "solution antenna modes" and "residual antenna modes," respectively. The penalty functional \mathcal{J} which defines the generalized inverse will be expressed in terms of principal antenna elements; the reduced functional will be shown to be a chi-squared variable with as many degrees of freedom as there are independent observations.

6.2 Canonical decomposition

It was shown in §5.5 that the infinite-dimensional problem of minimizing the penalty functional $\mathcal{J}[\psi]$ is reducible, using representer expansions, to the finite-dimensional problem of minimizing the penalty function $\mathcal{J}[\mathbf{b}]$, where \mathbf{b} is the vector of representer coefficients: see (5.3.1), (5.5.7), (5.5.12) and (5.5.13). The minimal coefficients $\hat{\mathbf{b}}$ are determined by

$$(\mathbf{R} + \mathbf{w}^{-1})\hat{\mathbf{b}} = \mathbf{h}, \qquad (6.2.1)$$

where \mathbf{R} is the representer matrix (5.5.10), \mathbf{w}^{-1} is the measurement error covariance matrix (see (5.6.3)) and \mathbf{h} is the measurement misfit for the first-guess streamfunction: see (5.5.11).

Suppose in the first instance that the data are perfect: $\mathbf{w}^{-1} = \mathbf{0}$. It is readily verified that the generalized inverse estimate $\hat{\psi}$ coincides with the data, which have therefore been interpolated. If no two

measurement functionals are identical, then, by construction, the M representers are linearly independent and so the symmetric represen- ter measurement matrix \mathbf{R} is positive-definite. Accordingly \mathbf{R} has an eigenvalue decomposition of the form

$$\mathbf{R} = \mathbf{Z}\mathbf{\Theta}\mathbf{Z}^* \qquad (6.2.2)$$

where $\mathbf{\Theta}$ is the diagonal matrix of real, positive eigenvalues: $\mathbf{\Theta} = \mathrm{diag}(\theta_1, \ldots, \theta_M)$, where $\theta_1 \geq \ldots \geq \theta_M > 0$, and \mathbf{Z} is a unitary matrix having the mth eigenvector of \mathbf{R} as its mth column: $\mathbf{Z} = (\mathbf{z}_1, \ldots, \mathbf{z}_M)$, where $\mathbf{R}\mathbf{z}_m = \theta_m\mathbf{z}_m$ $(1 \leq m \leq M)$. Hence the representer coefficients are given by

$$\hat{\mathbf{b}} = \mathbf{Z}\mathbf{\Theta}^{-1}\mathbf{Z}^*\mathbf{h} = \sum_{m=1}^{M} \mathbf{z}_m\theta_m^{-1}\mathbf{z}_m^*\mathbf{h} . \qquad (6.2.3)$$

Typically, θ_1 exceeds θ_M by orders of magnitude, and \mathbf{z}_M has much finer structure than does \mathbf{z}_1. Thus the solution of (6.2.1) for the representer coefficients $\hat{\mathbf{b}}$ would, in general, be ill-conditioned, with $\hat{\mathbf{b}}$ having fine structure of large amplitude. The corresponding sum of representers would have fine structure in space and time. It would be concluded that the components of the observing system, or "antenna elements," were too close in space or time to permit interpolation of the data.

Now suppose that the observations have uncorrelated data errors with common variance w^{-1}: that is, $\mathbf{w} = w\mathbf{I}_M$, where \mathbf{I}_M is the $M \times M$ unit matrix. Then the appropriate decomposition is

$$\mathbf{R} + \mathbf{w}^{-1} = \mathbf{Z}(\mathbf{\Theta} + w^{-1}\mathbf{I})\mathbf{Z}^* . \qquad (6.2.4)$$

The eigenvectors are still those of \mathbf{R}, but the eigenvalues become $\theta_1 + w^{-1}, \ldots, \theta_M + w^{-1}$. If w^{-1} is sufficiently large, then the range of the eigenvalues may be reduced to the extent that the solution of (6.2.1) for $\hat{\mathbf{b}}$ is well conditioned. On the other hand if the data errors were perfectly correlated: $w_{nm} \equiv w$ $(1 \leq n, m \leq M)$, then as will be seen in an example, the conditioning of (6.2.1) would not be improved even for significant values of w^{-1}. Returning to the well-conditioned case, there will in general be small-amplitude fine structure in $\hat{\mathbf{b}}$ leading in turn to small-amplitude, fine spatial and temporal structure in $\hat{\psi}$. Satisfactory results may be obtained in

either case, by discarding the contributions to $\hat{\mathbf{b}}$ due to the "noisy" eigenvectors associated with the least significant eigenvalues, namely those eigenvalues which are less than w^{-1}.

In the general case the decomposition is

$$\mathbf{P} = \mathbf{R} + \mathbf{w}^{-1} = \mathbf{Z}\boldsymbol{\Phi}\mathbf{Z}^* \tag{6.2.5}$$

for some unitary matrix \mathbf{Z} and some diagonal matrix $\boldsymbol{\Phi} =$ diag (ϕ_1, \ldots, ϕ_M) with $\phi_1 \geq \ldots \geq \phi_M > 0$. Note that unless \mathbf{R} and \mathbf{w} are simultaneously diagonizable, their eigenvectors are not those of \mathbf{P}. A satisfactory estimate of $\hat{\mathbf{b}}$ may be obtained if sufficiently many of the smaller eigenvalues are discarded, say

$$\hat{\mathbf{b}} \simeq \hat{\mathbf{b}}^G = \sum_{m=1}^{M'} \mathbf{z}_m \, \phi_m^{-1} \, \mathbf{z}_m^* \mathbf{h}, \tag{6.2.6}$$

where $M' \leq M$ and $w^{-1} > \phi_{M'+1} \geq \ldots \geq \phi_M$ for some representative data variance w^{-1}. It is clear that (6.2.6) is in effect the generalized or Moore-Penrose inverse solution of (6.2.1), after formal reduction of the rank of the system from M to M'.

It is now possible to define a more sparse antenna comprised just of the M' "principal elements" defined by the M' measurement functionals

$$\mathcal{L}'_1 = \mathbf{z}_1^*\boldsymbol{\mathcal{L}}; \; \ldots \; ; \mathcal{L}'_{M'} = \mathbf{z}_{M'}^*\boldsymbol{\mathcal{L}}. \tag{6.2.7}$$

Using the $M \times M'$ matrix $\mathbf{Z}' = (\mathbf{z}_1, \ldots, \mathbf{z}_{M'})$, these functionals may be expressed as a vector of length M':

$$\boldsymbol{\mathcal{L}}' = \mathbf{Z}'^*\boldsymbol{\mathcal{L}}. \tag{6.2.8}$$

The corresponding M'-vectors of representers and their residuals are respectively the "solution antenna modes" and "residual antenna modes":

$$\mathbf{r}' = \mathbf{Z}'^*\mathbf{r}, \qquad \mathbf{s}' = \mathbf{Z}'^*\mathbf{s}. \tag{6.2.9}$$

The corresponding data, and the first-guess data misfits, are the M'-vectors

$$\mathbf{d}' = \mathbf{Z}'^*\mathbf{d}, \qquad \mathbf{h}' = \mathbf{Z}'^*\mathbf{h}. \tag{6.2.10}$$

The original measurement error term, in the penalty functional \mathcal{J} defined by (5.3.1), may be replaced with a "principal measurement error" term:

$$\epsilon^* \mathbf{w} \epsilon \rightarrow \epsilon'^* \mathbf{w}' \epsilon' . \tag{6.2.11}$$

In (6.2.11), ϵ' is the M'-vectors of data misfits:

$$\epsilon' = \mathbf{Z}'^* \epsilon = \mathbf{d}' - \mathcal{L}'[\psi] , \tag{6.2.12}$$

while the $M' \times M'$ weight matrix is

$$\mathbf{w}' = (\mathbf{Z}'^* \mathbf{w}^{-1} \mathbf{Z}')^{-1} . \tag{6.2.13}$$

Then the modified penalty functional \mathcal{J}' is minimized by the generalized inverse estimate

$$\hat{\psi}' = \psi_F + \hat{\mathbf{b}}'^* \mathbf{r}' , \tag{6.2.14}$$

where the M'-vector of representer coefficients $\hat{\mathbf{b}}'$ satisfies

$$(\mathbf{R}' + (\mathbf{w}')^{-1}) \hat{\mathbf{b}}' = \mathbf{h}' . \tag{6.2.15}$$

In (6.2.15), \mathbf{R}' is the $M' \times M'$ representer matrix

$$\mathbf{R}' = \langle \mathbf{r}', \mathbf{r}'^* \rangle = \mathcal{L}'[\Gamma] \mathcal{L}'^* . \tag{6.2.16}$$

The reproducing kernel Γ is independent of the antenna. The system (6.2.15) is trivially solved since

$$\mathbf{P}' = \mathbf{R}' + (\mathbf{w}')^{-1} = \mathbf{Z}'^* (\mathbf{R} + \mathbf{w}^{-1}) \mathbf{Z}' = \mathbf{\Phi}' , \tag{6.2.17}$$

where $\mathbf{\Phi}'$ is the $M' \times M'$ diagonal matrix

$$\mathbf{\Phi}' = \mathrm{diag}(\phi_1, \ldots, \phi_{M'}) . \tag{6.2.18}$$

Indeed,

$$\hat{\mathbf{b}}' = (\mathbf{\Phi}')^{-1} \mathbf{h}' , \tag{6.2.19}$$

and consequently

$$\hat{\psi}' = \psi_F + \hat{\mathbf{b}}'^{*}\mathbf{r}' = \psi_F + \hat{\mathbf{b}}^{G*}\mathbf{r} \qquad (6.2.20)$$

where $\hat{\mathbf{b}}^G$ is defined by (6.2.6). In other words, the approximation (6.2.6) corresponds precisely to minimizing the functional \mathcal{J}' associated with the sparse antenna of principal elements \mathcal{L}', with the data \mathbf{d}' and with the data weight matrix \mathbf{w}'. The reduced penalty functional for the sparse antenna is given by

$$\widehat{\mathcal{J}}' = \mathbf{h}'^{*}(\boldsymbol{\Phi}')^{-1}\mathbf{h}'. \qquad (6.2.21)$$

A scaled, first-guess data misfit may be defined by

$$\mathbf{k}' = (\boldsymbol{\Phi})'^{-1/2}\mathbf{h}', \qquad (6.2.22)$$

in terms of which the reduced penalty functional is

$$\widehat{\mathcal{J}}' = \mathbf{k}'^{*}\mathbf{k}' = \mathrm{Tr}(\mathbf{k}'\mathbf{k}'^{*}). \qquad (6.2.23)$$

However, $\overline{\mathbf{h}'\mathbf{h}'^{*}} = \boldsymbol{\Phi}'$, so the mean of (6.2.23) is

$$\overline{\widehat{\mathcal{J}}'} = \mathrm{Tr}(\overline{\mathbf{k}'\mathbf{k}'^{*}}) = \mathrm{Tr}(\mathbf{I}_{M'}) = M'. \qquad (6.2.24)$$

Consequently $\widehat{\mathcal{J}}'$ may be regarded as a chi-squared variable with M' degrees of freedom:

$$\widehat{\mathcal{J}}' = \chi^2_{M'}. \qquad (6.2.25)$$

Strictly speaking, it is necessary that each component of \mathbf{k}' be normally distributed before stating (6.2.25). However, each component is a linear combination of the first-guess data misfit \mathbf{h}, which has vanishing mean and finite variance. Hence, according to the central limit theorem each component of \mathbf{k}' will be normally distributed, asymptotically as $M \to \infty$.

Exercise 6.2.1. Derive the error covariances for the posterior estimates $\hat{\psi}$, \hat{q}, \hat{a} and $\hat{\mathbf{e}}$. □

Figure 6.3.1. A simple streamfunction antenna for the quasi-geostrophic model of Chapter 5. Data are collected at five points in the doubly-periodic domain and at seven equi-spaced time intervals; thus the total number of measurements is $M = 35$.

6.3 Examples

i) A streamfunction antenna

A very simple antenna is shown in Fig. 6.3.1. The antenna measures the streamfunction governed by the quasi-geostrophic model of Chapter 5. Data are collected at each of the five points in the doubly periodic domain and at seven equi-spaced time intervals, thus $M = 35$. The forcing error is assumed to be delta-correlated in time, and statistically homogeneous in space:

$$Q(\mathbf{x}_1, t_1, \mathbf{x}_2, t_2) = \delta(t_1 - t_2) \sum_{\mathbf{n}} Q_{\mathbf{n}} e^{i\mathbf{n} \cdot (\mathbf{x}_1 - \mathbf{x}_2)}, \qquad (6.3.1)$$

where the forcing error variance spectrum is

$$Q_{\mathbf{n}} = Q_0 (|\mathbf{n}|^2 + \rho^{-2})^{-b/2}. \qquad (6.3.2)$$

The real constant Q_0 is chosen such that the total variance of the

Figure 6.3.2. Eigenvalues of the representer matrix \mathbf{R} ($M = 35$: 5 sites, 7 times) for deformation radii $\rho = 0.1$ (broken line) and $\rho = 10.0$ (solid line). Other parameters: see Table 6.3.1.

forcing error, that is,

$$Q(0) = \int_{t_1-0}^{t_1+0} Q(\mathbf{x}_1, t_1, \mathbf{x}_1, t_2)\, dt_2 = Q_0 \sum_{\mathbf{n}} (|\mathbf{n}|^2 + \rho^{-2})^{-b/2} \qquad (6.3.3)$$

takes a prescribed value. The initial error covariance is assumed to be the equilibrium streamfunction covariance in the absence of data, and is also homogeneous, that is,

$$A(\mathbf{x}_1, \mathbf{x}_2) = \sum_{\mathbf{n}} A_{\mathbf{n}} e^{i\mathbf{n}\cdot(\mathbf{x}_1 - \mathbf{x}_2)}, \qquad (6.3.4)$$

where the initial error variance spectrum is

$$A_{\mathbf{n}} = Q_{\mathbf{n}} (2\operatorname{Re} L_{\mathbf{n}})^{-1}. \qquad (6.3.5)$$

Detailed descriptions of the chosen variance spectra are shown in Table 6.3.1, together with values of model parameters. The 35 eigenvalues of the representer matrix \mathbf{R} are shown in Fig. 6.3.2 for two

choices of the deformation radius: $\rho = 0.1$ and $\rho = 10.0$. For the smaller value of ρ, the range of the eigenvalues is less than one order of magnitude, and thus \mathbf{R} is very well conditioned. For the larger value of ρ, the range exceeds two orders of magnitude: \mathbf{R} is much less well conditioned. On the other hand the largest eigenvalue for $\rho = 0.1$ is about equal to the thirteenth largest eigenvalue for $\rho = 10.0$. In fact none of the eigenvalues exceed 0.1 when $\rho = 0.1$. These results may be explained by considering first an antenna which takes just five measurements (streamfunction values at one time, at the five points shown in Fig. 6.3.1), and second by considering an antenna at one location taking measurement at the seven times indicated in Fig. 6.3.1.

Table 6.3.1
Parameters for the model (see (5.2.3) and (5.3.1)), and for the covariances (see (6.3.2)–(6.3.5)), used to obtain the spectra in Figs. 6.3.2–6.3.4. The value of the deformation radius ρ is either 0.1 or 10.0.

Model	$\beta = 1$, $D_0 = 0$, $A_H = 0.2$, $T = 10.0$
Covariances	$Q(0) = 0.0625$, $b = 6$, $w^{-1} = 0.01$

In the first case (see Fig. 6.3.3) the representer matrix has the values

$$R_{jk} = Q_0 (2A_H)^{-1} \sum_{\mathbf{n}} |\mathbf{n}|^{-2} (|\mathbf{n}|^2 + \rho^{-2})^{-b/2} e^{i\mathbf{n}\cdot(\mathbf{x}_j - \mathbf{x}_k)}, \qquad (6.3.6)$$

where the measurement sites are $\mathbf{x}_j (1 \leq j \leq 5)$. Note that R_{jk} is independent both of β and of the single measurement time. Recall from §5.6 that

$$R_{jk} = \overline{\psi(\mathbf{x}_j, t)' \, \psi(\mathbf{x}_k, t)'}, \qquad (6.3.7)$$

where $\psi(\mathbf{x}, t)'$ is the response of the model to the random forcing $q(\mathbf{x}, t)$ and to the random initial values $a(\mathbf{x})$. Again, these have zero means and covariances Q, A as in (6.3.1), (6.3.4). The value of Q_0 has been chosen so that $Q(0)$, the forcing variance, is independent of ρ. However, when $\rho = 0.1$ much more of the forcing variance is contained in higher wavenumbers (larger values of $|\mathbf{n}|$) than is the case when $\rho = 10$. Hence, as a consequence of the effects of diffusion,

Figure 6.3.3. Eigenvalues for the representer matrix \mathbf{R} ($M = 5$: 5 sites, 1 time (6.25)) for $\rho = 0.1$ (broken line) and $\rho = 10.0$ (solid line).

the solution variance R_{jj} is less when ρ is smaller. Since the sum of the eigenvalues of the matrix \mathbf{R} is also its trace, the average eigenvalue will be smaller when ρ is smaller. It is apparent from Fig. 6.3.3 that for $\rho = 0.1$ and $w = 0.1$, diffusion is so great that the signal-to-noise ratio $\phi_1 w$ is less than unity. The data have little influence on the inverse estimate $\hat{\psi}$, which is close to the first-guess ψ_F. When $\rho = 10$ the signal-to-noise ratio is an order of magnitude greater, and the data will significantly influence the inverse estimate.

 Examination of the complex Fourier series (6.3.6) also shows that \mathbf{R} is more strongly diagonally dominant when ρ is smaller, since the amplitudes of the terms with non-zero phases ($j \neq k$) are then greater, leading to more destructive interference in the summation. The better conditioning of \mathbf{R} for $\rho = 0.1$ is a consequence of this greater diagonal dominance. In other words, the variance spectrum of ψ' is flatter when ρ is smaller; measurements at sites separated by a fixed distance are correspondingly less correlated and so yield more information.

 The eigenvalues of \mathbf{R} for seven sequential measurements at one point are shown in Fig. 6.3.4. The values of the representer matrix are given by

Figure 6.3.4. Eigenvalues for the representer matrix \mathbf{R} ($M = 7$: 1 site $(0,0)$, 7 times) for $\rho = 0.1$ (broken line) and $\rho = 10.0$ (solid line).

$$R_{jk} = Q_0(2A_H)^{-1} \sum_{\mathbf{n}} |\mathbf{n}|^{-2}(|\mathbf{n}|^2 + \rho^{-2})^{-b/2} \exp\left\{ -A_H|\mathbf{n}|^2|t_j - t_k| \right.$$

$$\left. - i\beta n(|\mathbf{n}|^2 + \rho^{-2})^{-1}(t_j - t_k) \right\}, \qquad (6.3.8)$$

where $\mathbf{n} = (n, m)$. Now the values are dependent upon β, but not upon the position of the measurements. Again there is less stream-function variance when ρ is smaller, but \mathbf{R} is more strongly diagonally dominant. There is *less* destructive interference in (6.3.8) when ρ is smaller (thereby tending to reduce the diagonal dominance and so worsen the condition of \mathbf{R}), but the tendency towards diagonal dominance due to diffusion is much greater when ρ is smaller. Quite simply, the diffusive decorrelation time ρ^2/A_H is shorter when ρ is smaller, and so the measurements are in effect less frequent.

ii) An antenna for the equatorial waveguide

Interannual variability in the equatorial Pacific is marked by intermittent but very strong events known as El Niño. Sea-surface temperature (SST) departs from its annual cycle by as much as 4°C over a two-year period: see Philander (1990) and Fig. 6.3.5. Nevertheless the SST time series in such an interval have low information content, and may be represented as the sum of a few sinusoidal

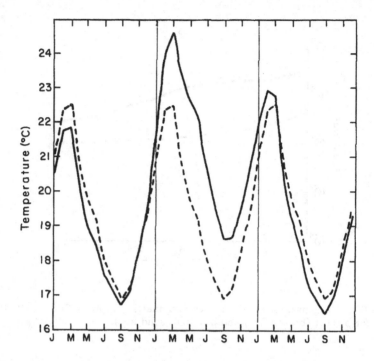

Figure 6.3.5. The seasonal sea-surface temperature changes (dashed line) and the changes during a typical El Niño episode (solid line) as measured along a ship track 100 km off the coast of South American between 3°S and 12°S (after Rasmussen & Carpenter, 1983).

oscillations in that two-year interval. Consequently it is possible to reduce greatly the computational task of combining observations with a linear dynamical model of equatorial circulation, by working in the frequency domain rather than in the time domain.

There is a widely used model, based on the linear, single-layer reduced-gravity shallow-water equations (Philander, 1990, p. 106). In the frequency domain these equations are

$$\begin{pmatrix} i\omega - \kappa & -f & c\frac{\partial}{\partial x} \\ f & i\omega - \kappa & c\frac{\partial}{\partial y} \\ c\frac{\partial}{\partial x} & c\frac{\partial}{\partial y} & +i\omega - \lambda \end{pmatrix} \begin{pmatrix} u \\ v \\ \theta \end{pmatrix} = \begin{pmatrix} F^u \\ F^v \\ F^\theta \end{pmatrix} + \begin{pmatrix} q^u \\ q^v \\ q^\theta \end{pmatrix}, \quad (6.3.9)$$

where x and y are zonal and meridional Cartesian coordinates, ω is the frequency, u and v are zonal and meridional velocity components, θ is an upper layer thickness divided by the time scale c/H, where H is the mean upper layer thickness and c is the phase speed of gravity waves

in the upper layer. The Coriolis parameter $f = f(y)$ is approximated by βy on the equatorial β-plane; F^u, F^v and F^θ are prior estimates of zonal forcing, meridional forcing and heating, respectively, while q^u, q^v and q^θ are the respective errors in those prior estimates. Finally, κ and λ are linear drag and Rayleigh damping coefficients respectively. The system (6.3.9) may be expressed as

$$\mathbf{Su} = \mathbf{F} + \mathbf{q}, \qquad (6.3.10)$$

where \mathbf{S} is a 3×3 matrix operator, while \mathbf{u}, \mathbf{F} and \mathbf{q} are 3-vector fields depending on $\mathbf{x} = (x, y)$ and on the frequency ω. Boundary conditions at simple zonal and meridional boundaries are discussed in Bennett (1990).

There are several observing systems in use in the equatorial Pacific: island tide gauges measuring mean sea-level changes associated with fluctuations in the upper-layer thickness (Wyrtki *et al.*, 1988) and ship-of-opportunity expendable bathythermograph (SOP XBT) sampling programs (McPhaden & Taft, 1984). The latter of these makes use of merchant marine shipping routes: see Fig. 6.3.6, which also shows the domain for the model (6.3.10). Upper-layer thickness is measured by dropping an XBT about every three hours: since a typical steaming speed is 20 knots, these samples are about 60 nautical miles or 100 km apart. Ships cross the equatorial region between 20°S and 20°N in about 5 days, so the cruise data may be considered synoptic when examining interannual variability. Each track is traversed one way or the other every two weeks, so interannual time scales are well resolved.

The XBT probes are not dropped at fixed locations. This is not a fundamental difficulty but is very inconvenient. It is much more desirable to have an observing system with a regular spatial structure. This could be achieved by smoothing the data collected along a track using some appropriate basis functions, and taking uniformly spaced samples from the smoothing function. However, the last step is unnecessary as the coefficients of the basis functions provide an orderly data set. The coefficients are calculated by integrating along the tracks. Various choices of path parameter are available, such as distance along the track, but latitude is equally satisfactory. Longitude is a linear function of latitude along the tracks, which are approximately straight lines on the equatorial β-plane. Interannual variability is largely confined to the latitude band from 5°S to 5°N, so an efficient set of basis functions would be similarly concentrated

Figure 6.3.6. Model domain for (6.3.10), and merchant marine ship routes labeled 1–13. G, Guam; M, Majuro; H, Honolulu; N, Noumea; S, Suva; PP, Pago-Pago; T, Tahiti; P, Panama (after Bennett, 1990).

in that band. A suitable basis is provided by Hermite functions of latitude, the independent variable being scaled by the equatorial internal deformation radius $(c/\beta)^{1/2}$. These function are also the meridional structure of eigenfunctions of the operator \mathbf{S} in (6.3.10), but this would not be so if \mathbf{S} included advection by a mean zonal flow with meridional shear. In such a case, a Hermite expansion of the solution of (6.3.10) would be a less efficient numerical technique than, say, finite differences. The point here is that the choice of basis functions for definition of an orderly data set need not be related to the method of solving the model equations. Finally, the data may be harmonically analyzed in time to yield complex Fourier coefficients at selected frequencies, such as $\omega = 0.25$, 0.5 and 1.0 cycles per year (cpy). In summary, the data are related to the "circulation field" \mathbf{u}

at frequency ω by

$$d_m = \mathcal{L}[\mathbf{u}] + \epsilon_m , \qquad (1 \le m \le M) , \qquad (6.3.11)$$

where d_m is a datum, \mathcal{L}_m is the measurement functional

$$\mathcal{L}_m[\mathbf{u}] = \omega \int_{T_k} \psi_j(\eta)\theta(\mathbf{X}_k(\eta),\omega)\, d\eta \qquad (6.3.12)$$

and ϵ_m is the measurement error. In (6.3.12), $\eta = y(c/\beta)^{-1/2}$, ψ_j is the Hermite function of order j, $\mathbf{X}_k(\eta) = (X_k(\eta), \eta(c/\beta)^{1/2})$ where $x = X_k(\eta)$ is the longitude-latitude relationship along the kth track, and T_k denotes the range of integration in η along the kth track. The measurement index m is related to the Hermite mode j and the track index n by $m = m(j,n) = jN_T + n$, where $0 \le j \le N_H - 1$, N_H being the number of Hermite functions and N_T is the number of tracks. Thus $1 \le m \le M = N_T \times N_H$. The integral in (6.3.12) is multiplied by frequency so that \mathcal{L}_m has the same units as the terms in (6.3.10), namely, those of acceleration. Note that the formal argument of the measurement functionals is the vector field \mathbf{u}; for this SOP XBT antenna the functionals yield integrated values of only the third component $\theta = \theta(\mathbf{x},\omega)$. Other antennas such as current meter arrays would of course only yield values of the first two components u and v.

A scalar-valued inner product may be defined as follows for all pairs of 3-vector fields $\mathbf{p}(\mathbf{x},\omega)$, $\mathbf{o}(\mathbf{x},\omega)$ satisfying the model boundary conditions:

$$\langle \mathbf{p}, \mathbf{o} \rangle = A^{-2} \int d\mathbf{x}_1 \int d\mathbf{x}_2 (\mathbf{Sp}(\mathbf{x}_1))^* \mathbf{W}(\mathbf{x}_1, \mathbf{x}_2, \omega)\mathbf{So}(\mathbf{x}_2) , \qquad (6.3.13)$$

where the symmetric 3×3 weight matrix \mathbf{W} is positive-definite and symmetric with respect to its spatial arguments; A is the area of the model domain. A penalty functional may be defined by

$$\mathcal{J}[\mathbf{u}] = \langle \mathbf{u} - \mathbf{u}_F, \mathbf{u} - \mathbf{u}_F \rangle + (\mathbf{d} - \mathcal{L}[\mathbf{u}])^* \mathbf{w}(\mathbf{d} - \mathcal{L}[\mathbf{u}]) , \qquad (6.3.14)$$

where \mathbf{u}_F is the response to the prior estimate of forcing: $\mathbf{Su}_F = \mathbf{F}$; \mathbf{d} and \mathcal{L} are the M-vectors of data and measurement functionals defined in (6.3.11), and \mathbf{w} is an $M \times M$ positive-definite symmetric weight

matrix. Then the 3-vector field $\hat{\mathbf{u}}(\mathbf{x}, \omega)$ which minimizes (6.3.14) provides the weighted least-squares best fit to the model and data. Corresponding to $\hat{\mathbf{u}}$ are model and measurement residuals $\hat{\mathbf{q}}$ and $\hat{\epsilon}$, respectively, which are optimal estimates of the errors \mathbf{q} and ϵ in the forcing first guess and data, respectively. It is readily shown that \mathcal{J} is minimized by a representer expansion in the form

$$\hat{\mathbf{u}} = \mathbf{u}_F + \sum_{m=1}^{M} \hat{b}_m \mathbf{r}_m \,, \qquad (6.3.15)$$

where $\mathbf{r}_m = \mathbf{r}_m(\mathbf{x}, \omega)$ is a 3-vector representer field, satisfying the model boundary conditions and having the property that

$$\langle \mathbf{r}_m, \mathbf{p} \rangle = \mathcal{L}_m[\mathbf{p}] \qquad (6.3.16)$$

for any 3-vector field \mathbf{p} also satisfying the model boundary conditions. The M-vector $\hat{\mathbf{b}}$ of representer coefficients yielding the minimum for \mathcal{J} satisfies

$$\mathbf{P}\hat{\mathbf{b}} = \mathbf{h} \,, \qquad (6.3.17)$$

where $\mathbf{h} = \mathbf{d} - \mathcal{L}[\mathbf{u}_F]$ is the data misfit for the first-guess solution, and \mathbf{P} is related to the $M \times M$ Hermitian positive-definite representer matrix \mathbf{R} by

$$\mathbf{P} = \mathbf{R} + \mathbf{w}^{-1} \,. \qquad (6.3.18)$$

The components of \mathbf{R} are given by

$$R_{nm} = \langle \mathbf{r}_n, \mathbf{r}_m \rangle \,. \qquad (6.3.19)$$

Exercise 6.3.1 Show that the expression (6.3.14) does define an inner product, provided that the model boundary conditions preclude eigensolutions. □

Exercise 6.3.2 Verify that the penalty functional \mathcal{J} defined by (6.3.14) is minimized by the representer expansion (6.3.15), (6.3.17).
□

Exercise 6.3.3 Show that the representer residuals are weighted solutions of adjoint model equations, in which the forcing consists of measurements of impulses. For definiteness assume simple model

boundary conditions, such as no normal flow across the bounding latitudes and longitudes. □

Exercise 6.3.4 It is always possible to express $\mathcal{L}[\mathbf{p}]$ as an integral of the form

$$\mathcal{L}[\mathbf{p}] = \int \mu(\mathbf{x})\,\mathbf{p}(\mathbf{x})\,d\mathbf{x}, \qquad (6.3.20)$$

which may be denoted $\mu \circ \mathbf{p}$. In (6.3.20), μ is the measurement kernel; for each \mathbf{x}, $\mu(\mathbf{x})$ is an $M \times 3$ matrix. It may involve Dirac delta-functions, as in the case of pointwise measurement, for example. The penalty functional \mathcal{J} may be expressed in terms of μ and the \circ notation as

$$\mathcal{J}[\mathbf{u}] = \big(\mathbf{S}(\mathbf{u} - \mathbf{u}_F)\big) \circ \big(\mathbf{S}(\mathbf{u} - \mathbf{u}_F)\big)$$
$$+ (\mathbf{d} - \mu \circ \mathbf{u})^* \, \mathbf{w}(\mathbf{d} - \mu \circ \mathbf{u}). \qquad (6.3.21)$$

Show that \mathcal{J} is least when $\mathbf{u} = \hat{\mathbf{u}}$, satisfying

$$\big(\mathbf{S}^\dagger \mathbf{W} \circ \mathbf{S} + \mu^* \, \mathbf{w}\mu\circ\big)\,(\hat{\mathbf{u}} - \mathbf{u}_F) = \mu^* \, \mathbf{w}\mathbf{h}, \qquad (6.3.22)$$

where \mathbf{S}^\dagger is an operator adjoint to \mathbf{S}, in a certain sense. Find that sense and construct \mathbf{S}^\dagger explicitly. Reconcile (6.3.22) with the representer solution (6.3.15). Hence rearrange the formal solution of (6.3.22) into a form depending on \mathbf{W}^{-1} and \mathbf{w}^{-1} rather than on \mathbf{W} and \mathbf{w}. Tarantola (1987) calls this the method of "total inversion." □

The representers $\mathbf{r}_m(1 \leq m \leq M)$ are determined by the measurement functionals \mathcal{L}_m, the model operator \mathbf{S} plus boundary conditions, and the functional inverse of the weighting function \mathbf{W}^{-1} in the inner product (6.3.14). A natural choice for \mathbf{W}^{-1} is the forcing error covariance \mathbf{Q}:

$$A^{-2} \int d\mathbf{x}_3 \mathbf{W}(\mathbf{x}_1, \mathbf{x}_3, \omega)\mathbf{Q}(\mathbf{x}_3, \mathbf{x}_2, \omega) = \delta(\mathbf{x}_1 - \mathbf{x}_2), \qquad (6.3.23)$$

where

$$\mathbf{Q}(\mathbf{x}_1, \mathbf{x}_2, \omega) = \overline{\mathbf{q}(\mathbf{x}_1, \omega)\mathbf{q}(\mathbf{x}_2, \omega)^*}, \qquad (6.3.24)$$

the overbar denoting an ensemble average. It is assumed that \mathbf{q} has zero mean. Likewise a natural choice for \mathbf{w} in \mathcal{J} is the matrix inverse of the measurement error covariance:

$$\mathbf{w}^{-1}(\omega) = \overline{\epsilon(\omega)\epsilon(\omega)^*}, \qquad (6.3.25)$$

Figure 6.3.7. Eigenvalues of $\mathbf{P} = \mathbf{R} + \mathbf{w}^{-1}$ for the TOGA SOP XBT antenna ($M = 78$). Solid line, $\mathbf{w}^{-1} = 0$ (perfect data); dashed line, $\mathbf{w}^{-1} = w^{-1}\mathbf{I}_M$ with $w = 5$ (uncorrelated data errors with the same variance); dotted line, $(\mathbf{w}^{-1})_{nm} = w^{-1}$ for $1 \leq n, m \leq M$ (correlated errors with the same variance) (after Bennett, 1990).

it being assumed that ϵ has zero mean.

In the calculations described below, \mathbf{Q} is a multiple of the 3×3 unit matrix: $\mathbf{Q}(\mathbf{x}_1, \mathbf{x}_2, \omega) = Q(\mathbf{x}_1, \mathbf{x}_2, \omega)\mathbf{I}_3$, and the covariance function Q is displacement-invariant: $Q(\mathbf{x}_1, \mathbf{x}_2, \omega) = Q(\mathbf{x}_1 - \mathbf{x}_2, \omega)$. Thus the first-guess forcing errors q^u, q^v and q^θ are uncorrelated, are statistically homogeneous and have the same covariance. Isotropy is not assumed: unless otherwise stated, the decorrelation length scales in the zonal and meridional directions are 32° and 4°, respectively. The power spectrum corresponding to $Q(\mathbf{x}, \omega)$ has algebraic decay for large wavenumber, as in (2.3.27), with order $m = 3$. Fixed parameters in \mathbf{S} are $\beta = 2.3 \times 10^{-11}$ m^{-1} s^{-1}, $\kappa = \lambda = 3.17 \times 10^{-8}$ s$^{-1} = 1$ year^{-1}.

The eigenvalues of \mathbf{P} are shown in Fig. 6.3.7 for $\omega = 1$ cpy and $c = 3.0$ m s^{-1}. Three cases are shown: $\mathbf{w}^{-1} = \mathbf{0}$ (perfect data), $\mathbf{w}^{-1} = w^{-1}\mathbf{I}_M$ (uncorrelated data errors with common variance $w^{-1} = 0.2$), and $(\mathbf{w}^{-1})_{nm} = w^{-1} = 0.2$ for $1 \leq n, m \leq M$ (perfectly correlated data errors with common variance). With $N_T = 13$ SOP tracks and $N_H = 6$ Hermite moments (100-km wavelength), there

is a total of $M = 78$ measurements. Due to roundoff there are a few negative eigenvalues of very small amplitude, and these have not been plotted. Clearly **P** is ill-conditioned if the data are perfect or if the data errors are perfectly correlated; the resulting interpolation or smoothing would be very sensitive to the data. Equally clearly **P** is well conditioned if the data have uncorrelated errors (with common variance $w^{-1} = 0.2$), in which case smoothing would not be sensitive to the data. Interpretation of the value of w^{-1} requires care: multiplication of the penalty functional by a positive constant K does not alter the optimal estimate \hat{u}; in particular the first-guess data misfit **h** is unchanged while the representers r_m and the matrix **P** are both reduced by a factor of K, so the linear combination of representers in (6.3.15) is invariant. It follows that the value of the data error variance w^{-1} is meaningful only in proportion to the value of the forcing error variance $Q(0)$. Moreover, by virtue of the formulation of \mathcal{J} as weighted sums of squares of residuals, it follows that $Q(0)$ and w^{-1} may be defined as error variances relative to typical terms in the model equations and typical data values, respectively. For example, with the parameters used here, $Q(0) \simeq 100$; thus $w^{-1}Q(0)^{-1} \simeq 0.002$, implying that as a prior estimate, the standard deviation of the relative data errors is about $1/22$ of the standard deviation of the first-guess forcing errors. At best, the latter would be no smaller than about 20–25% in the equatorial Pacific, so a very optimistic assumption of 1% relative data errors is being made. The crucial point here is that even with such a small estimate for the data errors, smoothing is well conditioned. The conditioning is improved as c decreases: see Fig. 6.3.8. At a fixed frequency ω there is a corresponding decrease in the wavelength c/ω of Kelvin waves and long Rossby waves at the equator; consequently the antenna is more sparse. Equally, increasing the frequency leads to better conditioning: see Fig. 6.3.9. In these latter two figures, it is the eigenvalues of **R** which are plotted; those of **P** for the case $\mathbf{w} = w\mathbf{I}$ may be inferred by adding w^{-1} to each eigenvalue of **R**. Alternatively the conditioning may be estimated in terms of the number of eigenvalues of **R** exceeding w^{-1}.

These eigenvalue "spectra" indicate that about one-half of the $M = 78$ measurements are contributing independent information, at the assumed level of data error. Examinations of the eigenvectors of **P** show that no measurements from the three tracks with southern terminals at Majuro contribute significant information. The results

Figure 6.3.8. Eigenvalues of **R** for the TOGA SOP XBT antenna ($M = 78$). Solid line, phase speed $c = 3.0$ m s^{-1}; dashed line, $c = 1.8$ m s^{-1}; dotted line, $c = 1.1$ m s^{-1}. All at frequency $\omega = 10^{-7}$ s$^{-1} \simeq 0.5$ cpy (after Bennett, 1990).

are simply explained: the Hermite expansion has been truncated at order 5 since the original SOP XBT data do not resolve scales less than about 100 km, and the amplitudes of the first 6 Hermite functions (with arguments scaled by $(c/\beta)^{1/2}$) are insignificant at latitudes more than 10° from the equator. Hence the first 6 Hermite moments along the three tracks north of Majuro are negligible in comparison with moments along tracks spanning 10°S to 10°N. The structure of the SOP XBT antenna is discussed further in Bennett (1990). An examination of the principal antenna elements, which are the representer combinations most stably estimated from the data, reveals mixtures of equatorial Kelvin-wave-like and long Rossby-wave-like structures. The principal antenna elements, which are *forced* solutions of the model, should be used in preference to individual, free waves when interpolating or smoothing data (Stommel, 1989).

iii) Tides

An analysis of an antenna for observing barotropic tides in the North Pacific is shown in Fig. 6.3.10 (Bennett, 1985). The elements in the antenna include pressure gauges on the ocean floor, and subsurface

Figure 6.3.9. Eigenvalues of **R** for TOGA SOP XBT antenna ($M = 78$). Solid line, frequency $\omega = 0.5 \times 10^{-7}$ s^{-1}($\simeq 0.25$ cpy); dashed line, $\omega = 1.0 \times 10^{-7}$ s^{-1}($\simeq 0.5$ cpy); dotted line, $\omega = 2.0 \times 10^{-7}$ s^{-1}($\simeq 1.0$ cpy). All at phase speed $c = 3.0$ m s^{-1} (after Bennett, 1990).

acoustic transducers engaged in reciprocal-shooting tomography. This acoustic technique measures the travel times of sound pulses between pairs of transducers. The difference in the times for travel in opposite directions yields an estimate of the water velocity along the path, averaged along the path. The dynamics of subtropical barotropic tides in a small region are again given by (6.3.9), with $f(y) \equiv f_0$, a constant, and $c = (gH)^{1/2}$, where g is the gravitational acceleration, H being the mean depth. Spectra of the representer matrix **R** are shown in Fig. 6.3.11. Clearly there is little to choose between two plausible dispositions of the pressure gauges. Less than half of the data are significant, given the indicated data error variance w^{-1}.

An antenna for coastal, barotropic semi-diurnal tides was examined by McIntosh (1987a); see also Bennett & McIntosh (1982) and McIntosh & Bennett (1984). The antenna elements were seven coastal tide gauges and one current meter mooring (see Fig. 6.3.12). The mooring yielded two horizontal velocity components, for a total of nine scalar measurements. The first guess for the model forcing vector was assumed to be exact; in effect the dynamics were given

Figure 6.3.10. North Pacific: open-ocean region A with boundary B. Open circles are acoustic transmitters/receivers. Solid lines are reciprocal shooting paths. Crosses and solid circles are, respectively, pressure transducers in the "square" and "diamond" arrays. Both arrays include the reciprocal shooting paths (after Bennett, 1985).

infinite weight. Similarly the rigid boundary condition at the coast was given infinite weight, but the radiation boundary conditions along the two parts of the open boundary were assumed to have uniform error variances. Details may be found in McIntosh (1984). The nine eigenvalues of the representer matrix are shown in Fig. 6.3.13. The complex eigenvectors for the six largest eigenvalues are shown as rotary stick plots in Fig. 6.3.14. The length of each stick is proportional to the amplitude of a complex component of an eigenvector; the rotation of the stick is the phase. The leading eigenvector (#1) is dominated by measurements #1, 2, 3, 4 and 7 (the tide gauges in the western part of Bass Strait). The second eigenvector is dominated by measurements #5, 6 and 7 (the tide gauges in the eastern part of the strait). Measurements #8 and 9 (the current meter mooring in the western approach) only contribute significantly to the last four eigenvalues, indicating that interpolating the tide gauge data together with the current meter data would be badly conditioned. McIntosh

Figure 6.3.11. Eigenvalues of the North Pacific representer matrix **R** ($M =$ 11). Squares denote the "square" array, diamonds denote the "diamond" array. Negative eigenvalues (shown in parentheses) indicate errors in the computation of **R** (after Bennett, 1985).

(1987a) concluded that, given the relatively coarse resolution of his numerical model (16.5 km), the mooring was too close to coastal tide gauges and would have yielded more independent information if deployed elsewhere. McIntosh also examined the sea-level components of the antenna modes; he interpreted both of the first two as pairs of opposed Kelvin waves, one traveling westward along the north coast and the other traveling eastward along the south coast. In the first mode the pair combines to form an amphidrome in the northeastern part of the strait; in the second mode there is an amphidrome in the northwest. The third mode was interpreted as a Poincaré wave tunneling in from the eastern approaches. The higher-order antenna modes could not be interpreted as simple combinations of simple wave solutions.

iv) Acoustic tomography

The preceding examples of "antenna design" are really just as-

Figure 6.3.12. Bass Strait bathymetry (m) and real and model coastlines. Numbered dots give the location of tide gauge data, and the numbered cross represents the two orthogonal components of current data (after McIntosh & Bennett, 1984).

Figure 6.3.13. Eigenvalues of the Bass Strait representer matrix \mathbf{R} ($M = 9$). Squares: exact semidiurnal tidal dynamics and rigid coasts (Proudman limit). Triangles: all constraints inexact (after McIntosh, 1984).

sessments of efficiencies of a few subjectively chosen antennae. Design proper would involve an algorithm for meeting an objective criterion of a good antenna configuration. The objective criterion used in the

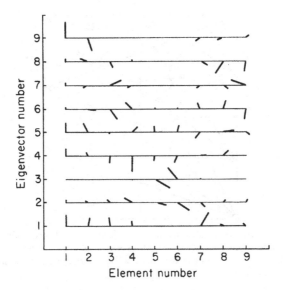

Figure 6.3.14. Complex elements of seven eigenvectors for the Bass Strait tidal model in the Proudman limit (after McIntosh, 1987a).

preceding examples is the degree of conditioning of the $M \times M$ representer matrix. The criterion has been expressed as the ratio of λ_M, the smallest eigenvalue of the matrix, to the largest eigenvalue λ_1. Maximizing the ratio by varying the parameters in the M measurement functionals $\mathcal{L}_1, \ldots, \mathcal{L}_M$ (such as the locations of point measurements) is clearly a highly nonlinear problem, even if the dynamics and measurement functionals are linear. Thus the efficiency of the minimizing algorithm itself is an important consideration.

Barth & Wunsch (1990) were able to obtain major improvements in the design of an acoustic tomography array by using the method of "simulated annealing" (Kirkpatrick, Gelatt & Vecchi, 1983) in order to maximize the conditioning of a simple inverse problem. Central to acoustic tomography is the travel time equation:

$$T = \int_{\Gamma} ds \, C^{-1}, \qquad (6.3.26)$$

where T is the time taken for the acoustic signal moving with a speed of sound $C(x, y, z)$ to travel a path Γ from a source to a receiver. A tomographic problem consists of determining C from measurements of T. If the reference speed of sound is taken to be C_0, then any

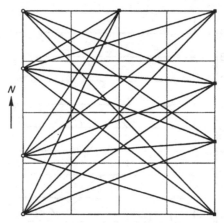

Figure 6.3.15. North Atlantic: original configuration of acoustic tomography transmitters (open circles) and receivers (solid squares). The 16 squares are regions in which the sound speed is assumed to be uniform (after Barth & Wunsch, 1990).

fluctuations in the speed of sound will produce fluctuations in the measured travel times, which can often be linearized to

$$\Delta T = - \int_{\Gamma} ds \, C_0^{-2} \Delta C \qquad (6.3.27)$$

(Barth & Wunsch, 1990). Breaking up the ocean into J squares or "pixels," in which the sound speed is constant, leads to a system of linear equations of the form

$$\mathbf{A} \, \Delta \mathbf{C} = \Delta \mathbf{T}, \qquad (6.3.28)$$

where the jth component of the vector $\Delta \mathbf{C}$ is the sound speed fluctuation in the jth pixel $(1 \le j \le J)$, the mth components of the vector $\Delta \mathbf{T}$ is the travel time fluctuation between the mth transmitter-receiver pair $(1 \le m \le M)$, while the components of the matrix \mathbf{A} are proportional to the length of the mth sound path in the jth square. Behringer et al. (1982) deployed four fixed sound transmitters and five fixed receivers in the North Atlantic $(M = 20)$; Barth & Wunsch (1990) represented the sound speed field with a 4×4 square grid $(J = 16)$: see Fig. 6.3.15, which also shows the original configuration of Behringer et al. (1982). As discussed in Chapter 1, the conditioning of the inverse problem (6.3.28) is given by the J singular values

$\lambda_1, \ldots, \lambda_J$ of \mathbf{A}, that is, the J eigenvalues of $\mathbf{A}^*\mathbf{A}$. Thus, the design strategy should be to locate the transmitters and receivers so that the ratio λ_J/λ_1 is maximized. In practice, Barth & Wunsch were able to obtain a greater value for the ratio by maximizing λ_J instead of the ratio itself.

The maximum of λ_J was sought using the method of "simulated annealing" (Kirkpatrick *et al.*, 1983). The method proceeds by comparing alternative configurations of transmitters and receivers. Given one configuration, a second configuration is generated by making random perturbations to the first. Then the quantity P_{12} is calculated, where

$$P_{12} = \exp\left\{ \left(\lambda_J^{(2)} - \lambda_J^{(1)}\right)/\theta \right\}. \qquad (6.3.29)$$

In (6.3.29), $\lambda_J^{(1)}$ and $\lambda_J^{(2)}$ are the smallest singular values for the first and second configurations, respectively, while θ is a positive "annealing parameter." If $P_{12} > 1$, then the second configuration is chosen. If $P_{12} < 1$, then the second configuration, yielding the *smaller* value, is chosen when $\xi < P_{12}$, where ξ is a random number uniformly distributed over the interval $0 \leq \xi \leq 1$. That is, the second configuration is chosen, with probability $P_{12} < 1$. The chosen configuration is then randomly perturbed again. This process is repeated until a satisfactorily large value of λ_J is found. Since the larger value is not always chosen, the ultimate choice should be compared with the "largest rejected larger value." Next, the annealing parameter θ is fractionally reduced and the search is repeated. Simulated annealing terminates when further reduction of θ fails to yield significantly larger values of λ. For further details, see Barth & Wunsch (1990). In particular the random perturbations to the configurations are scaled by θ: the search range is reduced as an apparent maximum for λ_J is approached. The optimal configuration is shown in Fig. 6.3.16. The singular value spectra for the original choice of Behringer *et al.* (1982) and the optimal choice of Barth & Wunsch are shown in Fig. 6.3.17. The largest values of λ_J for each value of the annealing parameter are shown in Fig. 6.3.18. A variety of related experiments were carried out by Barth & Wunsch. In one case, simulated annealing yielded a significantly larger value for λ_J than that obtained with many trials of a conventional nonlinear optimization algorithm based on gradient-descent methods. The cumulative computational times for the two approaches were comparable.

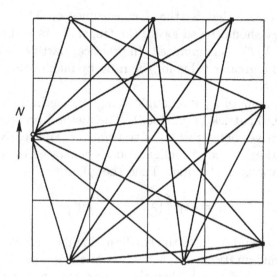

Figure 6.3.16. North Atlantic: the final optimal configuration, constrained to locations on the boundary, according to simulated annealing (after Barth & Wunsch, 1990).

Figure 6.3.17. North Atlantic: singular values of matrix **A** (see (6.3.26)) ($J = 16$) for the original configuration in Fig. 6.3.15 (open circles) and the final optimal configuration in Fig. 6.3.16 (solid squares) (after Barth & Wunsch, 1990).

This fascinating approach to array design, as its name suggests, derives from statistical physics. The selection criterion, based on

Figure 6.3.18. North Atlantic: output of simulated annealing for the acoustic tomography antenna originally configured as in Fig. 6.3.15. Each iterate on the abscissa corresponds to a reduction of the annealing parameter θ by a factor in the range 0.9–0.995 (after Barth & Wunsch, 1990).

the quantity P_{12} defined by (6.3.29), is the celebrated "importance sampling" scheme due to Metropolis *et al.* (1953). It may be proved that the probability of a particular configuration being selected (rejection of the "second" configuration implies acceptance of the "first" once more) is given by

$$P \propto \exp\{\lambda_J/\theta\}. \tag{6.3.30}$$

Importance sampling provides the basis for Monte Carlo simulations in statistical physics (Binder & Heermann, 1988), where the objective is the estimation of mean values of a state variable, call it λ_J, with distribution P. Sampling should approximate popularity. The search for the maximum λ_J is concentrated in the neighborhood of a local maximum, while allowing *some* searches which are initially downhill ($\xi < P_{12} < 1$) but which might lead eventually to the global maximum. The choice of (6.3.30) as the "concentration coefficient" is a matter of judgement: rational or other distributions might be more appropriate at various stages of the search if not entirely.

The broader issue in antenna design is the choice of objective criterion. Maximizing the conditioning of the appropriate matrix

minimizes the redundancy of the observing system, yet redundancy is a desirable property from a practical point of view. The issue reduces to the incremental cost of the antenna.

7

NONLINEAR QUASI-GEOSTROPHIC
MODELS

7.1 Introduction

An essential feature of the inverse theory described in the preceding chapters has been linearity, which allows explicit expressions for the solutions of the various inverse problems. The expressions permit complete analyses of the analytical and statistical properties of the estimates. The one striking exception has been the evolution equation for the state error covariance in the Kalman filter. The equation is nonlinear, being of matrix-Riccati form, as a consequence of the requirement for a sequential approximation to the fixed-interval smoother. However, the filter estimate of the state does depend linearly on the prior estimates of forcing and initial conditions, and on the data.

The linearity of an inverse theory may be violated in many ways.

i) Nonlinear dynamics

The complexity of fluid motion is fundamentally due to advection, which takes a nonlinear form in the Eulerian description of flow. While there are a few classes of interesting and important oceanic motions which are adequately described by linearized dynamics, or at most linearized advection, such as tides, coastal trapped waves, equatorial interannual variability and Sverdrup flow, the importance of advection is unavoidable for most oceanic circulation. The classical examples are the western boundary currents of the subtropical gyres, the highly variable extension regions of the boundary currents, the meandering

181

Antarctic Circumpolar Current, and equatorial mesoscale variability. The dynamics of a model may also be nonlinear as a consequence of turbulence closures expressed in terms of shear-dependent diffusivities. Ocean models which include chemical tracers or biological fields may be nonlinear owing to chemical reactions, or to predation or to grazing.

ii) Nonlinear measurements

Only linear measurement functionals have been considered thus far. A simple and important exception is provided by scatterometer measurement of wind speed, which has a nonlinear relationship to the wind stress vector driving a model. Thus, even though the stress vector may have a linear relationship to the circulation, a wind speed datum cannot be expressed as the value of a linear functional acting on the circulation.

iii) Boundary conditions

Regional models of ocean circulation may have boundaries through which there is a specified mass flow. Since the fluid velocity influences the propagation of disturbances, the number of conditions which should be specified at such open boundaries may depend upon the sign or the magnitude of the fluid velocity. This is a nonlinearity of a most unusual kind, especially if the number of boundary conditions also depends upon unspecified aspects of the circulation such as the stratification. The latter possibility arises in non-geostrophic models. For quasi-geostrophic models the number of boundary conditions is determined entirely by the specified normal flow at the boundary. Primitive-equation models of the "layered" variety have additional nonlinear boundary conditions related to those layer interfaces which rise to the ocean surface.

iv) Parameter estimation

The parameters such as the deformation radius and the eddy diffusivity in the linear quasi-geostrophic model of Chapter 5 were kept fixed in the search for best-fit estimates of the circulation. Only the prior estimates of the forcing and the initial values were varied. Had the parameters also been varied, the extremum would have been defined by a nonlinear system of equations. The system would have comprised the Euler-Lagrange equations, and, for each variable parameter, an equation expressing stationarity of the penalty functional with respect to that parameter. The nonlinearity of the system is

immediately evident since the variable parameters appear as coefficients in the dynamics.

v) Nonlinear estimation

The penalty functions and functionals considered in the preceding chapters are all quadratic in the errors in the prior estimates. Consequently the conditions at an extremum are all linear equations for the residuals or estimated errors. Derivatives of residuals may also have been penalized, but again in a weighted least-squares sense so that linearity is preserved. It may be argued on statistical grounds that certain residuals should be penalized in some other sense; for example if the residuals are not normally distributed, in which case minimization of a quadratic functional is not the estimator of maximum likelihood. Minimization of the maximum value of the residual field may be more appropriate, yielding a nonlinear optimization problem. The analysis used thus far, based on linear projections, may still be useful in such cases but is not so straightforward. In the absence of compelling statistical evidence of non-normality, the adoption of a non-quadratic penalty functional, with its associated nonlinear estimation problems, in not readily justified. Indeed, many such functionals may be bounded above by quadratic functionals (the so-called embedding theorems: see Adams (1975)).

This chapter examines nonlinearity solely due to advection. Iterative methods for solving the nonlinear Euler-Lagrange equations will be considered in detail.

7.2 A nonlinear quasi-geostrophic model

The model is a simple extension of the single-layer, reduced-gravity dynamics considered in §5.2, and now includes advection of the quasi-geostrophic potential vorticity $E\psi$:

$$\frac{E\partial\psi}{\partial t} + \varepsilon\mathbf{u}\cdot\boldsymbol{\nabla}E\psi + \frac{\partial\psi}{\partial x} = E(F+q) + DE\psi, \qquad (7.2.1)$$

where E is the elliptic operator (5.2.4), $\psi = \psi(\mathbf{x}, t)$ is the streamfunction, ε is a Rossby number, $\mathbf{u} = \mathbf{u}(\mathbf{x}, t) = \mathbf{e}_3 \times \boldsymbol{\nabla}\psi$ is the geostrophic velocity field, where $\mathbf{e}_3 = (0, 0, 1)$, EF is the prior estimate of the forcing, Eq is the error in that prior estimate, and D is the diffusion

operator $-D_0 + A_H \nabla^2$. The Rossby number ε is based on β, the constant northward gradient of the Coriolis parameter. That is,

$$\varepsilon = U(\beta H^2)^{-1}, \qquad (7.2.2)$$

where U is a velocity scale and H a length scale. Consequently the coefficient β in (5.2.3) is replaced by unity in (7.2.1).

Assuming that all fields are doubly periodic in x and y, the operator E is readily inverted, yielding the streamfunction evolution equation

$$\frac{\partial \psi}{\partial t} = -E^{-1} \varepsilon J(\psi, E\psi) + L\psi + F + q, \qquad (7.2.3)$$

where J is the Jacobian operator:

$$J(a, b) = \frac{\partial(a, b)}{\partial(x, y)} = a_x b_y - a_y b_x, \qquad (7.2.4)$$

while L is the linear operator

$$L = -E^{-1} \frac{\partial}{\partial x} + D. \qquad (7.2.5)$$

The prior estimate $\Psi = \Psi(\mathbf{x})$ and error $a = a(\mathbf{x})$ for the initial streamfunction are related to the streamfunction by (5.2.5), that is,

$$\psi(\mathbf{x}, 0) = \Psi(\mathbf{x}) + a(\mathbf{x}). \qquad (7.2.6)$$

The data \mathbf{d} and measurement errors ϵ are related to ψ by linear functionals \mathcal{L} as in (5.2.1), that is,

$$\mathbf{d} = \mathcal{L}[\psi] + \epsilon. \qquad (7.2.7)$$

It is assumed that all variables and parameters in (7.2.3)–(7.2.7) are dimensionless.

7.3 The generalized inverse

The streamfunction ψ_F is the response to the prior estimates F and Ψ for the forcing and initial values, respectively. Thus ψ_F satisfies

$$\frac{\partial \psi_F}{\partial t} = -E^{-1} \varepsilon J(\psi_F, E\psi_F) + L\psi_F + F, \qquad (7.3.1)$$

subject to

$$\psi_F(\mathbf{x}, 0) = \Psi(\mathbf{x}).$$ (7.3.2)

Now that (7.3.1) is nonlinear, ψ_F is not the sum of separate responses to F and to Ψ.

The measured values of ψ_F do not, in general, coincide with the data, that is, $\mathbf{h} = \mathbf{d} - \mathcal{L}[\psi_F]$ is not the null vector $\mathbf{0}$, and so posterior estimates $\hat{q} = \hat{q}(\mathbf{x}, t)$ and $\hat{a} = \hat{a}(\mathbf{x})$ will be made for the prior errors q and a in the forcing and the initial streamfunction, respectively. The corresponding posterior or inverse estimate $\hat{\psi}$ for the streamfunction will not be the sum of separate responses to \hat{q} and to \hat{a}, again as a consequence of the nonlinearity of (7.2.3). Just as in the linear model, $\hat{\psi}$ will be chosen so that the corresponding \hat{q}, \hat{a} and $\hat{\boldsymbol{\epsilon}}$ are minimal in a weighted least-squares sense over a fixed time interval $0 \leq t \leq T$. However, the penalty functional is now

$$\mathcal{N}[\psi] = (2\pi)^{-4} \int_0^T dt_1 \int_0^T dt_2 \int d\mathbf{x}_1 \int d\mathbf{x}_2 \, q(\mathbf{x}_1, t_1) \, W(\mathbf{x}_1, t_1, \mathbf{x}_2, t_2) \, q(\mathbf{x}_2, t_2)$$

$$+ (2\pi)^{-4} \int d\mathbf{x}_1 \int d\mathbf{x}_2 a(\mathbf{x}_1) \, V(\mathbf{x}_1, \mathbf{x}_2) \, a(\mathbf{x}_2) + \boldsymbol{\epsilon}^* \mathbf{w} \boldsymbol{\epsilon},$$ (7.3.3)

where W, V and \mathbf{w} are symmetric positive-definite weights. Note that $\mathcal{N}[\psi]$ differs from $\mathcal{J}[\psi]$ defined by (5.3.1), since ψ is now related to q by the nonlinear relation (7.2.3) instead of the linear relation (5.2.2).

Minimization of (7.3.3) cannot be conveniently achieved using representers, since \mathcal{N} is not related to an inner product such as (5.5.2). Recall that an inner product $\langle \theta, \phi \rangle$ must be linear in each of its arguments θ and ϕ. Moreover, it would no longer be convenient to express the minimum $\hat{\psi}$ as a sum of particular solutions as in (5.4.1). However, the calculus of variations may be used to show that if $\hat{\psi}(\mathbf{x}, t)$ is a local extremum of the penalty functional $\mathcal{N}[\psi]$, then $\hat{\psi}$ satisfies

$$\frac{\partial \hat{\psi}}{\partial t} = -E^{-1} \varepsilon J(\hat{\psi}, E\hat{\psi}) + L\hat{\psi} + F + Q \bullet \lambda,$$ (7.3.4)

where the adjoint variable $\lambda = \lambda(\mathbf{x}, t)$ now satisfies the adjoint equation (Tziperman & Thacker, 1989):

$$-\frac{\partial \lambda}{\partial t} = E\varepsilon J(\hat{\psi}, E^{-1}\lambda) - \varepsilon_2 J(E\hat{\psi}, E^{-1}\lambda) + L^{\dagger}\lambda$$

$$+ (2\pi)^2 \mathcal{L}^*_{(\mathbf{x}_2, t_2)} \big[\delta(\mathbf{x} - \mathbf{x}_2)\delta(t - t_2) \big] \mathbf{w}(\mathbf{d} - \mathcal{L}[\hat{\psi}]).$$ (7.3.5)

In (7.3.5), $\varepsilon_2 = 0$ or ε; it serves to switch in or out the second Jacobian which arises from functional variation of the advecting velocity $\mathbf{u} = \mathbf{e}_3 \times \nabla \psi$ implicit in (7.2.3). The subscripts on \mathcal{L} indicate that the vector of measurement functionals acts on the (\mathbf{x}_2, t_2) dependences of the delta-functions. In (7.3.4) the last term is

$$(Q \bullet \lambda)(\mathbf{x}, t) = (2\pi)^{-2} \int_0^T dt_2 \int d\mathbf{x}_2 \, Q(\mathbf{x}, t, \mathbf{x}_2, t_2) \, \lambda(\mathbf{x}_2, t_2) \,, \qquad (7.3.6)$$

where Q is the doubly periodic functional inverse of W: see (5.3.6). The calculus of variations also yields initial and final conditions:

$$\hat{\psi} = \Psi + A \circ \lambda \,, \qquad \text{at } t = 0 \,, \qquad (7.3.7)$$

and

$$\lambda = 0 \,, \qquad \text{at } t = T \,. \qquad (7.3.8)$$

In (7.3.7) the last term is

$$(A \circ \lambda)(\mathbf{x}) = (2\pi)^{-2} \int d\mathbf{x} \, A(\mathbf{x}) \lambda(\mathbf{x}, 0) \,, \qquad (7.3.9)$$

where A is the doubly periodic functional inverse of V: see (5.3.7). The operator L^\dagger differs from L only in that the sign of the first derivative is reversed:

$$L^\dagger = E^{-1} \frac{\partial}{\partial x} + D \,. \qquad (7.3.10)$$

The optimal estimates of q, a and ϵ are

$$\hat{q} = Q \bullet \lambda \,, \quad \hat{a} = A \circ \lambda \,, \quad \hat{\epsilon} = \mathbf{d} - \mathcal{L}[\hat{\psi}] \,. \qquad (7.3.11)$$

The Euler-Lagrange system of equations (7.3.4), (7.3.5), (7.3.7) and (7.3.8) is nonlinear even though the adjoint equation (7.3.5) and final condition (7.3.8) are linear in λ. The nonlinearity is evident in (7.3.4) and (7.3.6), which are coupled to (7.3.5) and (7.3.8), forming a two-point boundary-value problem in the interval $0 \le t \le T$.

7.4 A naive iteration scheme

The nonlinear, two-point boundary-value problem (7.3.4)–(7.3.9) may be approximated by a sequence of pairs of initial-value problems, one of which is nonlinear. The sequence of streamfunctions and adjoint variables $\{\psi_n, \lambda_n\}_{n=1,\infty}$ satisfies the iterated equations

$$\frac{\partial \hat{\psi}_n}{\partial t} = -E^{-1}\varepsilon J(\hat{\psi}_n, E\hat{\psi}_n) + L\hat{\psi}_n + F + Q \bullet \lambda_n \,, \qquad (7.4.1)$$

$$-\frac{\partial \lambda_n}{\partial t} = E\varepsilon J(\hat{\psi}_{n-1}, E^{-1}\lambda_n) - \varepsilon_2 J(E\hat{\psi}_{n-1}, E^{-1}\lambda_n) + L^\dagger \lambda_n$$
$$+ (2\pi)^2 \boldsymbol{\mathcal{L}}^*[\delta\delta]\mathbf{w}(\mathbf{d} - \boldsymbol{\mathcal{L}}[\hat{\psi}_{n-1}]) \,, \qquad (7.4.2)$$

$$\hat{\psi}_n = \Psi + A \circ \lambda_n \,, \qquad \text{at } t = 0 \,, \qquad (7.4.3)$$

$$\lambda_n = 0 \,, \qquad \text{at } t = T \,. \qquad (7.4.4)$$

That is, the streamfunction in the adjoint equation for λ_n is evaluated using the preceding iterate $\hat{\psi}_{n-1}$. The adjoint variable λ_n is obtained by integrating the linear equation (7.4.2) backward in time from $t = T$, subject to (7.4.4). Then $\hat{\psi}_n$ is obtained by integrating the nonlinear equation (7.4.1) forward in time from $t = 0$, subject to (7.4.3). Thus there are two model or model-like integrations per iteration.

It is instructive to attempt to prove that the sequence so constructed is convergent. This is very difficult, due to the complexity of the partial differential equations. Much can be learned by studying a prototype problem involving only ordinary linear differential equations, such as

$$\frac{d\hat{\psi}_n}{dt}(t) = L\hat{\psi}_n(t) + Q\lambda_n(t) \,, \qquad (7.4.5)$$

$$-\frac{d\lambda_n}{dt}(t) = L^*\lambda_n(t) + w(d(t) - \hat{\psi}_{n-1}(t)) \,, \qquad (7.4.6)$$

with

$$\hat{\psi}_n(0) = \Psi + A\lambda_n(0) \,, \qquad (7.4.7)$$
$$\lambda_n(T) = 0 \,.$$

Here Q, w, and A are positive constants, Ψ is a complex constant, while L is a complex constant with negative real part L_r. Consequently $\hat{\psi}_n(t)$ is a complex variable. The data $d(t)$ are available

continuously in the time interval $0 \le t \le T$. Defining a norm for the "streamfunction" iterate $\hat{\psi}_n(t)$ by

$$||\hat{\psi}_n||^2 = \int_0^T |\hat{\psi}_n(t)|^2 \, dt \,, \qquad (7.4.8)$$

it may be shown from (7.4.5)–(7.4.7) that

$$||\hat{\psi}_n - \hat{\psi}_{n-1}||^2 \le k ||\hat{\psi}_{n-1} - \hat{\psi}_{n-2}||^2 \,. \qquad (7.4.9)$$

The definition of the constant k in (7.4.9) is as follows. First define constants c_1 and c_2 by

$$c_1 = 2L_r + w \,, \qquad c_2 = 2L_r + Q \,. \qquad (7.4.10)$$

If $c_2 < 0$, then

$$k = w e^{c_1 T} |c_2|^{-1} (A + QT) \,, \qquad (7.4.11)$$

otherwise

$$k = w e^{(c_1 + c_2)T} c_2^{-2} (A c_2 + Q) \,. \qquad (7.4.12)$$

It is clear from (7.4.9) that the iterates $\hat{\psi}_n$ form a Cauchy sequence if $k < 1$. Such is the case only for sufficiently small values of the parameters w, A, Q, and T (except if c_1 is negative, in which case $k \to 0$ as $T \to \infty$). In general $k > 1$, and (7.4.9) *yields no conclusion.*

An inequality like (7.4.9) may be derived from the system (7.4.1)–(7.4.4), the constant k increasing with increasing Rossby number ε but otherwise depending on the other parameters much as in (7.4.10)–(7.4.12). Thus there is no theoretical evidence to suggest that the naive iteration converges, except for small Rossby numbers, small data weights, large initial weights, large forcing weights and short smoothing-time intervals. This is borne out by numerical experiments. The actual values of the parameters in the experiments will be given in §7.6.

7.5 A bounded iteration scheme

Extrema $\hat{\psi}$ of the penalty functional $\mathcal{N}[\psi]$ are solutions of a nonlinear, two-point boundary-value problem in the time interval

$0 \leq t \leq T$. The naive iteration scheme described in the preceding section yields a sequence of pairs of initial-value problems, one of the pair being nonlinear. However, analysis suggests that this scheme is convergent only in cases of little interest, such as data of low worth. Numerical experiments support the analysis.

There is an alternative iteration scheme which yields a sequence of linear two-point boundary-value problems. Each of these may be solved using representer expansions, and the solutions may be shown to form a bounded sequence, without restrictions on the parameters. For the class of functions under consideration (Sobolev spaces, which are reflexive and therefore locally sequentially compact: see Adams (1975)), it may then be inferred that the sequence has one or more convergent subsequences. Numerical experiments support this analysis provided that the viscous subrange is resolved and that the calculation of the coefficients in the representer expansion is well conditioned.

According to this alternative scheme, the iterations of the Euler-Lagrange equations for extrema $\hat{\psi}$ of the functional $\mathcal{N}[\psi]$ are

$$\frac{\partial \hat{\psi}_n}{\partial t} = -E^{-1}\varepsilon J(\hat{\psi}_{n-1}, E\hat{\psi}_n) + L\hat{\psi}_n + F + Q \bullet \lambda_n , \quad (7.5.1)$$

$$-\frac{\partial \lambda_n}{\partial t} = E\varepsilon J(\hat{\psi}_{n-1}, E^{-1}\lambda_n) - \varepsilon_2 J(E\hat{\psi}_{n-1}, E^{-1}\lambda_{n-1})$$

$$+ L^{\dagger}\lambda_n + (2\pi)^2 \mathcal{L}^*[\delta\delta]\mathbf{w}(\mathbf{d} - \mathcal{L}[\hat{\psi}_n]) , \quad (7.5.2)$$

$$\hat{\psi}_n = \Psi + A \circ \lambda_n , \qquad \text{at } t = 0 , \quad (7.5.3)$$

$$\lambda_n = 0 , \qquad \text{at } t = T . \quad (7.5.4)$$

In (7.5.1) and (7.5.2) the advecting velocities are evaluated using the previous streamfunction iterate $\hat{\psi}_{n-1}$, while the second Jacobian in (7.5.2), which arises from functional variation of the advecting velocity in (7.2.3), is evaluated entirely in terms of the previous iterates. This term will be denoted $\varepsilon_2 J_{n-1}$ subsequently. The measurements of the streamfunction in (7.5.2) are evaluated using the current iterate $\hat{\psi}_n$; this retains the coupling in the system which therefore remains a two-point boundary-value problem for $0 \leq t \leq T$. It is a linear problem and, due to evaluation of the second Jacobian as $\varepsilon_2 J_{n-1}$, of the same form as the linear Euler-Lagrange equations examined in Chapter 5. The latter equations are satisfied by the unique extremum of the quadratic penalty functional $\mathcal{J}[\psi]$. The equations (7.5.1)–(7.5.4) differ by inclusion of advection by a known velocity field, and

by the presence of the inhomogeneity $\varepsilon_2 J_{n-1}$ in the adjoint equation.

The solution strategy is closely based on that used in Chapter 5. A first-guess solution is found satisfying

$$\frac{\partial \psi_{F_n}}{\partial t} = -E^{-1}\varepsilon J(\hat{\psi}_{n-1}, E\psi_{F_n}) + L\psi_{F_n} + F + Q \bullet \lambda_{F_n} \quad (7.5.5)$$

$$\frac{-\partial \lambda_{F_n}}{\partial t} = E\varepsilon J(\hat{\psi}_{n-1}, E\lambda_{F_n}) - \varepsilon_2 J_{n-1} + L^\dagger \lambda_{F_n} , \quad (7.5.6)$$

$$\psi_{F_n} = \Psi + A \circ \lambda_{F_n} , \qquad \text{at } t = 0 , \quad (7.5.7)$$

$$\lambda_{F_n} = 0 , \qquad \text{at } t = T . \quad (7.5.8)$$

If $\varepsilon_2 = 0$ then $\lambda_{F_n} \equiv 0$. Otherwise there are now estimates $Q \bullet \lambda_{F_n}$ and $A \circ \lambda_{F_n}$ for the errors in F and Ψ, which again are the first guesses for the forcing and initial streamfunction, respectively,

Next, a vector of representers \mathbf{r}_n is constructed, satisfying

$$\frac{\partial \mathbf{r}_n}{\partial t} = -E^{-1}\varepsilon J(\hat{\psi}_{n-1}, E\mathbf{r}_n) + L\mathbf{r}_n + Q \bullet \boldsymbol{\alpha}_n , \quad (7.5.9)$$

$$-\frac{\partial \boldsymbol{\alpha}_n}{\partial t} = E\varepsilon J(\hat{\psi}_{n-1}, E\boldsymbol{\alpha}_n) + L^\dagger \boldsymbol{\alpha}_n + \boldsymbol{\mathcal{L}}[\delta\delta] , \quad (7.5.10)$$

$$\mathbf{r}_n = A \circ \boldsymbol{\alpha}_n , \qquad \text{at } t = 0 , \quad (7.5.11)$$

$$\boldsymbol{\alpha}_n = \mathbf{0} , \qquad \text{at } t = T . \quad (7.5.12)$$

The representers \mathbf{r}_n and their adjoints $\boldsymbol{\alpha}_n$ are advected by the velocity field for the previous iterate $\hat{\psi}_{n-1}$. Construction of the representers requires $2M$ model integrations. Note that (7.5.10) may be integrated backward in time, subject to (7.5.12); then (7.5.9) may be integrated forward in time from $t = 0$, subject to (7.5.11).

The nth iterate $\hat{\psi}_n$ may be expressed as

$$\hat{\psi}_n = \psi_{F_n} + \widehat{\mathbf{b}}_n^* \mathbf{r}_n , \quad (7.5.13)$$

where for each n, $\widehat{\mathbf{b}}_n$ is a vector of coefficients for the M representers \mathbf{r}_n. These coefficients satisfy the finite-dimensional linear system

$$\mathbf{P}_n \widehat{\mathbf{b}}_n = \mathbf{h}_n , \quad (7.5.14)$$

where

$$\mathbf{h}_n = \mathbf{d} - \boldsymbol{\mathcal{L}}[\psi_{F_n}] \quad (7.5.15)$$

is the data misfit for the nth first-guess streamfunction, while \mathbf{P}_n is given by

$$\mathbf{P}_n = \mathbf{R}_n + \mathbf{w}^{-1}\,, \qquad (7.5.16)$$

\mathbf{R}_n being the representer matrix at the nth iterate:

$$\mathbf{R}_n = \boldsymbol{\mathcal{L}}[\mathbf{r}_n^*]\,. \qquad (7.5.17)$$

It follows that

$$\mathbf{w}(\mathbf{d} - \boldsymbol{\mathcal{L}}[\hat{\psi}_n]) = \widehat{\mathbf{b}}_n\,; \qquad (7.5.18)$$

hence, as an alternative to evaluating $\hat{\psi}_n$ using (7.5.13), the nth iterate may be found by substituting (7.5.18) into (7.5.2). Then the two-point boundary value is decoupled, leaving a pair of initial-value problems as discussed in Chapter 5. In all, $2M + 4$ model integrations must be performed *per iterate*. This reduces to $2M + 3$ if $\varepsilon_2 = 0$.

The issue of convergence must be considered. The iterated equations may be used to prove that the iterates $\hat{\psi}_n$ form a Cauchy sequence for sufficiently small ϵ, w, and T (see, for example, the general methods in Hartman (1973, §XII.II.4)). An alternative approach will be taken instead, in which it is proved that the iterates $\hat{\psi}_n$ form a *bounded* sequence. This will be accomplished by exploiting the explicit expression (7.5.13) which yields the streamfunction iterate $\hat{\psi}_n$ as a linear combination of the first-guess iterate ψ_{F_n} and the representer iterates \mathbf{r}_n.

Consider the first-guess ψ_{F_n}, which satisfies (7.5.5)–(7.5.8). For the case $\varepsilon_2 = 0$, it follows that $\lambda_{F_n} \equiv 0$, and so each $\hat{\psi}_{F_n}$ has the same forcing F and initial values Ψ.

Now (7.5.5), after applying the operator E, yields the rate of change of potential vorticity $E\psi_{F_n} + \beta y$ due to advection, diffusion and forcing. Neither advection nor diffusion increases the range of values of a field, so it may be concluded that

$$\left|\psi_{F_n}(\mathbf{x}, t)\right| \leq \max\left|\Psi(\mathbf{x})\right| + T \max\left|F(\mathbf{x}, t)\right|. \qquad (7.5.19)$$

In particular this upper bound for $|\psi_{F_n}|$ is independent of the iteration index n; setting ε_2 to zero is crucial here. Similarly, a uniform upper bound may be derived for the representers \mathbf{r}_n, provided the measurement kernel $\boldsymbol{\mathcal{L}}^*[\delta\delta]$ is finite. This precludes measurements at isolated points or times, but in practice measurements must be interpreted as

local averages, for which the kernels are bounded. Next, the uniform boundedness of ψ_{F_n} implies that \mathbf{h}_n is bounded independently of n, and since $||\mathbf{P}_n^{-1}|| \leq ||\mathbf{w}||$ it follows that $\hat{\psi}_n$ is also bounded uniformly with respect to n.

Since $\hat{\psi}_n(\mathbf{x}, t)$ is a bounded sequence, for each (\mathbf{x}, t), it follows that there is at least one convergent subsequence $\hat{\psi}_{N_m}(\mathbf{x}, t)$, where $N_m \to \infty$ as $m \to \infty$ and $\hat{\psi}_{N_m}(\mathbf{x}, t) \to \hat{\psi}_{\#}(\mathbf{x}, t)$. Other subsequences may have different limits $\hat{\psi}_{\$}(\mathbf{x}, t)$, $\hat{\psi}_{\%}(\mathbf{x}, t)$ and so on. These limits may be badly behaved as functions of \mathbf{x} and t, and so may satisfy the nonlinear Euler-Lagrange equations only in a weak sense. However, the proof that $\hat{\psi}_n$ is uniformly bounded also establishes that $E\hat{\psi}_n$ is uniformly bounded. It follows that all second spatial derivatives of $\hat{\psi}_{\#}$ are at least square-integrable. Examination of (7.5.9)–(7.5.12) shows that boundedness of the measurement kernels ensures continuity in time of the representer adjoints $\boldsymbol{\alpha}_n$, and hence continuity of the first time derivative of the representers \mathbf{r}_n. Continuity of $\frac{\partial}{\partial t}\hat{\psi}_{\#}$ is an immediate consequence, assuming that the forcing F is also continuous in time.

The above results having been established, it is possible to prove by a "bootstrap" procedure that higher spatial and temporal derivatives of $\hat{\psi}_n$ are continuous, provided that F, Ψ, Q and A are correspondingly well-behaved functions. The procedure involves differentiating the model equations, and then finding bounds for the resulting nonconservative terms by invoking a delicate analytical inequality for solutions of elliptic equations. Recall that the relationship between streamfunction and vorticity is an elliptic equation, involving the operator E. For details of the procedure, see Bennett & Kloeden (1980, 1981a, 1982). The analytical properties of the iterates $\hat{\psi}_{n-1}$ are passed on to the limits of the convergent subsequences.

In summary, it has been shown that the bounded iteration scheme yields sequences of streamfunction estimates $\hat{\psi}_n$ possessing subsequences with well-behaved limits. There may be several such "limit points." The argument depends crucially on the assumption that $\varepsilon_2 = 0$, thereby suppressing the functional variation of the advecting velocity in the original quasi-geostrophic model. [With ε_2 set to zero, (7.5.1)–(7.5.4) form a system of linear Euler-Lagrange equations for the minimum of a quadratic penalty functional \mathcal{J}_n, which differs from the functional \mathcal{J} defined in (5.3.1) solely owing to the presence of a given advecting velocity field $\hat{\mathbf{u}}_{n-1} = \mathbf{e}_3 \times \boldsymbol{\nabla}\hat{\psi}_{n-1}$. In the language of

control theory, the case $\varepsilon_2 = 0$ defines an *extended smoother*. Applying the Gelfand & Fomin sweep algorithm to (7.5.1)–(7.5.4) would yield an *extended filter*: see §5.7.] The boundedness argument also depends on the availability of the representer expansion for $\hat{\psi}_n$. With the exception of ε_2, the proof of existence of limit points does not depend upon values of model parameters and inverse parameters. It would seem possible to provide uniqueness of the limit point only for very small values of ε, T and w. The difficulty in proving uniqueness for larger values is very similar to the previously mentioned problem of proving that the iterates $\hat{\psi}_n$ form a Cauchy sequence. For a complete analysis of convergence and uniqueness, see Hagelberg (1992).

7.6 An illustration

In this illustration of the preceding nonlinear inverse formulation and bounded iteration scheme, the observing system or antenna contains 128 elements as shown in Fig. 7.6.1. Streamfunction is measured at each of the 16 "moorings" found in one cell of the region of doubly periodic flow, and is also measured at each of 8 times in the interval $0 \leq t \leq T = 10$. Listed in Table 7.6.1 are prior estimates of the initial streamfunction and the streamfunction forcing, together with covariances of the error in the prior estimates and in the data.

Table 7.6.1			
Prior estimates			
$\Psi(x,y) = (\cos 2x + \cos y)/2$	prior estimate of initial streamfunction		
$F(x,y,t) = \cos(t/2)\Psi(x,y)$	prior estimate of streamfunction forcing		
$A(x_1,x_2) = A(x_1 - x_2)$	initial error covariance		
$A(x) = A_0 \sum\limits_{n}(\mathbf{n}	^2 + \rho^{-2})^{-3}$	red wavenumber
$\times\, (2A_H	\mathbf{n}	^2)^{-1}e^{i\mathbf{n}\cdot\mathbf{x}}$	spectrum, $A(x)$
$Q(x_1,t_1,x_2,t_2) = Q(x_1 - x_2)\delta(t_1 - t_2)$	forcing error covariance		
$Q(x) = Q_0 \sum\limits_{n}(\mathbf{n}	^2 + \rho^{-2})^{-3}e^{i\mathbf{n}\cdot\mathbf{x}}$	red wavenumber spectrum, $Q(x)$
$\mathbf{w}^{-1} = w^{-1}\mathbf{I}_M$	uncorrelated measurement errors		
$Q(0) = 0.0625 \qquad A(0) = 0.01$ $w^{-1} = 0.01$	error covariances		

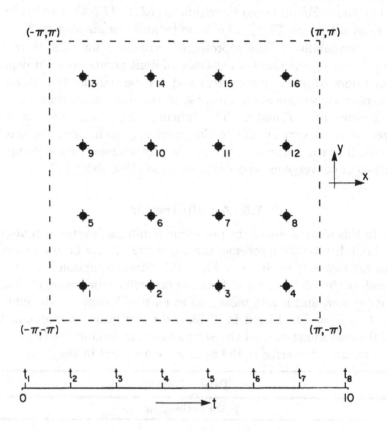

Figure 7.6.1. A simple 128-element antenna: 16 "moorings" $(x_1 - x_{16})$ within a single cell of a doubly-periodic flow $(x_1 = (-3\pi/4, -3\pi/4);\ \Delta x = \Delta y = \pi/2)$, and 8 measurement times $(t_1 - t_8)$ within the interval $0 \le t \le 10$ $(t_1 = 0.1\,;\ \Delta t = 1.4)$ (after Bennett & Thorburn, 1992).

Pseudo-random initial errors and forcing errors were generated consistently with these statistics, and added to the prior estimates. The "true" streamfunction resulting from these fields was measured; then suitably scaled pseudo-random errors were added to the measurements, yielding the synthetic data. The input for the inverse calculation consisted of the prior initial and forcing estimates, the synthetic data, and prior estimates of the error covariances. These last estimates were correct in this first experiment, that is, they took the

forms given in Table 7.6.1. Other model parameters are listed in Table 7.6.2. In formal terms, the assumptions (5.6.1)–(5.6.3) for the first and second moments of errors, together with the prior estimates for these moments shown in Table 7.6.1, comprise a *null hypothesis* about the ocean model. In this first experiment, the hypothesis was correct.

Table 7.6.2. (a) Dimensional parameters		
$\tilde{\rho}$	3×10^4 m = 30 km	length scale (internal deformation radius)
$2\pi\rho$	1.88×10^5 m = 188 km	spatial period
\tilde{U}	2×10^{-2} m s^{-1}	velocity scale
$\tilde{\tau} = (\tilde{\beta}\tilde{\rho})^{-1}$	1.6×10^6 s \cong 19 d	time scale
$T\tilde{\tau}$	190 d	smoothing interval
\tilde{D}_0	0	linear drag coefficient
\tilde{A}_H	54 m^2 s^{-1}	eddy diffusivity

Table 7.6.2. (b) Dimensionless parameters		
ρ	1.0	internal deformation radius
T	10.0	smoothing interval
$D_0 = \tilde{D}_0\tilde{\tau}$	0.0	linear drag coefficient
$A_H = \tilde{A}_H(\tilde{\beta}\tilde{\rho}^3)^{-1}$	0.2	eddy viscosity
$\epsilon = \tilde{U}(\tilde{\beta}\tilde{\rho}^2)^{-1}$	2.0	Rossby number
Re $= \tilde{\rho}\tilde{U}\,\tilde{A}_H^{-1} = \epsilon A_H^{-1}$	10.0	Reynolds number

The setting for the inverse problem is shown in Fig. 7.6.2, in terms of time series of the streamfunction at one mooring ($\mathbf{x} = \mathbf{x}_5$). The rough curve is the true streamfunction $\psi(\mathbf{x}_5, t)$, obtained by solving the nonlinear model (7.2.3), (7.2.6) with a pseudo-random sample for q and an independent pseudo-random sample for a. The 128 data values differ from the true streamfunction owing to 128 independent pseudo-random measurement errors ϵ. The error bars indicate prior estimates of standard errors. The smooth curve is the prior streamfunction estimate $\psi(\mathbf{x}_5, t)$, obtained by solving the nonlinear model (7.3.1), (7.3.2). Synoptic charts or "maps" of the true streamfunction $\psi(\mathbf{x}, 5)$ and the prior estimate $\psi_F(\mathbf{x}, t_5)$ (where $t_5 = 5$ is a measurement time) are shown in Figs. 7.6.3 and 7.6.4, respectively. The objective of the

Figure 7.6.2. Rough curve: times series for the true streamfunction $\psi(\mathbf{x}_5, t)$ at the mooring $\mathbf{x} = \mathbf{x}_5 = (-3\pi/4,\ -\pi/4)$; 8 data values with standard errors $\pm w^{-1/2} = \pm 0.1$. Smooth curve: time series for the prior streamfunction estimate $\psi_F(\mathbf{x}_5, t)$ (after Bennett & Thorburn, 1992).

Figure 7.6.3. Map of the true streamfunction $\psi(\mathbf{x}, t_5)$. The 16 moorings are indicated (after Bennett & Thorburn, 1992).

inverse calculation is to obtain an improved fit to the data, initial conditions and dynamics, thereby leading, presumably, to a better

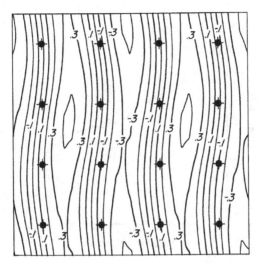

Figure 7.6.4. Map of the prior streamfunction estimate $\psi_F(\mathbf{x}, t_5)$ (after Bennett & Thorburn, 1992).

Figure 7.6.5. Solid line, left-hand scale: prior penalty functional \mathcal{J}_n. Broken line, right hand scale: posterior penalty functional $\widehat{\mathcal{J}}_n$ $(0 \leq n \leq 25)$ (after Bennett & Thorburn, 1992).

fit to the streamfunction away from the antenna. The sequences of prior and posterior values of the sequences of quadratic penalty functionals \mathcal{J}_n, associated with the sequence of linear Euler-Lagrange equations (7.5.1)–(7.5.4), are shown in Fig. 7.6.5. Note the order-of-magnitude reduction in values. Note also that the posterior sequence

Figure 7.6.6. Broken line: eigenvalues of \mathbf{R}_0 (no advection). Solid lines: eigenvalues of \mathbf{R}_n (advection by \mathbf{u}_{n-1}) for $n = 10$, 20, 25, 30. There are $M = 128$ eigenvalues for each matrix (after Bennett & Thorburn, 1992).

Figure 7.6.7. Rough curve: times series of the true streamfunction $\psi(\mathbf{x}_5, t)$. The data at \mathbf{x}_5 are also shown. Smooth curve: time series of the posterior or inverse streamfunction estimate $\hat{\psi}(\mathbf{x}_5, t)$. The smooth curve actually has discontinuously turning tangents at data times. Dashed curve: strong constraint estimate $\tilde{\psi}(\mathbf{x}_5, t)$. Compare with Fig. 7.6.2 (after Bennett & Thorburn, 1992).

has a minimum at $n = 13$, yet the iteration is continued. The reason is that the streamfunction ψ_{12} for the advecting velocity in (7.5.1) is significantly different from ψ_{13}. A larger posterior penalty obtains in the limit, but the dynamics are then consistent with the nonlinear model (7.2.3). Sequences of the 128 eigenvalues of the representer matrix \mathbf{R}_n are shown in Fig. 7.6.6. Their limited range indicates that the inverse is well conditioned. Time series of streamfunctions at the mooring $\mathbf{x} = \mathbf{x}_5 = (-3\pi/4, -\pi/4)$ are shown in Fig. 7.6.7. The true streamfunction $\psi(\mathbf{x}_5, t)$ and the data are as before: see Fig. 7.6.2. However, the prior estimate $\psi_F(\mathbf{x}_5, t)$ is now replaced with the inverse estimate $\hat{\psi}(\mathbf{x}_5, t)$, which provides an improved fit to the data, and in fact to the true streamfunction. As indicated in Fig. 7.6.5, a unique limit is obtained for practical purposes at $n = 20$. A map of the inverse estimate $\hat{\psi}(\mathbf{x}, t_5)$ is shown in Fig. 7.6.8; compare with the true streamfunction $\psi(\mathbf{x}, t_5)$ and prior estimate $\psi_F(\mathbf{x}, t_5)$ shown in Figs. 7.6.3 and 7.6.4, respectively. The inverse estimate is clearly a substantial improvement on the prior. Maps of prior and posterior error variances Γ and $C_{\hat{\psi}\hat{\psi}}$ (see §5.6) may be found in Bennett & Thorburn (1992). These moments are conditional upon the advecting velocity \mathbf{u} taking the value $\hat{\mathbf{u}}$. Alternative statistical linearizations lead to unbounded overestimates of error variance, owing to the linear barotropic instability mechanism (Pedlosky, 1987, Chapter 7).

The dotted curve in Fig. 7.6.7 and the map in Fig. 7.6.9 show a *strong-constraint* inverse estimate $\tilde{\psi}$. That is, the prior forcing error covariance Q was assumed to vanish identically even though the pseudo-random sample q for the forcing error was scaled consistently with Q as given in Table 7.6.1. The strong-constraint estimate $\tilde{\psi}$ is clearly inferior to the *weak-constraint* estimate $\hat{\psi}$, which was based on a valid null hypothesis (correct prior estimate for Q). In this experiment, $\tilde{\psi}$ is little better than the prior estimate ψ_F.

It is shown in §6.2 that the posterior or reduced penalty functional $\widehat{\mathcal{J}}$ is a chi-squared variable, with as many degrees of freedom as there are independent measurements. The same argument may be used here for each reduced functional $\widehat{\mathcal{J}}_n$ following, again, a statistical linearization about the inverse $\hat{\psi}$. A probability function for $\widehat{\mathcal{J}}_{20}$ is shown in Fig. 7.6.10 for an experiment involving only 15 measurements, and based on 33 samples. Each sample is generated by realizations of the random errors in the initial conditions, the forcing and the measurements. One such sample has just been presented in detail. It may be seen that χ^2_{13} provides a better fit than does χ^2_{15}, but the

Figure 7.6.8. Map of the inverse streamfunction estimate $\hat{\psi}(\mathbf{x}, t_5)$. Compare with Figs. 7.6.3 and 7.6.4 (after Bennett & Thorburn, 1992).

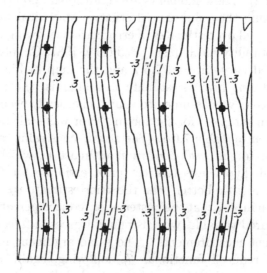

Figure 7.6.9. Map of the strong constraint estimate $\tilde{\psi}(\mathbf{x}, t_5)$. Compare with Figs. 7.6.3, 7.6.4 and 7.6.8 (after Bennett & Thorburn, 1992).

Kolmogorov-Smirnov test gives only a 70% probability that the former hypothesis is correct. Clearly, it is difficult to confirm a correct null hypothesis. Another set of experiments was performed in which the

Figure 7.6.10. Solid line: sample probability function for the posterior or reduced penalty functional $\widehat{\mathcal{J}}$ (correct null hypothesis). Short dashed line: probability function for χ^2_{13}. Long dashed line: probability function for χ^2_{15}.

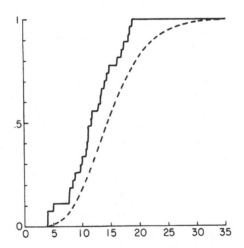

Figure 7.6.11. Solid line: sample probability function for the posterior or reduced penalty functional $\widehat{\mathcal{J}}$ (incorrect null hypothesis). Dashed line: probability function for χ^2_{15}.

true streamfunction forcing variance was 0.125, while the prior estimate was 0.0625. All other parameters were correctly estimated. The resulting probability function for $\widehat{\mathcal{J}}_\infty$ is shown in Fig. 7.6.11; there is only a 6% probability that the distribution is χ^2_{15}. Clearly,

it is relatively easy to reject an incorrect hypothesis. However, these results indicate that an error variance has been overestimated.

These examples, based on the synthetic antenna in Fig. 7.6.1 and priors in Table 7.6.1, serve to illustrate the feasibility of bounded iteration on the nonlinear Euler-Lagrange equations. A real application, with open boundary conditions, is described in Chapter 8.

8

OPEN-OCEAN MODELING:
QUASI-GEOSTROPHY

8.1 Introduction

The conventional introduction to partial differential equations uses examples from classical physics: potentials, diffusion and waves. The partial differential equations (pde's) only describe the local shape of solution surfaces in space or space-time; a unique, global determination of the solution requires that some of its values be provided at least on certain faces of the domain in space or space-time. That is, initial and boundary conditions are needed. The choice of initial conditions is usually simple, with causality being the guide. The thermodynamic or mechanical state of the system must be completely specified at an instant. Boundary conditions, on the other hand, make manifest the nature of the rest of the universe. In particular the familiar conditions of Dirichlet, Neumann and Robin (Jackson, 1962) indicate both the type of surrounding medium and the time interval of interest. For example, the Dirichlet condition for the diffusion equation reveals the presence of a reservoir of so great a capacity that its level does not change as the reservoir is emptied or filled by diffusive exchange with the system of interest. On the other hand, the homogeneous Neumann condition indicates that the surrounding medium is so effectively insulating that no significant diffusion takes place during the time interval of interest. The Robin or mixed condition is no such idealization of a simple property. By specifying a diffusive flux proportional to the state of the system, the

condition models a more complex transfer process, involving both the system and the rest of the universe.

The essential function of the boundary conditions described above is to describe the physical properties of a well-defined thermodynamic or mechanical interface. A different type of boundary condition was introduced by Sommerfeld (1952); it characterizes the wave field far from a source in an unbounded medium. The field may be represented everywhere as a sum of incoming and outgoing waves of all wavelengths. The Sommerfeld condition asserts that at each wavelength, the incoming radiation vanishes asymptotically at large distance from the source. In mathematical terms, the condition expresses the vanishing of the Wronskian of the far field and the outgoing wave. That is, at each wavelength the far field is linearly independent of the incoming wave.

A completely new kind of boundary condition was introduced by Charney, Fjörtoft & von Neumann (1950). This remarkable condition holds at a boundary which is not an interface between one system and another. Instead, the boundary merely defines a subregion of a single system. The need for such a condition is in part a consequence of the advent of powerful digital computers, but it is partly owing to the limit to their power which influences the selection of the subregion (another influence is the availability of boundary and initial data). On the other hand, the boundary condition and its properties exist independently of any particular scheme of numerical computation and must be studied in the context of continuum mechanics. A review of classical fluid-dynamic boundary conditions is an appropriate beginning.

8.2 Classical boundary conditions of plane hydrodynamics

Consider the motion of an unforced, inviscid, incompressible fluid in the plane. The vorticity evolves according to the conservation law

$$\omega_t + \mathbf{u} \cdot \nabla \omega = 0, \qquad (8.2.1)$$

where $\omega = \omega(\mathbf{x}, t)$ is the component of vorticity normal to the plane:

$$\omega = \mathbf{e}_3 \cdot \nabla \times \mathbf{u}. \qquad (8.2.2)$$

Since the plane velocity field $\mathbf{u} = \mathbf{u}(\mathbf{x}, t)$ is solenoidal:

$$\nabla \cdot \mathbf{u} = 0, \qquad (8.2.3)$$

Figure 8.2.1. Simply connected region R with rigid boundary B. The unit outward normal on B is **n**, while the unit tangent is **s**.

there will, in simply connected domains, be a scalar streamfunction field $\psi = \psi(\mathbf{x}, t)$ such that

$$\mathbf{u} = \mathbf{e}_3 \times \boldsymbol{\nabla}\psi\,, \tag{8.2.4}$$

and consequently

$$\nabla^2\psi = \omega\,. \tag{8.2.5}$$

The system (8.2.1)–(8.2.5) is readily integrated in an unbounded region. (The following conventions will be used here: a *domain* for a pde is the set of points in space-time where a solution is sought; a *region* is the spatial projection of a *domain*.) Given an initial field of vorticity, concentrated in a particular part of an unbounded region, a streamfunction may be constructed by solving the elliptic equation (8.2.5) subject to the assumption that the streamfunction vanishes asymptotically at great distances from the concentration of vorticity. The velocity field may then be derived from the streamfunction using (8.2.4). The time derivative of the vorticity follows from (8.2.1), and so the vorticity field may be continued for times $t \geq 0$.

Integration of the system is also simple in rigidly bounded, simply-connected regions. Assume that there is no flow across the boundary defined by the simple closed curve B:

$$\mathbf{u} \cdot \mathbf{n} = 0, \qquad \text{on } B\,, \tag{8.2.6}$$

where **n** is the unit outward normal on B (see Fig. 8.2.1). By virtue of (8.2.4), it follows that the streamfunction ψ is at most a function of time on B:

$$\mathbf{s} \cdot \boldsymbol{\nabla}\psi = -\mathbf{n} \cdot \mathbf{e}_3 \times \boldsymbol{\nabla}\psi = 0\,, \tag{8.2.7}$$

where $\mathbf{s} = \mathbf{e}_3 \times \mathbf{n}$ is the unit tangent on B. It may be assumed that ψ vanishes on B, since adding a spatially uniform function of time to

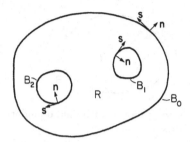

Figure 8.2.2. Multiply-connected region R with rigid boundaries B_0, B_1, B_2, ...;
again the unit outward normal is \mathbf{n} and the unit tangent is \mathbf{s}.

ψ would not alter the velocity field \mathbf{u}. That is, given an initial
vorticity field, the streamfunction may be constructed inside B by
solving (8.2.5) subject to the Dirichlet condition

$$\psi = 0, \qquad \text{on } B. \tag{8.2.8}$$

If the region R is multiply connected (Fig. 8.2.2), then in general

$$\psi = \psi_k(t), \qquad \text{on } B_k, \qquad (k = 0, 1, \ldots, K). \tag{8.2.9}$$

The streamfunction $\psi(\mathbf{x}, t)$ may be expressed in terms of the Green's
function $\mathcal{G}(\mathbf{x}, \mathbf{x}')$:

$$\psi(\mathbf{x}, t) = \int_R \mathcal{G}(\mathbf{x}, \mathbf{x}')\omega(\mathbf{x}', t)\, d\mathbf{x}' + \sum_{k=0}^{K} \psi_k(t) \int_{B_k} \frac{\partial \mathcal{G}}{\partial n'}(\mathbf{x}, \mathbf{x}')\, ds' \tag{8.2.10}$$

where $d\mathbf{x}'$ is an element of area in R, while ds' is an element of length
on the boundary B_k, and $(\partial \mathcal{G}/\partial n') = \mathbf{n}' \cdot \boldsymbol{\nabla}'\mathcal{G}(\mathbf{x}, \mathbf{x}')$ is the outward
normal gradient of \mathcal{G} with respect to the variable \mathbf{x}'. Kelvin's theorem
(Pedlosky, 1987, Chapter 2) may be used to determine the boundary
streamfunction values $\psi_k(t)$. Starting from the momentum equations

$$\mathbf{u}_t + \mathbf{u} \cdot \boldsymbol{\nabla}\mathbf{u} = -\rho^{-1}\boldsymbol{\nabla}p, \tag{8.2.11}$$

where p is pressure and ρ the constant density, the theorem proves
that the circulation around any circuit C moving with the fluid is
constant:

$$\frac{d}{dt} \int_C \mathbf{u} \cdot d\mathbf{s} = 0. \tag{8.2.12}$$

Hence, owing to the rigidity of the boundary, the circulation around each boundary component is steady:

$$\int_{B_k} \frac{\partial \psi}{\partial n}(s,t)\, ds = \Gamma_k, \qquad (k = 0, \ldots, K), \qquad (8.2.13)$$

where each Γ_k is a constant. Substituting (8.2.10) into (8.2.13) yields a system of $K + 1$ linear equations for the $K + 1$ unknowns $\psi_k (k = 0, 1, \ldots, K)$. The constants $\Gamma_k (k = 0, 1, \ldots, K)$ are determined from the initial velocity field $\mathbf{u}(\mathbf{x}, 0)$. The flux between any two points on separate boundary pieces B_k and B_j has the constant value $\psi_k - \psi_j$.

Exercise 8.2.1 Define the Green's function \mathcal{G} used above. □

8.3 Charney-Fjörtoft-von Neumann boundary condition

Only initial values of vorticity are required for the integration of the vorticity conservation law (8.2.1) in a rigidly bounded region. This may be seen by expressing (8.2.1) in characteristic form. The characteristics are also particle paths. A fluid particle observed at position \mathbf{x} at time t will pass through the position \mathbf{X} at time r, provided

$$\frac{\partial}{\partial r}\mathbf{X}(\mathbf{x}, t|r) = \mathbf{u}\big(\mathbf{X}(\mathbf{x}, t|r),\, r\big), \qquad (8.3.1)$$

and

$$\mathbf{X}(\mathbf{x}, t|t) = \mathbf{x}. \qquad (8.3.2)$$

It will be convenient to use Lagrangian notation such as

$$\mathbf{u}(\mathbf{x}, t|r) \equiv \mathbf{u}(\mathbf{X}(\mathbf{x}, t|r), r) \qquad (8.3.3a)$$

and

$$\omega(\mathbf{x}, t|r) \equiv \omega(\mathbf{X}(\mathbf{x}, t|r), r) \qquad (8.3.3b)$$

to denote field values observed at time r, following a particle which passes through position \mathbf{x} at time t. The Lagrangian derivative of

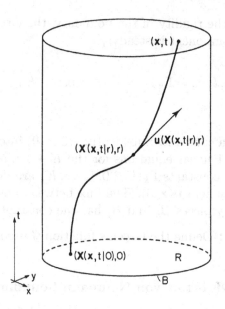

Figure 8.3.1. Space-time domain in which the spatial region R has rigid boundary B. Hence the particle passing through position \mathbf{x} at time t, and also through $\mathbf{X} = \mathbf{X}(x, t|r)$ at time r, must have been at $\mathbf{X}(\mathbf{x}, t|0)$ inside R at time 0.

vorticity is

$$
\begin{aligned}
\frac{\partial}{\partial r}[\omega(\mathbf{x}, t|r)] &= \frac{\partial}{\partial r}\left[\omega(\mathbf{X}(\mathbf{x}, t|r), r)\right] \\
&= \frac{\partial \mathbf{X}}{\partial r} \cdot \nabla\omega(\mathbf{X}, r) + \omega_t(\mathbf{X}, r) \\
&= \mathbf{u}(\mathbf{X}, r) \cdot \nabla\omega(\mathbf{X}, r) + \omega_t(\mathbf{X}, r) \\
&= 0 .
\end{aligned}
\tag{8.3.4}
$$

That is, vorticity is conserved following the motion. Consequently the vorticity field at time $t > 0$ may be determined by tracing particle paths back to the initial instant:

$$
\begin{aligned}
\omega(\mathbf{x}, t) &= \omega(\mathbf{x}, t|t) = \omega(\mathbf{x}, t|0) \\
&= \omega(\mathbf{X}(\mathbf{x}, t|0), 0) \\
&= \omega_I\big(\mathbf{X}(\mathbf{x}, t|0)\big) ,
\end{aligned}
\tag{8.3.5}
$$

where $\omega_I = \omega_I(\mathbf{x})$ is the initial vorticity field. See Fig. 8.3.1. Entirely rigid boundaries ensure that all fluid particles must have been inside R at $t = 0$.

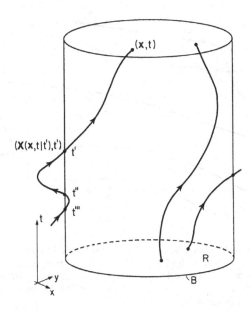

Figure 8.3.2. Space-time domain in which the spatial region R has an open boundary B. The particle passing through position x at time t may have crossed the boundary at earlier times t', t'', t''', ..., the latest crossing taking place at $\mathbf{X} = \mathbf{X}(\mathbf{x}, t|t')$.

On the other hand, if there are parts of the boundary across which particles may freely pass, then the initial vorticity alone will not suffice to determine the vorticity at subsequent times. Such "open" boundaries may be present in order to limit the size of R for computational economy, or they may represent the extent of an observational program. In either case, such parts of the boundary are merely sets of points in space, without any physical significance. Returning to the problem of determining the vorticity field at position \mathbf{x} and at time $t > 0$, there are now two possibilities: either the particle passing through position \mathbf{x} at time t was inside R for all $t \geq 0$, in which case (8.3.5) holds, or else the particle entered R at some previous time t' where $0 < t' < t$. The particle may have departed and reentered R several times t'', t''', etc.; let t' be the latest time of entry. See Fig. 8.3.2. In this case the vorticity is given by

$$\omega(\mathbf{x}, t) = \omega(\mathbf{x}, t|t) = \omega(\mathbf{x}, t|t') = \omega(\mathbf{X}(\mathbf{x}, t|t'), t') . \qquad (8.3.6)$$

Consequently ω is determined if values of vorticity $\omega_{IN} = \omega_{IN}(\mathbf{x}, t)$

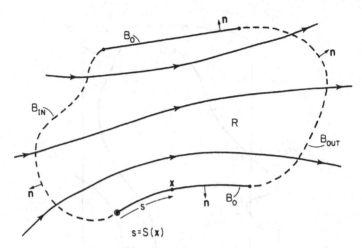

Figure 8.3.3. Region R with boundary B. The latter is comprised of an inflow part B_{IN} where $\mathbf{u} \cdot \mathbf{n} < 0$, a rigid part B_0 where $\mathbf{u} \cdot \mathbf{n} = 0$ and an outflow part B_{OUT} where $\mathbf{u} \cdot \mathbf{n} > 0$. The arc length around B, measured from an arbitrarily chosen origin on B, is given by $s = S(\mathbf{x})$, where \mathbf{x} is a point on B. Streamlines indicate the instantaneous flow.

are provided on those parts of the open boundary where flow is inward:

$$\mathbf{u} \cdot \mathbf{n} < 0, \qquad \text{on } B_{IN}, \tag{8.3.7}$$

and then

$$\omega(\mathbf{x}, t) = \omega_{IN}(\mathbf{X}(\mathbf{x}, t|t'), t'). \tag{8.3.8}$$

If ω_{IN} is expressed as a function of time and of a contour parameter on B_{IN} such as arc length s, then (8.3.8) may in turn be expressed in the form

$$\omega(\mathbf{x}, t) = \omega_{IN}(S(\mathbf{x}, t|t'), t), \tag{8.3.9}$$

where $S(\mathbf{x}, t|t') \equiv S(\mathbf{X}(\mathbf{x}, t|t'))$, and the relationship $s = S(\mathbf{x})$ gives the value of the contour parameter at any point \mathbf{x} on \mathbf{B}.

The incompressibility of the flow ensures that if there is inflow at some parts of an open boundary, then there must simultaneously be outflow at other parts:

$$\int_B \mathbf{u} \cdot \mathbf{n} \, ds = \int_R \boldsymbol{\nabla} \cdot \mathbf{u} \, d\mathbf{x} = 0. \tag{8.3.10}$$

Hence the boundary B may be partitioned (see Fig. 8.3.3) into inflow parts B_{IN}, rigid parts B_0, and outflow parts B_{OUT}:

$$B = B_{IN} \cup B_0 \cup B_{OUT}, \tag{8.3.11}$$

defined by

$$\mathbf{u} \cdot \mathbf{n} < 0, \quad \text{on } B_{IN}, \qquad (8.3.12a)$$

$$\mathbf{u} \cdot \mathbf{n} = 0, \quad \text{on } B_0, \qquad (8.3.12b)$$

$$\mathbf{u} \cdot \mathbf{n} > 0, \quad \text{on } B_{OUT}. \qquad (8.3.12c)$$

To make the partition (8.3.11)–(8.3.12) it is sufficient to know the boundary values of the streamfunction, by virtue of (8.2.4). However, for multiply-connected regions such boundary information may not be prescribed *a priori*. Suppose B_k is a simple closed contour forming part of the open boundary. Then in general

$$\int_{B_k} \frac{\partial \psi}{\partial s} \, ds = -\int_{B_k} \mathbf{u} \cdot \mathbf{n} \, ds \neq 0; \qquad (8.3.13)$$

consequently \mathbf{u} cannot be determined solely by a continuous streamfunction field. Instead, \mathbf{u} is determined by a continuous velocity potential field $\chi = \chi(\mathbf{x}, t)$ as well as a continuous streamfunction field $\psi = \psi(\mathbf{x}, t)$:

$$\mathbf{u} = \boldsymbol{\nabla}\chi + \mathbf{e}_3 \times \boldsymbol{\nabla}\psi. \qquad (8.3.14)$$

Nevertheless, it is still the case that

$$\mathbf{e}_3 \cdot \boldsymbol{\nabla} \times \mathbf{u} = \nabla^2 \psi = \omega, \qquad (8.3.15)$$

and as a consequence of the incompressibility condition (8.2.3),

$$\boldsymbol{\nabla} \cdot \mathbf{u} = \nabla^2 \chi = 0. \qquad (8.3.16)$$

The normal velocity on B is given now by

$$\mathbf{u} \cdot \mathbf{n} = \frac{\partial \chi}{\partial n} - \frac{\partial \psi}{\partial s}. \qquad (8.3.17)$$

Suppose that the normal velocity is prescribed on B, that is,

$$\mathbf{u} \cdot \mathbf{n} = \nu, \quad \text{on } B, \qquad (8.3.18)$$

where $\nu = \nu(s,t)$ is a given function of time and arc length measured around B. (The latter coordinate may be defined piecewise on complex boundaries.) Before proceeding, it is necessary to establish that the velocity field \mathbf{u} inside R is uniquely determined by the vorticity field ω, by the values ν_k for the normal component of velocity on each boundary part B_k, and by the circulation Γ_k around each boundary part B_k. Uniqueness is readily established by considering the difference between two velocity fields having the same vorticities in R, the same normal components on B, and the same circulations around each part of B. The difference field \mathbf{U} is irrotational, has a vanishing normal component on B which becomes in effect a rigid boundary, and has no circulation around any part of B. Consequently \mathbf{U} may be expressed in terms of a streamfunction Ψ which is harmonic and, by virtue of the analysis in §8.2, must vanish everywhere on B. It follows that $\Psi \equiv 0$ and hence $\mathbf{U} \equiv \mathbf{0}$, which establishes the sought-after uniqueness of \mathbf{u} given ω, ν_k and $\Gamma_k (k = 0, 1, \ldots, K)$.

On the other hand, the decomposition (8.3.14) is not unique; clearly χ and ψ are defined only up to the addition of any harmonic fields θ and ϕ, respectively, satisfying

$$\frac{\partial \theta}{\partial n} - \frac{\partial \phi}{\partial s} = 0, \qquad \text{on } B. \tag{8.3.19}$$

An example of a construction for χ and ψ is as follows. Let $\delta = \delta_k(t)$ be the net outflow across the boundary part of B_k:

$$\delta_k = \int_{B_k} \nu_k \, ds. \tag{8.3.20}$$

Then let χ be the unique solution of the Neumann problem

$$\nabla^2 \chi = 0, \qquad \text{in } R, \tag{8.3.21}$$

and

$$\frac{\partial \chi}{\partial n} = \delta_k, \qquad \text{on } B_k. \tag{8.3.22}$$

Actually χ is undefined up to an additive constant, which may be set to zero without loss of generality. Now define

$$\nu_k^+(s,t) = \nu_k(s,t) - \delta_k(t) P_k^{-1}, \qquad (k = 0, 1, \ldots, K), \tag{8.3.23}$$

where P_k is the perimeter of B_k. Then

$$\int_{B_k} \nu_k^+ \, ds = 0, \qquad (k = 0, 1, \dots, K), \qquad (8.3.24)$$

and the boundary streamfunction values

$$\psi_k(s, t) = \int_{s_k}^{s} \nu_k^+(s', t) \, ds' + \phi_k(t) \qquad (8.3.25)$$

are continuous around B_k. The lower terminal s_k in (8.3.25) may be assigned the value $\sum_{j=1}^{k-1} P_j$, that is, the net arc length of preceding parts of B. The values of the functions $\phi_k(t)$ may be determined from the specified circulations $\Gamma_j (j = 0, 1, \dots, K)$ as in §8.2. A continuous streamfunction field ψ is obtained by solving the Dirichlet problem

$$\nabla^2 \psi = \omega, \qquad \text{in } R, \qquad (8.3.26)$$

and

$$\psi = \psi_k, \qquad \text{on } B_k, \quad (k = 0, 1, \dots, K). \qquad (8.3.27)$$

The construction of **u** is far less complicated in a simply-connected region (Fig. 8.3.3). The net outflow δ must vanish at all times, and so there is no need to introduce a velocity potential χ. Moreover, it is sufficient to define the boundary streamfunction as

$$\psi_B(s, t) = \int_0^s \nu(s', t) \, ds. \qquad (8.3.28)$$

It is not necessary to specify the circulation around the boundary. Further analysis will be restricted to this simple problem for which the model equations, initial conditions and boundary conditions are, once again for convenience,

$$\omega_t + \mathbf{u} \cdot \nabla \omega = G, \qquad (8.3.29)$$

$$\nabla^2 \psi = \omega, \qquad (8.3.30)$$

and

$$\mathbf{u} = \mathbf{e}_3 \times \nabla \psi, \qquad (8.3.31)$$
$$\text{in } R \text{ for } t \geq 0,$$
$$\omega(\mathbf{x}, 0) = \omega_I(\mathbf{x}), \qquad (8.3.32)$$
$$\text{in } R \text{ at } t = 0,$$
$$\omega(\mathbf{x}, t) = \omega_{IN}(S(\mathbf{x}), t), \qquad (8.3.33)$$
$$\text{for } \mathbf{x} \text{ on } B_{IN} \text{ and } t \geq 0,$$

and

$$\psi(\mathbf{x}, t) = \psi_{IN}(S(\mathbf{x}), t) \qquad (8.3.34)$$
$$\text{for } \mathbf{x} \text{ on } B \text{ and } t \geq 0.$$

In (8.3.33) and (8.3.34), $s = S(\mathbf{x})$ again defines arc length on B. The system (8.3.29)–(8.3.34) will be referred to subsequently as the CFvN problem. Note that a source term $G = G(\mathbf{x}, t)$ has now been included in the vorticity equation (8.3.29). The conservation law (8.3.5) is accordingly replaced with

$$\omega(\mathbf{x}, t) = \omega_I\big(\mathbf{X}(\mathbf{x}, t|0)\big) + \int_0^t G(\mathbf{X}(\mathbf{x}, t|r), r)\, dr \qquad (8.3.35)$$

for particles passing through position \mathbf{x} at time t, after having originated in R. A similar modification applies to (8.3.9):

$$\omega(\mathbf{x}, t) = \omega_{IN}(S(\mathbf{x}, t|t'), T) + \int_{t'}^t G(\mathbf{X}(\mathbf{x}, t|r), r)\, dr \qquad (8.3.36)$$

for particles entering R at time t'. According to (8.3.35) and (8.3.36), the vorticity is the sum of an initial or boundary value, plus the "integrated Lagrangian history of G" (Kraichnan, 1965).

Exercise 8.3.1 Complete the proof that $\mathbf{U} \equiv 0$, that is, complete the proof of the uniqueness theorem discussed above. □

Exercise 8.3.2 Verify (8.3.35) and (8.3.36). □

8.4 Well-posedness

Courant & Hilbert (1962, p. 227) propose that "A mathematical problem which is to correspond to physical reality should satisfy the following basic requirements:

(1) The solution must exist.

(2) The solution should be uniquely determined.

(3) The solution should depend continuously on the data (requirement of stability)."

According to Courant & Hilbert, "the first requirement expresses the logical condition that not too much, i.e., no mutually contradictory properties, is demanded of the solution." This interpretation clearly excludes multiply-valued solutions. The spirit of requirement (3) clearly implies that (1) may be interpreted further to exclude non-classical solutions, that is, solutions for which the terms in the mathematical problem such as derivatives are either discontinuous or unbounded. With this interpretation, any problem satisfying requirements (1)–(3) may be described as classically well posed. As Courant & Hilbert remark, there are many profound mathematical problems in physics which are "improperly posed." They cite nonlinear phenomena and quantum theory, and go on to remark that "the advent of powerful numerical methods have shown that properly-posed problems are by far not the only ones which appropriately reflect real phenomena." They further remark (p. 231) that "Improperly posed problems of great potential importance in numerical analysis so far have not been reached by the mainstream of active research. Thus, for initial value problems where the initial data are only approximately known from observations, approximate methods, e.g., finite difference methods, are used to construct the solution. Then the question arises: How can later observations be used to improve and extend the computed solution? Progress concerning such problems could be of the greatest value, for example, for mathematical weather prediction." Evidently, contemporary research in data assimilation, and inverse methods in general, had been anticipated for some time by Courant & Hilbert. Indeed, their remarks go beyond the immediate objective of the current section, namely, determining the well-posedness of the CFvN problem simply as a "forward" model. As to the smoothness of solutions, it might be reasonable to represent fronts and tornadoes as discontinuities or singularities in the solution. Of course, such phenomena are actually continuous and bounded; a front, for example, is a region of finite width in which small-scale phenomena such as turbulent dissipation and strong vertical motions are of particular importance. Nevertheless, the width would be less than the smallest scale for which quasi-geostrophic dynamics would be valid, and so the frontal region would have to be regarded as vanishingly small. The

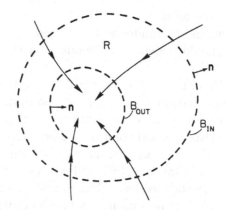

Figure 8.4.1. Multiply-connected region R with an entirely open boundary B. The normal velocity is uniformly bounded away from zero on both B_{IN} and B_{OUT}; that is, the Judovich condition (8.4.1) is satisfied.

question arises: can discontinuities or singularities arise spontaneously in the solutions of a model such as (8.3.29)–(8.3.34), assuming that the forcing and initial values are smooth? Consider first the case of an unbounded region. Wolibner (1933) has shown that in any time interval $0 \leq t \leq T$ there is a smooth solution of the initial-value problem (8.3.29)–(8.3.32), the gradients of vorticity having bounds depending upon T and the gradients of G and ω_I. These upper bounds become very large as $T^2|G|$ and $T|\omega_I|$ increase, but the bounds remain finite. Thus, for a given value of T, the numerical approximation of the solution is simply a matter of adequate resolution. Eventually the length scales would become so small that quasi-geostrophic dynamics would be violated, but the numerical scheme would still be providing a mathematically consistent and accurate approximation to the model. Were a discontinuity to develop, conventional numerical approximation methods such as finite-difference schemes would no longer be consistent, regardless of their resolution. The Wolibner result is readily modified to simply- or multiply-connected regions with rigid boundaries, or with open boundaries provided the normal velocity is smooth and is uniformly bounded away from zero (Judovich, 1966). That is, the result holds provided there is a positive constant γ such that

$$|\mathbf{u} \cdot \mathbf{n}| \geq \gamma > 0 \qquad (8.4.1)$$

on $B_{IN} \cup B_{OUT}$ for all $t \geq 0$. The analytical example provided by Charney, Fjörtoft & von Neumann (1950) meets the Judovich con-

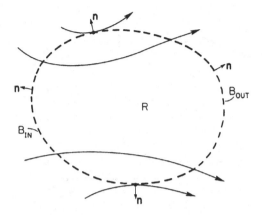

Figure 8.4.2. Simply-connected region R, with an entirely open and smooth boundary B. The Judovich condition (8.4.1) cannot be met at any time if the flow is incompressible. There must be points on the boundary where the flow is tangential to the boundary.

dition (8.4.1). The example involves flow on the surface of a section of a circular cylinder. The flow is uniformly inward at one end of the section, and uniformly outward at the other. The configuration is topologically equivalent to flow in the plane, in an annular region: see Fig 8.4.1.

Exercise 8.4.1 Sketch multiply-connected and simply-connected regions such that Judovich's condition (8.4.1) could hold on the open boundaries. □

The Judovich condition (8.4.1) cannot be met in simply-connected regions with entirely open boundaries described by smooth contours, as shown in Fig. 8.4.2. Moreover, the condition must be expected to break down at least some of the time in simply-connected regions with open boundaries described by piecewise smooth contours, as shown in Fig. 8.4.3. Since this latter configuration is the one of most common interest, it is essential to determine whether the Judovich condition is a necessary, as well as sufficient, condition for well-posedness. In other words, if the condition is violated, will the CFvN problem be ill-posed?

It is immediately evident from the vorticity equation (8.2.1) that continuation of ω off an open boundary may not always be possible. Suppose that the open boundary is locally smooth and has a large radius of curvature, relative to the scales of the boundary vorticity.

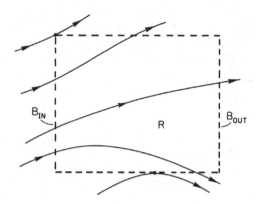

Figure 8.4.3. Simply-connected region R, with an entirely open and piecewise-smooth boundary B. The flow may be, but need not be, tangential to the boundary at any time.

Then (8.2.1) may be approximated by

$$\omega_t + \mathbf{u} \cdot \mathbf{n}\, \omega_n + \mathbf{u} \cdot \mathbf{s}\, \omega_s = 0, \qquad (8.4.2)$$

where \mathbf{n} and \mathbf{s} are respectively the outward unit normal and the unit tangent on B, while subscripts denote partial derivatives in normal and tangential coordinates (n, s). Rearranging in favor of the normal derivative yields

$$\omega_n = \frac{-(\omega_t + \mathbf{u} \cdot \mathbf{s}\, \omega_s)}{\mathbf{u} \cdot \mathbf{n}}. \qquad (8.4.3)$$

If the normal velocity $\mathbf{u}\cdot\mathbf{n}$ changes sign on a part of the open boundary, then continuation of ω off the boundary will only be possible if the numerator of (8.4.3) changes sign simultaneously with $\mathbf{u}\cdot\mathbf{n}$. These sign changes occur at points separating B_{IN} from B_{OUT}. Boundary data are specified on B_{IN}, so the vorticity derivatives in the numerator may be evaluated, at least as one-sided limits:

$$\omega_n = -\frac{(\partial\omega_{IN}/\partial t) + \mathbf{u} \cdot \mathbf{s}\,(\partial\omega_{IN}/\partial s)}{\mathbf{u} \cdot \mathbf{n}}. \qquad (8.4.4)$$

However, the tangential velocity $\mathbf{u}\cdot\mathbf{s}$ is not specified on B, so *a priori* it cannot be known if ω_n is finite and hence if continuation off the open boundary is possible. It may happen that $\mathbf{u} \cdot \mathbf{n}$ vanishes at a point on the open boundary without changing sign. In such a case,

the numerator in (8.4.4) may be evaluated as a two-sided limit, or else not at all.

The difficulty described above is not limited to behavior at the boundary. Information about interior behavior may be gained by examining the solution representations (8.3.35) and (8.3.36). Suppose that \mathbf{x} is some point inside R, such that $\mathbf{X}(\mathbf{x}, t|r)$ also lies inside R for all r: $0 \leq r \leq t$. That is, the fluid particle passing through position \mathbf{x} at time T was inside R throughout the time interval $0 \leq r \leq t$. If the velocity field \mathbf{u} is bounded, then the particle paths \mathbf{X} are continuous, so for sufficiently small $\Delta\mathbf{x}$ it will be the case that $\mathbf{X}(\mathbf{x} + \Delta\mathbf{x}, t|r)$ also lies inside R for all r: $0 \leq r \leq t$. Consequently,

$$\Delta\omega = \omega(\mathbf{x} + \Delta\mathbf{x}, t) - \omega(\mathbf{x}, t) \cong \Delta\mathbf{x} \cdot \mathbf{J} \cdot \boldsymbol{\nabla}\omega_I\big(\mathbf{X}(\mathbf{x}, t|0)\big)$$

$$+ \int_0^t \Delta\mathbf{x} \cdot \mathbf{J} \cdot \boldsymbol{\nabla}G\big(\mathbf{X}(\mathbf{x}, t|r), r\big)\, dr\,, \tag{8.4.5}$$

where the 2×2 Jacobian matrix \mathbf{J} is defined by

$$J_{jk} = \frac{\partial X_k}{\partial x_j}\,. \tag{8.4.6}$$

Hence ω will be a smooth function of position at time t, provided \mathbf{X}, ω_I and G are smooth functions of position at previous times. In particular, no restriction is placed on the *values* of \mathbf{X}, ω_I, G, or their gradients. A similar argument relates the continuity of ω_t to the smoothness of \mathbf{X}, ω_I and G. Now suppose \mathbf{x} lies inside R, but that $\mathbf{X}(\mathbf{x}, t|t')$ lies on B_{IN} at some previous time $t' = t'(\mathbf{x}, t)$. For $\Delta\mathbf{x}$ sufficiently small, $\mathbf{X}\big(\mathbf{x} + \Delta\mathbf{x}, t|t'(\mathbf{x} + \Delta\mathbf{x}, t)\big)$ also lies on B_{IN}. For the sake of simplicity alone, consider the special case $G \equiv 0$. Then (8.3.36) yields

$$\Delta\omega = \omega\,(\mathbf{x} + \Delta\mathbf{x},\, t) - \omega\,(\mathbf{x},\, t)$$

$$= \frac{\partial \omega_{IN}}{\partial s} \left\{ \boldsymbol{\nabla}\, S \cdot \mathbf{J} + \boldsymbol{\nabla}\, S \cdot \frac{\partial \mathbf{X}}{\partial r} \cdot \boldsymbol{\nabla}\, t' \right\} \cdot \Delta\mathbf{x}$$

$$+ \frac{\partial \omega_{IN}}{\partial t}\, \boldsymbol{\nabla}\, t' \cdot \Delta\mathbf{x}\,. \tag{8.4.7}$$

Suppose the boundary is defined by the equation

$$b(\mathbf{x}) = 0\,, \qquad \mathbf{x} \in B_{IN}\,. \tag{8.4.8}$$

Then

$$b\big(\mathbf{X}(\mathbf{x}+\Delta\mathbf{x},\,t|t'(\mathbf{x}+\Delta\mathbf{x},\,t))\big) = b\big(\mathbf{X}(\mathbf{x},\,t|t'(\mathbf{x},\,t))\big) = 0,\quad (8.4.9)$$

and so

$$\boldsymbol{\nabla} b\cdot\left\{\mathbf{J}+\frac{\partial\mathbf{X}}{\partial r}\,\boldsymbol{\nabla} t'\right\}\cdot\Delta\mathbf{x}=0.\qquad (8.4.10)$$

Now $\boldsymbol{\nabla} S=\mathbf{s}$, which is the unit tangent on the boundary, and without loss of generality it may be assumed that $\boldsymbol{\nabla} b=\mathbf{n}$, which is the unit normal.

Combining (8.4.7), (8.4.10) and (8.3.1) yields

$$\Delta\omega=\frac{\partial\omega_{IN}}{\partial t}\,\mathbf{s}\cdot\mathbf{J}\cdot\Delta\mathbf{x}-(\mathbf{u}\cdot\mathbf{n})^{-1}\left\{\frac{\partial\omega_{IN}}{\partial t}+\mathbf{u}\cdot\mathbf{s}\,\frac{\partial\omega_{IN}}{\partial s}\right\}\mathbf{n}\cdot\mathbf{J}\cdot\Delta\mathbf{x}.$$

$$(8.4.11)$$

This result, due to Judovich (1966), relates the *smoothness* of the interior vorticity to *values* of velocity, vorticity and vorticity gradients on the open boundary. In particular, since it cannot be known *a priori* if the tangential velocity takes the value

$$\mathbf{u}\cdot\mathbf{s}=-\left(\frac{\partial\omega_{IN}}{\partial s}\right)^{-1}\frac{\partial\omega_{IN}}{\partial t}\qquad (8.4.12)$$

whenever and wherever $\mathbf{u}\cdot\mathbf{n}=0$ on the open boundary, it must be concluded that the CFvN problem is not well posed in open regions where the Judovich condition (8.4.1) cannot be guaranteed.

8.5 Discontinuities, wavefronts, the β-effect and characteristics

i) Discontinuities

The preceding analysis indicates that the solution of the CFvN problem may have discontinuities in the vorticity field. Should there be a path such as

$$\mathbf{x}=\boldsymbol{\xi}(p,\,t),\qquad (8.5.1)$$

where p is a path parameter, across which the solution $\omega(\mathbf{x},\,t)$ has a simple discontinuity, then it is readily shown that

$$\boldsymbol{\xi}_t=\mathbf{u}(\boldsymbol{\xi},\,t)\qquad (8.5.2)$$

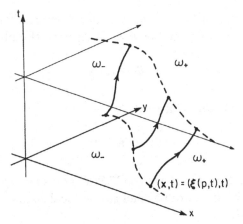

Figure 8.5.1. Paths $\mathbf{x} = \boldsymbol{\xi}\,(p,\,t)$ in the plane, for two values of t, across which the vorticity changes discontinuously from $\omega_-(p,t)$ to $\omega_+(p,t)$. The path parameter p might be arc length. The path moves with the flow; that is, vorticity discontinuities are "attached" to fluid particles.

(e.g., Courant & Hilbert, 1962). That is, discontinuities in the vorticity field are attached to fluid particles: see Fig. 8.5.1. Note that if (8.5.1) is expressed in the form

$$P(\mathbf{x},\, t) = 0\,, \qquad (8.5.3)$$

where $P = \mathbf{x} - \boldsymbol{\xi}$, then (8.5.2) becomes

$$P_t + \mathbf{u} \cdot \boldsymbol{\nabla} P = 0\,. \qquad (8.5.4)$$

ii) Wavefronts

Now suppose that the advecting velocity field $\mathbf{u}(\mathbf{x},\, t)$ appearing in (8.2.1) is prescribed, independently of the unknown field ω. Then (8.2.1) is a linear equation. Seeking a simple wave solution of the form

$$\omega = \omega_0\, e^{i\phi(\mathbf{x},t)}\,, \qquad (8.5.5)$$

where ω_0 is a constant, yields the equation

$$\phi_t + \mathbf{u} \cdot \boldsymbol{\nabla}\phi = 0 \qquad (8.5.6)$$

for the phase function. So when discontinuities are present, they occur on the surfaces of constant phase. In particular, if $\mathbf{u}(\mathbf{x},\, t)$ is

"frozen" about a local (vector) value, then there is a local plane wave
of frequency θ and wavenumber $\boldsymbol{\kappa} = (\kappa, \lambda)$ with the phase function

$$\phi = \theta t + \boldsymbol{\kappa} \cdot \mathbf{x} \,, \tag{8.5.7}$$

and (8.5.6) is the dispersion relation

$$\theta + \mathbf{u} \cdot \boldsymbol{\kappa} = 0 \,. \tag{8.5.8}$$

That is, in space-time the normals $(\nabla P,\, P_t)$ on surfaces of discontinu-
ity of the vorticity field are parallel to the local space-time wavenum-
ber vectors $(\boldsymbol{\kappa}, \theta)$.

iii) The β-effect
 It may now be seen that the inclusion of the β-effect in a baro-
tropic ocean model fails to render the CFvN problem well posed. The
vorticity equation becomes

$$\omega_t + \mathbf{u} \cdot \nabla \omega + \beta v = 0 \,, \tag{8.5.9}$$

where $v = \psi_x$ is the meridional velocity, while β is as usual the local
value of the meridional gradient of the Coriolis parameter. Consider
first the case of plane Rossby waves in the "frozen velocity field"
approximation. The dispersion relation is

$$\theta = -\mathbf{u} \cdot \boldsymbol{\kappa} + \beta \kappa \, |\boldsymbol{\kappa}|^{-2} \,. \tag{8.5.10}$$

In particular, (8.5.8) is recovered in the limit as $|\boldsymbol{\kappa}| \to \infty$, that is, the
β-effect has *no* effect on the shortest Rossby waves.
 Now consider the case of a discontinuous vorticity field which is
a weak solution of (8.5.9) (Courant & Hilbert, 1962). Provided ω is
bounded, the velocity field \mathbf{u} is continuous, and so the presence of
the β-term in (8.5.9) does not alter the condition (8.5.2) governing
the propagation of vorticity discontinuities. The normal $(\nabla P,\, P_t)$ on
the surface of discontinuity still obeys (8.5.4), which indicates that
the normal is parallel to the wavenumber vector $(\boldsymbol{\kappa}, \theta)$ in the high-
wavenumber limit $|\boldsymbol{\kappa}| \to \infty$ (Courant & Hilbert, 1962). Hence the
propagation of discontinuities in the CFvN problem is not influenced
by the ability of long Rossby waves to disperse from local disturbances
in the vorticity field. The β-effect will be included in the subsequent

examples of ill-posed CFvN problems. The effect facilitates the construction of boundary data, using combinations of plane Rossby wave solutions and other simple flow fields.

iv) Characteristics

The conservation law for vorticity, expressed as the partial differential equation (8.3.29), may also be expressed as a family of ordinary differential equations. The independent variable is the time r for which a fluid particle has been in motion. Again, the particle paths are given by (8.3.1); on these paths (8.3.29) becomes

$$\frac{\partial}{\partial r}\omega(\mathbf{x}, t|r) = G(\mathbf{x}, t|r) \qquad (8.5.11)$$

subject to an initial condition at $r = 0$, or at $r = t'(\mathbf{x}, t)$ in the case of a particle entering the region R on the open boundary B_{IN}. Thus the particle paths (8.3.1) are the characteristics of the first-order, hyperbolic equation (8.3.29) (Courant & Hilbert, 1962). The characteristic direction is the space-time vector $(\mathbf{u}, 1)$. For the case of a boundary fixed in space, the Judovich condition (8.4.1) states that nowhere on the boundary does the characteristic direction lie in the boundary. Any boundary point where (8.4.1) does not hold, that is, where

$$\mathbf{u} \cdot \mathbf{n} = 0, \qquad (8.5.12)$$

will be referred to as a characteristic point. The dependent variable, in this case the vorticity ω, may not be freely specified on a boundary interval containing a characteristic point, since the hyperbolic equation (8.3.29) reduces to a local constraint on the vorticity. Here the constraint involves the unknown component of velocity tangential to the boundary. It is this imposition of an implicit local constraint which renders the CFvN problem ill-posed when characteristic points are present.

Exercise 8.5.1 Rigid boundaries are uniformly characteristic; that is, they are composed entirely of characteristic points. Explain why such boundaries do not contribute to ill-posedness. □

Exercise 8.5.2 Explain the relationship between the tangential velocity at a characteristic point, and the characteristic form (8.5.11) of the vorticity equation. □

There is at least one time in the CFvN problem when the local constraint is determinate. At the initial instant $t = 0$, the stream-function and hence both components of velocity are specified. Thus, should there be a characteristic point on the open boundary, it is possible to verify that the specified boundary vorticity satisfies the constraint. Indeed, there is a consistency condition everywhere on B_{IN}. In the event that the constraint is met at the characteristic point initially, the question arises as to whether the condition is met at later times.

To simplify the discussion, let the region R be the right half-plane defined by $x > 0$, with the entire y-axis as an open boundary. Then the initial data are

$$\omega(x,\, y,\, 0) = \omega_I(x,\, y)\,, \tag{8.5.13}$$

while the boundary data are

$$\psi(0,\, y,\, t) = \psi_B(y,\, t) \tag{8.5.14}$$

for $0 \leq t$ and $-\infty < y < \infty$, together with

$$\omega(0,\, y,\, t) = \omega_{IN}(y,\, t) \tag{8.5.15}$$

for $0 \leq t$ and for all $y \in B_{IN}$, that is, for all y such that

$$u_B = \frac{-\partial \psi_B}{\partial y} > 0\,. \tag{8.5.16}$$

It is assumed that ω and ψ vanish as $x \to \infty$ or $|y| \to \infty$.

The initial streamfunction has the representation

$$\psi(\mathbf{x},\, 0) = \int \int_R \mathcal{G}(\mathbf{x}, \mathbf{x}')\, \omega_I\, (\mathbf{x}')\, d\mathbf{x}$$
$$+ \int_{-\infty}^{\infty} \frac{\partial \mathcal{G}}{\partial x'} (\mathbf{x},\, 0,\, y')\, \psi_B\, (y',\, 0)\, dy'\,, \tag{8.5.17}$$

where \mathcal{G} is the Green's function for Dirichlet's problem in a half-space. In (8.5.17), $\mathbf{x} = (x,\, y)$ and $\mathbf{x}' = (x',\, y')$. In particular, the velocity components on the boundary are

$$u(0,\, y,\, 0) = -\frac{\partial}{\partial y}\, \psi(0,\, y,\, 0) = -\frac{\partial}{\partial y}\, \psi_B(y,\, 0) \tag{8.5.18}$$

and

$$v(0, y, 0) = \frac{\partial}{\partial x} \psi(0, y, 0). \tag{8.5.19}$$

The data consistency conditions may now be given explicitly. In addition to the simple condition of continuity of vorticity,

$$\omega_{IN}(0, y, 0) = \omega_I(0, y), \tag{8.5.20}$$

which must hold for $y \in B_{IN}$, there is the condition of continuity of the time derivative of vorticity,

$$\frac{\partial}{\partial t} \omega_{IN}(0, y, 0) + u(0, y, 0) \frac{\partial \omega_I}{\partial x}(0, y)$$

$$+ v(0, y, 0) \frac{\partial}{\partial y} \omega_{IN}(0, y, 0) + \beta v(0, y, 0)$$

$$= G(0, y, 0), \tag{8.5.21}$$

which must also hold for $y \in B_{IN}$.

Characteristic points, should they exist, have the coordinates $(0, y_0(t))$, where

$$u(0, y_0(t), t) = -\frac{\partial}{\partial y} \psi_B(y_0(t), t) = 0 \tag{8.5.22}$$

for $0 \leq t$. At such points the local consistency condition for the boundary vorticity is

$$Z(t) = \frac{\partial \omega_{IN}}{\partial t}(0, y_0(t), t) + v(0, y_0(t), t) \frac{\partial \omega_{IN}}{\partial y}(0, y_0(t), t)$$

$$+ \beta v(0, y_0(t), t) - G(0, y_0(t), t) = 0 \tag{8.5.23}$$

for $0 \leq t$. Now the initial tangential velocity $v = v(0, y_0(0), 0)$ is related to the initial vorticity $\omega_I = \omega_I(\mathbf{x})$, and to the initial boundary streamfunction $\psi_B = \psi_B(y, 0)$, through (8.5.17) and (8.5.19), so the initial local consistency condition

$$Z(0) = 0 \tag{8.5.24}$$

may be verified directly. However, well-posedness requires that $Z(t) = 0$ for all $t \geq 0$; if Z has a convergent Taylor series expansion for small t, then a necessary condition for well-posedness is

$$Z^{(m)}(0) \equiv \frac{d^m}{dt^m} Z(0) = 0 , \qquad (0 \leq m < \infty) . \qquad (8.5.25)$$

An examination of (8.5.23) shows that each condition in (8.5.25) may be explicitly verified. Consider $m = 1$. Evaluation of $Z^{(1)}(0)$ requires, *inter alia*, values for $y_0^{(1)}(0)$ and $v_t(0, y_0(0), 0)$. The former is obtained by differentiating (8.5.22) with respect to t, and evaluating the result at $t = 0$. The latter is obtained by noting that for $t \geq 0$, the streamfunction may be represented as

$$\psi(\mathbf{x}, t) = \int \int_R \mathcal{G}(\mathbf{x}, \mathbf{x}') \, \omega(\mathbf{x}, t) \, d\mathbf{x}$$
$$+ \int_{-\infty}^{\infty} \frac{\partial \mathcal{G}}{\partial x'}(\mathbf{x}, 0, y') \psi_B(y', t) \, dy' . \qquad (8.5.26)$$

Differentiating (8.5.26) with respect to t, and evaluating the result at $t = 0$, where

$$\omega_t(\mathbf{x}, 0) = -\mathbf{u}(\mathbf{x}, 0) \cdot \nabla \, \omega_{IN}(\mathbf{x}) - \beta v(\mathbf{x}, 0) + G(\mathbf{x}, 0) , \qquad (8.5.27)$$

yields $\psi_t(\mathbf{x}, 0)$ and hence $v_t(\mathbf{x}, 0)$. By induction, $Z^{(m)}(0)$ may be evaluated for all $m \geq 0$. Thus the well-posedness of the CFvN problem may be established for small times $t \geq 0$, but only after verifying an infinite number of conditions (Bennett & Kloeden, 1978).

The classical definition of well-posedness stated in §8.4 requires not only that a classical solution exist, but that it be unique and continuously dependent upon the data. The latter two requirements may be addressed simultaneously. Suppose that $\omega^{(1)}$ and $\omega^{(2)}$ are two solutions of the CFvN problem, having differing initial values $\omega_I^{(1)}$ and $\omega_I^{(2)}$ respectively. For simplicity's sake, suppose that both solutions have the same inflow boundary vorticity ω_{IN}, and that both correspond to the same boundary streamfunction ψ_B. Judovich (1966) and Sundström (1969) have shown that

$$\int_R \left| \omega^{(1)}(\mathbf{x}, t) - \omega^{(2)}(\mathbf{x}, t) \right|^2 \, d\mathbf{x}$$
$$\leq \exp[tK(t)] \int_R \left| \omega_I^{(1)}(\mathbf{x}) - \omega_I^{(2)}(\mathbf{x}) \right|^2 \, d\mathbf{x} , \qquad (8.5.28)$$

where

$$K(t) = 2\lambda \, \max|\nabla\omega|. \qquad (8.5.29)$$

In (8.5.29), λ is a length scale which depends upon the geometry of the region R, but not upon the flow. The maximum for the vorticity gradient is taken over the entire time interval $0 \leq r \leq t$ and the entire region R. If $K(t)$ is finite, then (8.5.28) implies that for fixed values of t, $\omega^{(1)}(\mathbf{x}, t) = \omega^{(2)}(\mathbf{x}, t)$ when $\omega_I^{(1)}(\mathbf{x}) = \omega_I^{(2)}(\mathbf{x})$, that is, the solution is unique. It also implies that $\omega^{(1)}(\mathbf{x}, t) \to \omega^{(2)}(\mathbf{x}, t)$ as $\omega_I^{(1)}(\mathbf{x}) \to \omega_I^{(2)}(\mathbf{x})$: the solution depends continuously upon the initial data. The argument is readily extended to include unique and continuous dependence upon ω_{IN} and ψ_B. The problem, of course, is that the finiteness of $K(t)$ cannot be assured if the possibility of characteristic points cannot be excluded.

Discrete analogues of (8.5.28) have been developed by Sundström (1969) and Fix (1975) for finite-difference and finite-element approximations, respectively. The discrete analogue of $K(t)$ is necessarily finite, but it need not be bounded as numerical resolution is refined. Indeed, the existence of characteristic points would preclude boundedness (Bennett & Kloeden, 1981b).

v) Examples

Consider the following scalar fields $\phi(\mathbf{x}, t)$ and $\zeta(\mathbf{x}, t)$:

$$\phi(x, y, t) = \sin\left(x + y + \frac{1}{2}\beta t\right) + \sin(x + \beta t), \qquad (8.5.30)$$

$$\zeta(\mathbf{x}, t) = \nabla^2 \phi(\mathbf{x}, t). \qquad (8.5.31)$$

Each of the two plane waves in (8.5.30), (8.5.31) is an exact finite-amplitude solution of (8.5.9), but their sum is not such a free solution. Nor is the pair resonant. Choosing initial and boundary data for streamfunction ψ and vorticity ω from values of ϕ and ζ, respectively, leads to continuity of $\omega(\mathbf{x}, 0)$ at B_{IN}, but not to continuity of $\omega_t(\mathbf{x}, 0)$ at B_{IN}. In particular, $Z(0) \neq 0$ at characteristic points. Numerical integration of the CFvN problem in a square region with fully open boundaries having length of $\approx 2\pi$ leads, in a time of $O(\beta^{-1})$, to spatial vorticity gradients which depend sensitively upon numerical resolution and numerical method (Miller & Bennett, 1988).

Figure 8.5.2. Finite-element integration of the CFvN problem. Left panels: streamfunction at times $t = 3.2$, $t = 6.4$ and $t = 9.6$. Right panels: vorticity at matching times (after Miller & Bennett, 1988).

Next consider the scalar fields

$$\phi(x,\, y,\, t) = \sin\left(x + y + \frac{1}{2}\beta t\right)$$

$$+ \mu \, \sin\left(\frac{1}{2}\beta t\right) \left[\frac{x - 2X}{2X}\right] \left[\frac{y - 2Y}{2Y}\right], \quad (8.5.32)$$

$$\zeta(\mathbf{x},\, t) = \nabla^2 \phi(\mathbf{x},\, t). \qquad\qquad\qquad (8.5.33)$$

The region is the interior of the rectangle $|x| \le X/2$, $|y| \le Y/2$. The plane wave in (8.5.32–33) is again an exact, free, finite-amplitude solution of (8.5.9). The bilinear perturbation is harmonic in space, and so does not contribute either to initial or inflow vorticity. The sum of the two fields in (8.5.32–33) is not an exact, free solution of (8.5.9). Choosing initial and boundary data for ψ and ω from ϕ and ζ, respectively, it is again the case that $\omega(\mathbf{x}, 0)$ is continuous at B_{IN}. However, it is now the case that $\omega_t(\mathbf{x}, 0)$ is also continuous at B_{IN}, in particular, $Z(0) = 0$ at characteristic points. On the other hand, $\omega_{tt}(\mathbf{x}, 0)$ is not continuous, and, in particular, $Z^{(1)}(0) \ne 0$ at characteristic points.

Numerical results obtained by Miller & Bennett (1988) using a finite-element method are shown in Fig. 8.5.2. The three left and three right panels show contours of streamfunction and vorticity, respectively, at dimensionless times $t = 3.2$, 6.4 and 9.6 units. Some other dimensionless parameter values are $\beta = 1$, $X = Y = 8$ and $\mu = 4$, while $\Delta x = \Delta y = 8/32$. See Miller & Bennett (1988) for further details. As time evolves, the vorticity field develops very large gradients, appearing first at characteristic points. Increasing the resolution by reducing Δx, Δy and Δt by factors of 2 and 4 has little influence on the velocity field, but leads to severe degradation of the vorticity. Shown in Fig. 8.5.3 are sections of the vorticity, normally inward from the characteristic point marked on Fig. 8.5.2 at time $t = 6.4$. The three sections correspond to the three levels of spatial resolution. The local vorticity field is roughly doubling as the element size is halved, since the finite-element equivalent of the advection approximation ($u \, \Delta\omega/\Delta x$) is roughly doubling.

Figure 8.5.3. Sections of vorticity, normal to the boundary point marked with an arrow in Fig. 8.5.2. Sections (b) and (c) show the effects of doubling and quadrupling the resolution used in (a).

8.6 Fixes

i) Friction

The vorticity conservation law for plane viscous incompressible flow is

$$\omega_t + \mathbf{u} \cdot \nabla \omega = G + \nu \nabla^2 \omega, \qquad (8.6.1)$$

where ν is the kinematic viscosity, and the other symbols have the same meaning as in the preceding sections. In particular, the kinematic relations (8.2.2)–(8.2.5) still hold. Whereas (8.2.1) is a hyperbolic equation, (8.6.1) is parabolic so long as ν is positive. Consequently, a natural boundary condition for (8.6.1) is the Dirichlet condition, namely, specification of vorticity all around. Such a condition would be an overspecification on rigid boundaries where there is no slip, but is admissible on open boundaries where only the normal velocity or the streamfunction is specified. In particular, according to (8.6.1), vorticity diffuses away from the boundary, both outwards and inwards, and so vorticity must be specified on inflow and outflow boundaries. Thus, suitable boundary conditions for (8.6.1) in a simply-connected region having an entirely open boundary are

$$\psi = \psi_B, \qquad \text{on } B_{IN} \cup B_{OUT}, \qquad (8.6.2)$$

$$\omega = \omega_B, \qquad \text{on } B_{IN} \cup B_{OUT}. \qquad (8.6.3)$$

Initial values of vorticity must also be specified, as in (8.3.32). Lady-zhenskaya (1969, pp. 167, 185) has proved that the resulting initial-boundary-value problem is well posed, with smooth data leading to smooth solutions which depend continuously upon the data. In particular, discontinuities cannot develop at points on the boundary where the flow is locally tangential. However, boundary layers can develop. Indeed, if the viscosity is small, or more precisely if the Reynolds number is large, then the solution outside the boundary layer is the inviscid solution. The latter will in general have vorticity discontinuities across the paths of particles which enter the region at characteristic points. The viscous boundary layers patch across these discontinuities. That is, the layers will be found in the interior of the region. They represent no real physical phenomena, and exist in the solution solely due to the ill-posedness of the inviscid flow problem. Thus, in a real physical sense, the inclusion of viscosity does not resolve the ill-posedness of the CFvN problem.

ii) Filtering

Discontinuities in the vorticity field lead to fine structure in numerical solutions. The structure may be suppressed by the application of spatial low-pass filters during numerical integration. In one of their numerical experiments, Miller & Bennett (1988) used a fourth-order Shapiro filter (Shapiro, 1970) to smooth the vorticity field after every time step. The results are shown in Fig. 8.6.1; they may be compared with results of an unfiltered but otherwise identical experiment shown in Fig. 8.5.2. The smoothed vorticity field is certainly a subjective improvement upon the unsmoothed field. There are some large-scale differences between the two vorticity fields at time $t = 9.6$, while the two streamfunction fields have lost all resemblance by that time. Miller & Bennett (1988) concluded that, since there is no physical basis for preferring either solution, it is necessary to assimilate data in the interior of the region during the numerical integration, in order to maintain the physical reality of the solution.

iii) Fudging

Suppose that the ocean is an incompressible inviscid fluid, and that the circulation is truly planar. Suppose also that the initial and boundary data for a model have been collected by exact measurement of the ocean circulation, the measurements being taken with infinitely

Figure 8.6.1. As for Fig. 8.5.2, excepting for the use now of a low-pass filter to smooth the vorticity field after each time step (after Miller and Bennett, 1988).

fine resolution. Finally, suppose that the model can be integrated exactly, that is, without truncation or round-off. In such ideal circumstances the ill-posedness of the CFvN problem does not yield discontinuous vorticity fields, provided the true circulation does not possess vorticity discontinuities. The consistency condition (8.5.23) will always be met, since the model integration always recovers the true circulation. Discontinuities in the vorticity can only be present if introduced in the boundary or initial data; recall that the classical existence theorems (§8.4) for both infinite and rigidly bounded domains preclude the development of vorticity discontinuities in a finite time, given smooth data.

In the real world, of course, the model, the data and the numerical integration are all inaccurate, and so vorticity discontinuities must arise. Nevertheless the preceding argument suggests that if the inaccuracies are not too great, then a perturbation of the data might lead to a consistent solution free of discontinuities. Once perturbations are admitted, there is no need to restrict the specification of vorticity to the inflow boundary; vorticity may be specified all around the open boundary. Discontinuities may arise at the outflow boundary: they cannot be advected into the region, but they may excite upstream-propagating Rossby waves. A residual $Z = Z(s, t)$ may be defined all around the open boundary:

$$Z = \frac{\partial}{\partial t} \omega + \mathbf{u} \cdot \nabla \omega + \beta v - G, \qquad (8.6.4)$$

$$\text{on } B_{IN} \cup B_{OUT}.$$

As before, Z can be evaluated only at time $t = 0$, yet it must vanish for all time $t \geq 0$ if a smooth solution is to be obtained. On the other hand, partial derivatives of Z of all orders with respect to time t may be evaluated at $t = 0$. Suppose that initially the vorticity ω is continuous at the open boundary, but $Z(s, 0) \neq 0$. At time $\Delta t > 0$, there will be a discontinuity at the open boundary of strength $\Delta t\, Z(s, 0)$, unless the boundary vorticity data are perturbed by that amount, the perturbed data ω'_B being

$$\omega'_B(s, \Delta t) = \omega_B(s, \Delta t) - \Delta t\, Z(s, 0). \qquad (8.6.5)$$

There is no need to perturb the boundary streamfunction data. Numerical experiments show that such a simple fudging scheme, applied

at every time step, can keep the vorticity field relatively smooth for a few advective time scales. However, the scheme suffers from the same defect as filtering. There is no physical basis for accepting the results; the inherent unpredictability of real fluid motion (smooth but sensitive dependence upon initial conditions) leads to solutions without even large-scale resemblance to those obtained using unperturbed data. Once again it may be concluded that additional data must be assimilated into the interior of the model region, in order to ensure that the solution bears some resemblance to reality.

8.7 Inverse formulation (I)

Resolution of the ill-posedness of the CFvN problem requires perturbations to the boundary data, and would also appear to require the assimilation of data within the model region. Both of these needs may be met by formulating the CFvN problem, together with an interior data set, as an inverse problem as in §7.3. The proliferation of details such as boundary residuals for the streamfunction and vorticity, and the specifications of their respective weights, greatly complicates the problem. In this preliminary formulation, the boundary streamfunction data will be presumed exact. A systematic notation for weights will be introduced. The region R will be assumed simply connected, having an entirely open boundary $B = B_{IN} \cup B_{OUT}$. The β-effect will be neglected.

To begin, several residuals are introduced:

$$g = \omega_t + \mathbf{u} \cdot \nabla \omega - G, \tag{8.7.1}$$
$$\text{in } R \text{ for } 0 \le t \le T,$$

$$\phi = \omega - \omega_I, \tag{8.7.2}$$
$$\text{in } R \text{ at } t = 0,$$

$$\theta = \omega - \omega_B, \tag{8.7.3}$$
$$\text{on } B \text{ for } 0 \le t \le T,$$

and

$$\epsilon = \mathbf{d} - \mathcal{L}[\psi]. \tag{8.7.4}$$

In (8.7.1)–(8.7.4), G, ω_I, ω_B and \mathbf{d} are prior estimates, respectively, of the vorticity source, the initial vorticity, the boundary vorticity and the measured streamfunction. In (8.7.4), \mathcal{L} is an M-dimensional

vector of linear measurement functionals. The data depend upon values of ψ in R for $0 \leq t \leq T$. No residual is allowed in the boundary streamfunction, so

$$\psi = \psi_B , \qquad (8.7.5)$$

$$\text{on } B \text{ for } 0 \leq t \leq T .$$

In addition there are exact kinematic relationships between the vorticity, velocity and streamfunction:

$$\omega = \nabla^2 \psi , \qquad (8.7.6)$$

and

$$\mathbf{u} = \mathbf{e}_3 \times \boldsymbol{\nabla} \psi , \qquad (8.7.7)$$

$$\text{on } R \text{ for } 0 \leq t \leq T .$$

A penalty functional $\mathcal{J} = \mathcal{J}[\psi]$ may be defined by

$$\mathcal{J}[\psi] = \int_0^T dt_1 \int_0^T dt_2 \int_R d\mathbf{x}_1 \int_R d\mathbf{x}_2 \, g(\mathbf{x}_1, t_1) \, W_{gg}(\mathbf{x}_1, t_1, \mathbf{x}_2, t_2)$$

$$\times \, g(\mathbf{x}_2, t_2) + \int_R d\mathbf{x}_1 \int_R d\mathbf{x}_2 \, \phi(\mathbf{x}_1) \, W_{\phi\phi}(\mathbf{x}_1, \mathbf{x}_2) \, \phi(\mathbf{x}_2)$$

$$+ \int_0^T dt_1 \int_0^T dt_2 \int_B ds_1 \int_B ds_2 \, \theta(s_1, t_1) \, W_{\theta\theta}(s_1, t_1, s_2, t_2) \, \theta(s_2, t_2)$$

$$+ \, \boldsymbol{\epsilon}^* \, \mathbf{w} \, \boldsymbol{\epsilon} . \qquad (8.7.8)$$

The weights W_{gg}, $W_{\phi\phi}$, $W_{\theta\theta}$ and \mathbf{w} appearing in (8.7.8) are positive-definite and symmetric. In an obvious shorthand, \mathcal{J} becomes

$$\mathcal{J} = g \bullet W_{gg} \bullet g + \phi \circ W_{\phi\phi} \circ \phi$$

$$+ \, \theta \bullet W_{\theta\theta} \bullet \theta + \boldsymbol{\epsilon}^* \, \mathbf{w} \, \boldsymbol{\epsilon} . \qquad (8.7.9)$$

The solid circles in (8.7.9) denote integration over space and time, while the open circles denote integration over space alone. The exact or strong constraints (8.7.5)–(8.7.7) may be appended to \mathcal{J} using

Lagrange multipliers. The calculus of variations leads, after elim-
inating the Lagrange multipliers, to the following system of Euler-
Lagrange equations for extrema of \mathcal{J}. Let ν denote the weighted
optimal estimate of the model residual g, that is, let

$$\nu \equiv W_{gg} \bullet \hat{g}. \tag{8.7.10}$$

Then ν satisfies

$$-\nu_t - \hat{\mathbf{u}} \cdot \boldsymbol{\nabla} \nu - \nabla^{-2} J(\nu, \hat{\omega}) = \nabla^{-2} \boldsymbol{\mathcal{L}}[\delta\delta]^* \, \mathbf{w} \Big(\mathbf{d} - \boldsymbol{\mathcal{L}} [\hat{\psi}] \Big), \tag{8.7.11}$$
$$\text{in } R \text{ for } 0 \le t \le T,$$
$$\nu = 0, \tag{8.7.12}$$
$$\text{in } R \text{ at } t = T,$$
$$\nu = W_{\phi\phi} \circ (\hat{\omega} - \omega_I), \tag{8.7.13}$$
$$\text{in } R \text{ at } t = 0,$$

and

$$\nu \, \hat{\mathbf{u}} \cdot \mathbf{n} + W_{\theta\theta} \bullet (\hat{\omega} - \omega_B) = 0, \tag{8.7.14}$$
$$\text{on } B \text{ for } 0 \le t \le T,$$

where \mathbf{n} is the outward unit normal on B. The operator ∇^{-2} in
(8.7.11) denotes the inverse of the Laplacian operator ∇^2 in R, subject
to the homogeneous Dirichlet boundary condition on B (vanishing
solution values). Note that the normal velocity $\hat{\mathbf{u}} \cdot \mathbf{n}$ is prescribed
exactly on B by virtue of (8.7.5). The fields $\hat{\omega}$, $\hat{\mathbf{u}}$ and $\hat{\psi}$ are the
optimal estimates of ω, \mathbf{u} and ψ. In particular,

$$\hat{\omega}_t + \hat{\mathbf{u}} \cdot \boldsymbol{\nabla} \hat{\omega} = G + Q_{gg} \bullet \nu, \tag{8.7.15}$$

where Q_{gg} is the inverse of W_{gg}:

$$Q_{gg} \bullet W_{gg} = \delta\delta, \tag{8.7.16}$$

the right-hand side of (8.7.16) being a product of delta-functions in
space and time. The final condition for (8.7.11) is given by (8.7.12),
while the initial condition for (8.7.15) is derived from (8.7.13):

$$\hat{\omega} = \omega_I + Q_{\phi\phi} \circ \nu, \tag{8.7.17}$$
$$\text{in } R \text{ at } t = 0,$$

where $Q_{\phi\phi}$ is the inverse of $W_{\phi\phi}$. The boundary conditions for (8.7.11) and (8.7.15) are obtained from (8.7.14). Respectively, the conditions are

$$\nu = -(\hat{\mathbf{u}} \cdot \mathbf{n})^{-1} W_{\theta\theta} \bullet (\hat{\omega} - \omega_B),\qquad(8.7.18)$$
$$\text{on } B_{OUT} \text{ for } 0 \le t \le T,$$

and

$$\hat{\omega} = \omega_B + Q_{\theta\theta} \bullet (\nu\, \hat{\mathbf{u}} \cdot \mathbf{n}),\qquad(8.7.19)$$
$$\text{on } B_{IN} \text{ for } 0 \le t \le T,$$

where $Q_{\theta\theta}$ is the inverse of $W_{\theta\theta}$.

The case in which no vorticity data are provided on the outflow boundary B_{OUT} may be recovered from the above equations in the limit $W_{\theta\theta} \to 0$ on B_{OUT} (worthless outflow data): then (8.7.18) is simply

$$\nu = 0,\qquad(8.7.20)$$
$$\text{on } B_{OUT} \text{ for } 0 \le t \le T.$$

The strong-constraint $(W_{gg} \to \infty)$, worthless-outflow-data $(W_{\theta\theta} \to 0)$, worthless-initial-data $(W_{\phi\phi} \to 0)$ limit of the Euler-Lagrange equations is given in Moore (1991). Finally, it should be noted that the inverse streamfunction estimate satisfies

$$\nabla^2 \hat{\psi} = \hat{\omega},\qquad(8.7.21)$$
$$\text{in } R \text{ for } 0 \le t \le T,$$

subject to

$$\hat{\psi} = \psi_B,\qquad(8.7.22)$$
$$\text{on } B \text{ for } 0 \le t \le T,$$

while

$$\hat{\mathbf{u}} = \mathbf{e}_3 \times \boldsymbol{\nabla}\, \hat{\psi},\qquad(8.7.23)$$
$$\text{in } R \text{ for } 0 \le t \le T.$$

The system of Euler-Lagrange equations for extrema of \mathcal{J} may be expressed as a single equation of the form

$$\left(\frac{\partial}{\partial t} + \hat{\mathbf{u}} \cdot \boldsymbol{\nabla}\right) W_{gg} \bullet \left(\frac{\partial}{\partial t} + \hat{\mathbf{u}} \cdot \boldsymbol{\nabla}\right) \hat{\omega} = \cdots , \qquad (8.7.24)$$

$$\text{in } R \text{ for } 0 \le t \le T ,$$

subject to the initial condition

$$W_{\phi\phi} \circ (\hat{\omega} - \omega_I) - W_{gg} \bullet (\hat{\omega}_t + \hat{\mathbf{u}} \cdot \boldsymbol{\nabla}\,\hat{\omega} - G) = 0 , \quad (8.7.25)$$

$$\text{in } R \text{ at } t = 0 ,$$

the final condition

$$W_{gg} \bullet (\hat{\omega}_t + \hat{\mathbf{u}} \cdot \boldsymbol{\nabla}\,\hat{\omega} - G) = 0 , \qquad (8.7.26)$$

$$\text{in } R \text{ at } t = T ,$$

and the boundary condition

$$(\hat{\mathbf{u}} \cdot \mathbf{n})\, W_{gg} \bullet (\hat{\omega}_t + \hat{\mathbf{u}} \cdot \boldsymbol{\nabla}\,\hat{\omega} - G) + W_{\theta\theta} \bullet (\hat{\omega} - \omega_B) = 0 \,(8.7.27)$$

$$\text{on } B \text{ for } 0 \le t \le T .$$

This is a second-order boundary-value problem in R for $0 \le t \le T$, with mixed boundary conditions on B for $0 \le t \le T$ and on R at $t = 0, T$. Consequently, there is no overspecification of vorticity at outflow. It might be argued that there is no such overspecification in the CFvN problem either, since vorticity is not specified at outflow in that problem. However, vorticity data are in practice available both at inflow and outflow, if at all, and are likely to be of comparable quality all around the boundary. Thus the inverse formulation has the benefit of outflow vorticity data as well as interior data, should the latter be available. Moreover, in the inverse formulation the outflow data and interior data can affect the upstream estimate \hat{g} for the dynamical residual. For example, the residual might be an error in an estimate of the curl of the wind stress over the Gulf Stream near Cape Hatteras. Observations of velocity downstream, say near the Grand Banks, would have been influenced by the true wind stress at Cape Hatteras, and so it is desirable that the data assimilation scheme should allow

the downstream observations to contribute to the estimate of the upstream error in the wind stress curl. Sequential schemes such as nudging or the Kalman filter allow for no such upstream influence. The concepts of upstream and downstream should, of course, be modified to allow for Rossby wave propagation, but the argument is essentially unaltered.

The inverse formulation does not resolve the difficulties of the CFvN problem at characteristic points. It may be seen from (8.7.27) that $W_{\theta\theta} \bullet (\hat{\omega} - \omega_B) \to 0$ near such points, but such behavior may be inconsistent with (8.7.24), depending upon the local value of the unspecified tangential velocity $\hat{\mathbf{u}} \cdot \mathbf{s}$, regardless of the smoothing properties of the inverse weights Q_{gg}, $Q_{\phi\phi}$ and $Q_{\theta\theta}$.

Further difficulties with (8.7.18) must be expected at characteristic points when integrating the adjoint equation (8.7.11).

Exercise 8.7.1 Repeat the inverse formulation for a simple wave equation as in Exercise 5.3.2, but subject now to a boundary condition at $x = 0$:

$$u(0, t) = B(t) + b(t), \qquad (8.7.28)$$

where B is a prior estimate of the boundary values, and b the prior error in the boundary values. Let the spatial interval of interest be $0 \le x \le X$, containing the M measurement sites X_m, $1 \le m \le M$. Express the inverse estimate \hat{u} using an appropriately defined set of representers. □

Exercise 8.7.2 Derive the Euler-Lagrange equations for the extrema of the penalty functional \mathcal{J} defined by (8.7.8), subject to the constraints (8.7.5)–(8.7.7). □

Exercise 8.7.3 Repeat the preceding exercise, but allowing for residuals in the boundary data for the streamfunction. □

8.8 Inverse formulation (II)

The preceding inverse formulation of the CFvN problem failed to resolve the ill-posedness because the formulation failed to make explicit the requirement for a continuous vorticity field. Such a requirement can be imposed in a least-squares formulation by penalizing the square of various derivatives of the vorticity. The underlying principle is based on embedding theorems (Adams, 1975, p. 98) which establish that for functions of two real variables, square-integrability of the

second derivatives of the function ensures the (Hölder-) continuity of the function itself. Square-integrability of third derivatives would ensure (Hölder-) continuity of the first derivatives. These embedding theorems are very conservative; in practice, smooth behavior can be ensured without reference to such high-order derivatives (see §2.3). This suggests that a regularization of the inverse CFvN problem might be achieved if the penalty functional \mathcal{J}, defined by (8.7.8), were augmented as follows:

$$\mathcal{K} = \mathcal{J} + C \int_0^T dt \int_R d\mathbf{x} |\nabla \omega|^2 , \qquad (8.8.1)$$

where C is a positive constant. As before, \mathcal{K} may be regarded as a functional of the streamfunction ψ, by appending the kinematic relationships (8.7.5)–(8.7.7) using Lagrange multipliers. Once again it will be assumed for simplicity that $\beta = 0$, and that the streamfunction boundary data are exact. The region is as in §8.7. If $\hat{\psi}$ is an extremum of \mathcal{K}, with associated vorticity $\hat{\omega}$, velocity $\hat{\mathbf{u}}$ and weighted vorticity residual ν as in (8.7.10), then (8.7.11) becomes

$$-\nu_t - \hat{\mathbf{u}} \cdot \nabla \nu - \nabla^{-2} J(\nu, \hat{\omega}) - C\nabla^2 \hat{\omega}$$
$$= \nabla^{-2} \mathcal{L}[\delta\delta]^* \mathbf{w}(\mathbf{d} - \mathcal{L}[\hat{\psi}]) , \qquad (8.8.2)$$
$$\text{in } R \text{ for } 0 \leq t \leq T ,$$

the boundary condition (8.7.14) becomes

$$\nu \, \hat{\mathbf{u}} \cdot \mathbf{n} + W_{\theta\theta} \bullet (\hat{\omega} - \omega_B) + C \, \mathbf{n} \cdot \nabla \hat{\omega} = 0 , \qquad (8.8.3)$$
$$\text{on } B \text{ for } 0 \leq t \leq T ,$$

while (8.7.10), (8.7.12), (8.7.13) and (8.7.15) are unchanged.

The system of Euler-Lagrange equations for extrema of \mathcal{K} may be expressed as a single equation of the form

$$\left[\left(\frac{\partial}{\partial t} + \hat{\mathbf{u}} \cdot \nabla \right) W_{gg} \bullet \left(\frac{\partial}{\partial t} + \hat{\mathbf{u}} \cdot \nabla \right) + C\nabla^2 \right] \hat{\omega} = \cdots , \qquad (8.8.4)$$
$$\text{in } R \text{ for } 0 \leq t \leq T ,$$

subject to the initial condition (8.7.25), the final condition (8.7.26) and the boundary condition

$$\left(\hat{\mathbf{u}} \cdot \mathbf{n}\right) W_{gg} \bullet \left(\hat{\omega}_t + \hat{\mathbf{u}} \cdot \boldsymbol{\nabla} \hat{\omega}\right) + W_{\theta\theta} \bullet \left(\hat{\omega} - \omega_B\right)$$
$$+ C \mathbf{n} \cdot \boldsymbol{\nabla} \hat{\omega} = 0, \qquad (8.8.5)$$
$$\text{on } B \text{ for } 0 \le t \le T.$$

Consider, for example, the case of uniform, uncorrelated weighting for the dynamical residual:

$$W_{gg}(\mathbf{x}_1, t_1, \mathbf{x}_2, t_2) = W\delta(\mathbf{x}_1 - \mathbf{x}_2)\,\delta(t_1 - t_2), \qquad (8.8.6)$$

where W is a constant. Then (8.8.4) is an elliptic problem in spacetime for $\hat{\omega}$, with mixed boundary conditions. In particular, the structure of the problem is not described by characteristics such as particle paths, and so the issue of locally characteristic boundaries does not arise. Thus, the ill-posedness of the original CFvN problem would appear to be resolved.

A value must be chosen for the constant C. As (8.8.1) is formulated, C is a dimensional quantity. If the choice is large relative to the weights in \mathcal{J}, then minimization of \mathcal{K} will be achieved with relatively large residuals in \mathcal{J}. If a relatively small value is chosen, then extrema of \mathcal{K} will in some sense be close to extrema of \mathcal{J}, and yet smooth. However, there will in general be boundary layers near characteristic points, with width δ given by

$$\delta^2 = C\,\ell^2\,W^{-1}\,v^{-2}, \qquad (8.8.7)$$

where ℓ is the length scale for the local boundary vorticity, W is the magnitude of W_{gg} and v is the magnitude of the tangential velocity.

In conclusion, augmenting the penalty functional \mathcal{J} with a further penalty for gradients of vorticity does yield a formal regularization of the CFvN problem, but in a manner which would not appear to possess any advantages over the simple fixes described in §8.6. The corresponding system of Euler-Lagrange equations (8.8.2), (8.8.3) *et seq.* would be extremely difficult to solve.

Exercise 8.8.1 Derive (8.8.2)–(8.8.3). Devise an iterative solution scheme as in §7.5. Reduce the linearized Euler-Lagrange equations to initial-value problems using the sweep algorithm. Find a statistical

model for the resulting equations, initial conditions and boundary conditions. Assess the computational cost of implementing the algorithm, including operation count and storage requirements. □

8.9 Representer solutions

The inverse formulation (I) described in §8.7 does not resolve the problem of characteristic points on open boundaries, but it does provide a procedure for assimilating both interior and outflow data. Setting aside temporarily the problem of characteristic points, an iterative scheme for solving the Euler-Lagrange system will now be described. As in §7.5, the chosen iteration scheme linearizes the system in such a way that each iterate is the Euler-Lagrange system for a linear inverse problem. Specifically, the advecting velocity \hat{u} appearing in (8.7.11) and (8.7.15) is evaluated using the preceding iterate, while the Jacobian term in (8.7.11) is evaluated entirely in terms of the preceding iterate. Once again the solution of the linear Euler-Lagrange system may be expressed in terms of a first guess or prior estimate, plus a linear combination of representers. Attention here will be restricted to the representer expansion, and in particular the treatment of the outflow boundary condition (8.7.18). For simplicity, subscripts denoting the level of iteration will be suppressed. Then the adjoint variable iterates such as ν_n, for example, will simply be denoted by ν, with the single exception that the advecting velocity \hat{u}_{n-1} will be denoted by \hat{u} without any ambiguity occurring.

Two sets of representers must be constructed. The first set has one member for each measurement functional \mathcal{L}_m, $1 \leq m \leq M$. Vector notation is convenient; the adjoint representer vector $\boldsymbol{\alpha}$ satisfies the final-value problem

$$\boldsymbol{\alpha}_t + \hat{u} \cdot \nabla \boldsymbol{\alpha} = -\nabla^{-2}\mathcal{L}[\delta\delta]\,, \qquad (8.9.1)$$

$$\text{in } R \text{ for } 0 \leq t \leq T\,,$$

subject to the final condition

$$\boldsymbol{\alpha} = \mathbf{0}\,, \qquad (8.9.2)$$

$$\text{in } R \text{ at } t = T\,,$$

and the boundary condition

$$\boldsymbol{\alpha} = \mathbf{0}\,, \qquad (8.9.3)$$

$$\text{on } B_{OUT} \text{ for } 0 \leq t \leq T\,.$$

The representer vector **r** satisfies the initial-value problem

$$\mathbf{r}_t + \hat{\mathbf{u}} \cdot \nabla \mathbf{r} = Q_{gg} \bullet \boldsymbol{\alpha}, \qquad (8.9.4)$$
$$\text{in } R \text{ for } 0 \leq t \leq T,$$

subject to the initial condition

$$\mathbf{r} = Q_{\phi\phi} \circ \boldsymbol{\alpha}, \qquad (8.9.5)$$
$$\text{in } R \text{ at } t = 0,$$

and the boundary condition

$$\mathbf{r} = -Q_{\theta\theta} \bullet \left[(\hat{\mathbf{u}} \cdot \mathbf{n}) \, \boldsymbol{\alpha} \right], \qquad (8.9.6)$$
$$\text{on } B_{IN} \text{ for } 0 \leq t \leq T.$$

This *finite* set of representers is needed to resolve the coupling between the adjoint and physical variables in the adjoint differential equation (8.7.11). The second set of representers has one member for each point on the outflow boundary, at each moment of time. The representer adjoints are denoted $\alpha(\mathbf{x}, t, s', t')$, where s' is a path parameter on the outflow boundary, such as path length. These variables satisfy the final-value problem

$$\alpha_t + \hat{\mathbf{u}} \cdot \nabla \alpha = 0, \qquad (8.9.7)$$
$$\text{in } R \text{ for } 0 \leq t \leq T,$$
$$\alpha = 0, \qquad (8.9.8)$$
$$\text{in } R \text{ at } t = T,$$
$$\alpha = (\hat{\mathbf{u}} \cdot \mathbf{n})^{-1} \delta(s - s') \, \delta(t - t'), \qquad (8.9.9)$$
$$\text{on } B_{OUT} \text{ for } 0 \leq t \leq T.$$

Note that the outflow boundary may vary in time. The representers are denoted $\mathbf{r}(\mathbf{x}, t, s', t')$, and satisfy the initial-value problem

$$r_t + \hat{\mathbf{u}} \cdot \nabla r = Q_{gg} \bullet \alpha, \qquad (8.9.10)$$
$$\text{in } R \text{ for } 0 \leq t \leq T,$$
$$r = Q_{\phi\phi} \circ \alpha, \qquad (8.9.11)$$
$$\text{in } R \text{ at } t = 0,$$
$$r = -Q_{\theta\theta} \bullet \left[(\hat{\mathbf{u}} \cdot \nabla)\alpha \right], \qquad (8.9.12)$$
$$\text{on } B_{IN} \text{ for } 0 \leq t \leq T.$$

This *continuously infinite* set of representers resolves the coupling between the adjoint and physical variables in the adjoint outflow boundary condition (8.7.18).

The general linear combination of representers is now of the form

$$\sigma(\mathbf{x},\, t) = \mathbf{r}(\mathbf{x},\, t)^* \mathbf{b} + \int_{B_{OUT}} ds' \int_0^T dt'\, r(\mathbf{x},\, t,\, s',\, t')\, b(s',\, t')\,, \quad (8.9.13)$$

or in compact notation,

$$\sigma = \mathbf{r}^* \mathbf{b} + r \bullet b\,. \quad (8.9.14)$$

The coefficient vector \mathbf{b} and coefficient function $b = b(s',\, t')$ are again determined by solving an inhomogeneous linear system. The inhomogeneities are the differences between the interior data \mathbf{d} and the measured values of the prior estimate, as well as the differences between the outflow boundary data ω_{OUT} and the outflow boundary values of the prior estimate. Thus the interior and outflow data, both of which overspecify the original CFvN problem, are treated on equal terms by the representer expansion.

Exercise 8.9.1 Define an inner product $<,>$ and a class of stream-functions ψ such that

$$< \mathbf{r},\, \psi > = \mathcal{L}[\psi] \quad (8.9.15)$$

and

$$< r',\, \psi > = \nabla^2\, \psi(s',\, t')\,, \quad (8.9.16)$$

where \mathbf{r} is the representer vector satisfying (8.9.1)–(8.9.6), while $r' = r(\mathbf{x},\, t,\, s',\, t')$ is a member of the continuous set of representers satisfying (8.9.7)–(8.9.12) for points $(s',\, t')$ on B_{OUT}. □

Exercise 8.9.2 Derive the infinite linear system of equations for the representer coefficients \mathbf{b} and b in (8.9.13). Verify that the linear operator is symmetric and positive-definite. □

The formal constructions outlined in §8.7 and §8.8 indicate that the CFvN problem may be combined with interior data and vorticity data on the outflow boundary, using least-squares inverse methods. Further, the nonlinear Euler-Lagrange system may be solved iteratively, each iterate being solved in turn using suitably defined representers. Several difficulties remain. First, the inverse formulation

does not resolve the ill-posedness of the CFvN problem. Second, the representer adjoint variable for outflow vorticity would appear to be ill-defined at characteristic points: see (8.9.9). Third, a continuously infinite-dimensional linear system of equations must be solved in order to find the representer coefficients.

It would appear that the ill-posedness has no resolution which is both acceptable on rigorous mathematical grounds and physically consistent. In practice, some "fix" such as filtering must be found, along the lines described in §8.6. Choosing fixes of this kind is often a matter of personal taste. There is, however, no prospect of proving classical convergence of the iterative solutions of the nonlinear Euler-Lagrange system, once a digital filter is chosen.

The undefined nature of $\alpha(\mathbf{x}, t, s', t')$ when (s', t') is a characteristic point, and the issue of having to solve infinite systems of equations for the representer coefficients, may both be avoided after considering, in reality, the distribution of boundary data.

In any numerical approximation scheme the continuous set of boundary points is lumped into a finite set. Thus the system of linear equations for the representer coefficients is finite, in computational practice. Nevertheless, there may be many outflow grid points in a high-resolution model: as many as 200 should be expected in a 100×100 grid, for example. This would add greatly to the task of computing the representers, but almost certainly would contribute far less to the rank of the representer or measurement matrix, since the grid, by definition, resolves the field with high accuracy. Next, consider the availability of vorticity data. In practice, real vorticity data are very difficult to collect in the ocean and would only be available at a limited number of boundary points, if at all: see, for example, Freeland & McIntosh (1989). Values of boundary vorticity may be available from a basin-scale, coarse-resolution model within which the open-ocean model is nested; in that case the availability of the data would be restricted to those points common to the fine and coarse grids. In other words, the data would only be available in practice on a sparse subset of the boundary grid points in the open-ocean, fine-resolution model. The data could, of course, be interpolated to the finer grid, but that would be precisely the wrong strategy. Instead, those vorticity data available at the outflow boundary should be recognized for what they are and treated accordingly, which is to say, no differently from streamfunction or other data in the interior of the open-ocean region. The outflow vorticity data may

have been formally assigned to isolated points on the boundary, but
in reality the data are averages either over an array of current meter
moorings, or else over a cell in a coarse-resolution model. Thus, in
each case an outflow vorticity datum should be defined by a linear
measurement functional having a kernel supported by a patch of cells
in the high-resolution model. As a consequence the outflow boundary
value (8.9.9) for the adjoint representer variable would be a kernel of
finite diameter, rather than a "bare" Dirac delta-function. Notice that
not only has the continuously infinite set of representers defined by
(8.9.7)–(8.9.12) been replaced by a finite extension to the finite set of
representers satisfying (8.9.1)–(8.9.6), but the difficulty with the ill-
defined nature of $\alpha(\mathbf{x}, t, s', t')$ when (s', t') is locally characteristic
(see (8.9.9)) is no longer present. It is worth considering why the
inflow data should not be treated similarly, as they are also likely in
practice to be available only at isolated points. The argument against
such treatment is perhaps most clearly made once it is recognized that
the role of inflow data is the same as that of initial data, namely, to
provide information at the upstream ends of characteristics. Thus,
if a characteristic crosses the inflow boundary between two vorticity
data points, it would be necessary to estimate the vorticity at the
crossing point. Interpolation would be the obvious choice, that is to
say, a continuous distribution of inflow vorticity is required.

It must be conceded that the averaging kernel for an outflow
datum may well lie partly outside the open-ocean region as originally
defined. A time-dependent deformation of the outflow boundary,
engulfing the supports of outflow vorticity data kernels, is conceivable.
However, it is likely that such a boundary would, as it performs its
peristaltic motions, stimulate similar but unphysical disturbances in
the flow fields themselves. Heeding the advice of George C. Marshall
("Don't fight the problem"), it is suggested that kernels which overlap
the outflow boundary be redefined one-sidedly: see Fig. 8.9.1.

A reader who has followed Chapter 8 to this stage could be
excused for feeling that Marshall's admonition might well have been
heeded much sooner, perhaps at the end of §8.6, especially as the
ill-posedness of the original CFvN problem is not resolved by the
formulations and analysis of §8.7 *et seq.* The justification which is
offered for the analysis is that problems should be examined in as
much depth as possible before adopting compromise solutions. In any
event an inverse formulation of the CFvN problem is desirable so that
all available data may be fully exploited, and not only to prevent

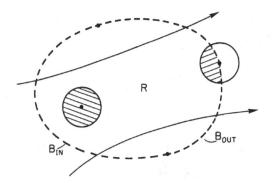

Figure 8.9.1. Hatched areas denote supports for vorticity data kernels. Parts which lie outside the region are discarded.

"fixes" from diverting the solution too far from reality. Moreover, the ill-posedness due to characteristic points is a relatively delicate concept compared to the gross ill-posedness of the open-ocean primitive-equation models considered in Chapter 9. A careful consideration of the former is a prudent if not strictly necessary step towards resolving the latter. Prudence also suggests that the methods discussed thus far should be tested with real data.

8.10 An atmospheric example

When testing an inverse method, it is essential that shortcomings of the method should not be confused with inadequacies of the data. Considering the difficulty of obtaining oceanic data which are high in quality and comprehensive in coverage, it is appropriate to take advantage of certain atmospheric data sets which possess both of these attributes. The original work of Charney, Fjörtoft & von Neumann (1950) involved the limited-area integration of a plane hydrodynamic model of the atmosphere. Operational numerical weather prediction now routinely uses global, multilevel primitive-equation models of great complexity, but it has been found that the CFvN model is as effective as more complex models at predicting the steering flows for tropical cyclones (Holland *et al.*, 1990). Some very extensive real data sets are available for the simpler model. Consequently an inverse formulation of the atmospheric limited-area problem will now be described in some detail, and preliminary results will be shown.

Figure 8.10.1. Domain for the TCM–90 forecasting experiments, shown in a Mercator projection. The 83 × 83 rectangular computational grid (not shown) has a resolution of approximately 93 km.

The region of interest is shown in Fig. 8.10.1. It was the subject of an intensive observational program during August and September 1990, known as the Tropical Cyclone Motion or TCM–90 Experiment (Elsberry, 1990). The flow field of interest here is the geostrophic velocity **u** at a pressure of 500 mbar (about 5.5 km altitude). The vertical velocity w has its extremum near this pressure surface, so the horizontal velocity is approximately solenoidal and hence has a streamfunction ψ. The reduced-gravity, quasi-geostrophic potential vorticity balance is

$$\zeta_t + m\left(u\zeta_x + v\zeta_y\right) = Z + z, \qquad (8.10.1)$$

where x and y are Mercator coordinates. The map factor m is a function of latitude y alone (Haltiner & Williams, 1980). The quasi-geostrophic potential vorticity ζ is related to the streamfunction by

$$m^2\left(\psi_{xx} + \psi_{yy}\right) - \rho^{-2}\,\psi + f(y) = \zeta. \qquad (8.10.2)$$

In (8.10.2), ρ is the internal deformation radius. The coefficient ρ^{-2} is also known as the Cressman correction (Cressman, 1958). The

Figure 8.10.2. Time line for generalized inversion of the tropical cyclone model. Initial data are available at $t = -24$ in the form of TCM–90 analyses. Boundary data are available for $-24 \leq t \leq 0$, also from TCM–90 analyses. For $0 \leq t \leq 48$, the boundary data are provided by a global forecast of relatively coarse resolution. The inverse takes advantage of additional TCM–90 data, at $t = -12$ hours and at $t = 0$ hours (as indicated by arrows). The dates refer to TC "Abe."

kinematic relationship between the velocity and streamfunction is

$$\mathbf{u} = (u, v) = m \left(-\psi_y, \, \psi_x\right). \qquad (8.10.3)$$

In (8.10.1), Z is a prior estimate of sources of potential vorticity, while z is a residual or error in the prior estimate.

The basic data consist of initial values for the streamfunction:

$$\psi (\mathbf{x}, \, 0) = \psi_I (\mathbf{x}) + \phi(\mathbf{x}), \qquad (8.10.4)$$

together with boundary values for the streamfunction:

$$\psi (s, \, t) = \psi_B (s, \, t) + \tau (s, \, t), \qquad (8.10.5)$$

and inflow boundary values for the vorticity:

$$\zeta (s, \, t) = \zeta_{IN} (s, \, t) + \theta (s, \, t), \qquad (8.10.6)$$

where

$$\mathbf{u} \cdot \mathbf{n} < 0. \qquad (8.10.7)$$

In (8.10.4)–(8.10.6) ϕ, τ and θ are unknown residuals.

The objective in this atmospheric example differs from the illustration of §7.6. The latter is a *hindcast*, that is, data in the interval $0 \leq t \leq T$ are used to improve the model solution within that interval. Here the example is a *forecast*: data in the interval -24 hours $\leq t \leq 0$ hours are used to improve the model solution in the subsequent interval $0 \leq t \leq +48$ hours. See Fig. 8.10.2.

Thus, the time interval of interest is -24 hours $\leq t \leq +48$ hours. In addition to the basic data set (8.10.4)–(8.10.6), data are available at $t = -T_l (l = 0, 1)$, where $T_k = k\Delta T$. The subinterval ΔT has the value 12 hours, that is, data are available at 0 and -12 hours. These data are all principal components for the streamfunction. That is, the measurement functionals are of the form

$$\mathcal{L}[\psi] = \int da \, \psi(\mathbf{x}, t) \, h(\mathbf{x}), \qquad (8.10.8)$$

where $h(\mathbf{x})$ is an eigenfunction of the spatial covariance of ψ. In practice ψ is defined on a finite-difference grid; the integral in (8.10.8) is approximated by a sum, and h is an eigenvector of the covariance matrix for ψ. Suppose that P_1 principal components are retained; then in all there are $2P_1$ data of this kind. Further data are available in the form of P_2 locally averaged values of vorticity at outflow. Then the combined number of measurements is $P = 2P_1 + P_2$. A single index $p = p(l, p_1, p_2)$ may be defined: for the principal components,

$$1 \leq p = lP_1 + p_1 \leq 2P_1, \qquad (8.10.9),$$
$$(0 \leq l \leq 1 \leq p_1 \leq P_1),$$

while for the locally averaged boundary vorticity,

$$2P_1 + 1 \leq p = 2P_1 + p_2 \leq 2P_1 + P_2 = P, \qquad (8.10.10)$$
$$(1 \leq p_2 \leq P_2).$$

The P linear measurement functionals \mathcal{L}_m form the components of a vector functional \mathcal{L}; the P data values are then given by

$$\mathbf{d} = \mathcal{L}[\psi] + \boldsymbol{\epsilon}, \qquad (8.10.11)$$

where $\boldsymbol{\epsilon}$ is the vector of prior measurement errors.

The various residuals are listed below:

$z(\mathbf{x}, t)$ - model residual: see (8.10.1)

$\phi(\mathbf{x})$ - initial streamfunction residual: see (8.10.4)

$\tau(s, t)$ - boundary streamfunction residual: see (8.10.5)

$\theta(s, t)$ - inflow vorticity residual: see (8.10.6)

ϵ - data residual: see (8.10.11)

A quadratic penalty functional may be defined by

$$J[\psi] = z \bullet W_{zz} \bullet z + \phi \circ W_{\phi\phi} \circ \phi$$
$$+ \tau \bullet W_{\tau\tau} \bullet \tau + \theta \bullet W_{\theta\theta} \bullet \theta + \epsilon^* \, \mathbf{w} \, \epsilon \, . \quad (8.10.12)$$

For example, the first term is

$$z \bullet W_{zz} \bullet z = \int_{-24}^{48} dt_1 \int_{-24}^{48} dt_2 \int_R da_1 \int_R da_2 \, z(\mathbf{x}_1, t_1)$$
$$\times \, W_{zz}(\mathbf{x}_1, t_1, \mathbf{x}_2, t_2) \, z(\mathbf{x}_2, t_2) \, , \quad (8.10.13)$$

while the third is

$$\tau \bullet W_{\tau\tau} \bullet \tau = \int_{-24}^{48} dt_1 \int_{-24}^{48} dt_2 \int_B ds_1 \int_B ds_2 \, \tau(s_1, t_1) \, W_{\tau\tau}(s_1, t_1, s_2, t_2) \, \tau(s_2, t_2) \, .$$
$$(8.10.14)$$

In (8.10.8) and (8.10.13), the area element has the "true" area on the sphere: $da = m^{-2} \, dxdy$, while the arc length ds has the "true" length on the sphere: $ds = m^{-1}\sqrt{dx^2 + dy^2}$. The "true" streamfunction gradient is $\nabla\psi = m \, (\psi_x, \psi_y)$, while the "true" divergence is $\nabla \cdot \mathbf{u} = m^2 \, \{(u/m)_x + (v/m)_y\}$, and the "true" vorticity is $\mathbf{e}_3 \cdot \nabla \times \mathbf{u} = m^2 \, \{(v/m)_x - (u/m)_y\}$. The weights in (8.10.12) are all symmetric and positive-definite. The residuals are related to the streamfunction ψ via their defining equations (8.10.1)–(8.10.11).

The calculus of variations yields the following Euler-Lagrange system of equations for extrema of J:

$$E\lambda = -\nabla \left(\mathbf{e}_3 \times \nu\nabla\hat\zeta \right) + \mathcal{L}\left[\delta\delta\right]^* \mathbf{w} \left(\mathbf{d} - \mathcal{L}[\hat\psi] \right) , \quad (8.10.15)$$
$$\text{in } R \text{ for } -24 \le t \le 48 \, ,$$

subject to

$$\lambda = 0 \, , \quad (8.10.16)$$
$$\text{on } B \text{ for } -24 \le t \le 48 \, ,$$

where E is the elliptic operator

$$E = \nabla \cdot \nabla - \rho^{-2} \, ; \qquad (8.10.17)$$

$$-\nu_t - \nabla \cdot (\hat{\mathbf{u}} \nu) = \lambda \, , \qquad (8.10.18)$$

$$\text{in } R \text{ for } -24 \leq t \leq 48 \, ,$$

subject to

$$\nu = 0 \, , \qquad (8.10.19)$$

$$\text{in } R \text{ at } t = 48$$

and

$$\nu \mathbf{n} \cdot \hat{\mathbf{u}} = 0 \, , \qquad (8.10.20)$$

$$\text{on } B_{OUT} \text{ for } -24 \leq t \leq 48 \, ;$$

$$\hat{\zeta}_t + \hat{\mathbf{u}} \cdot \nabla \hat{\zeta} = Z + Q_{zz} \bullet \nu \, , \qquad (8.10.21)$$

$$\text{in } R \text{ for } -24 \leq t \leq 48 \, ,$$

subject to

$$\hat{\zeta} = \zeta_I + Q_{\phi\phi} \circ \nu \, , \qquad (8.10.22)$$

$$\text{in } R \text{ at } t = -24 \, ,$$

and

$$\hat{\zeta} = \zeta_{IN} - Q_{\theta\theta} \bullet (\nu \mathbf{n} \cdot \hat{\mathbf{u}}) \, , \qquad (8.10.23)$$

$$\text{on } B_{IN} \text{ for } -24 \leq t \leq 48 \, ;$$

$$E\hat{\psi} = \hat{\zeta} - f(y) \, , \qquad (8.10.24)$$

$$\text{in } R \text{ for } -24 \leq t \leq 48 \, ,$$

subject to

$$\hat{\psi} = \psi_B + Q_{\tau\tau} \bullet \{ \mathbf{n} \cdot \nabla \lambda + \mathbf{n} \cdot \mathbf{e}_3 \times \nu \nabla \hat{\zeta} \} \, , \qquad (8.10.25)$$

$$\text{on } B \text{ for } -24 \leq t \leq 48 \, ,$$

and finally

$$\hat{\mathbf{u}} = \mathbf{e}_3 \times \nabla \hat{\psi} \, , \qquad (8.10.26)$$

$$\text{in } R \text{ for } -24 \leq t \leq 48 \, .$$

In (8.10.24), f is the Coriolis parameter. In (8.10.21), for example, Q_{zz} is the functional inverse of the weight W_{zz}:

$$\int_{-24}^{48} dt_2 \int_R da_2 \, Q_{zz}(\mathbf{x}_1, t_1, \mathbf{x}_2, t_2) \, W(\mathbf{x}_2, t_2, \mathbf{x}_3, t_3)$$

$$= \delta(\mathbf{x}_1 - \mathbf{x}_3) \, \delta(t_1 - t_3), \qquad\qquad (8.10.27)$$

while in (8.10.22) $Q_{\phi\phi}$ is defined by

$$\int_R da_2 \, Q_{\phi\phi}(\mathbf{x}_1, \mathbf{x}_2) \, W_{\phi\phi}(\mathbf{x}_2, \mathbf{x}_3) = \delta(\mathbf{x}_1 - \mathbf{x}_3). \qquad (8.10.28)$$

As before, the penalty functional \mathcal{J} is defined in terms of weights, but with the apparent exception of the data weight \mathbf{w} only the inverse weights need to be prescribed in order to find the extrema of \mathcal{J}. Note that $m_{yy} = O(R_e^{-1}) \ll \rho^{-2}$, where R_e is the earth's radius. Consequently, E is a negative-definite operator and may be inverted, subject to the Dirichlet boundary conditions (8.10.16) or (8.10.25).

Exercise 8.10.1 Derive the Euler-Lagrange system (8.10.15) through (8.10.26) for extrema of the penalty functional \mathcal{J} defined by (8.10.12). □

Exercise 8.10.2 Devise an iterative scheme which replaces the nonlinear Euler-Lagrange system (8.10.15)–(8.10.26) with a sequence of linear systems, each of which is the Euler-Lagrange system for a linear inverse problem. □

Exercise 8.10.3 Show that the solution of each linear system in Exercise 8.10.2 may be expressed as a sum of a prior estimate and a finite linear combination of representers. In particular define an inner product and prove that the representers do in fact represent the measurement functionals in the sense of the inner product. □

Exercise 8.10.4 Reformulate the inverse problem in the preceding section, admitting residuals in the diagnostic relations (8.10.2) and (8.10.3). □

A tropical cyclone forecast at $t = 48$, provided by a generalized inversion in the interval $-24 \le t \le 48$, is shown in Fig. 8.10.3, together with some comparison fields. The forecast is for Tropical Cyclone "Abe," at 1100 GMT on August 31, 1990. Shown in Fig. 8.10.3a is

Figure 8.10.3. Streamfunction fields for Tropical Cyclone "Abe" valid at 1100 GMT on August 31, 1990 (corresponds to $t = 48$ hours in Fig.8.10.2. (a) Standard forecast, using the TCM-90 vorticity analysis at $t = 0$ hours as the initial condition. (b) Verifying analysis, that is, TCM-90 streamfunction analysis at $t = 48$ hours. (c) Inverse streamfunction estimate $\hat{\psi}$ at $t = 48$ hours. It is significantly superior to the standard forecast of Abe, in location and intensity. Solid lines: positive values, broken lines: negative values; contour interval: 0.48×10^7 m^2 s^{-1}.

the standard streamfunction forecast at $t = 48$, using the TCM–90 vorticity analysis at $t = 0$ (1100 GMT on August 29) as the initial condition. The verifying streamfunction analysis is shown in Fig. 8.10.3b. This is the TCM–90 streamfunction analysis valid at 1100 GMT on August 31. The inverse streamfunction estimate $\hat{\psi}(\mathbf{x}, 48)$ is shown in Fig. 8.10.3c. It is significantly superior to the standard forecast, in terms of Abe's location and intensity, owing to the additional streamfunction data included at $t = -12$ and at $t = 0$ (2300 GMT on August 28 and 1100 GMT on August 29, respectively). For further details, see Bennett et al. (1992).

8.11 Transient tracers

The inverse problem for steady tracers discussed in §1.8 involved estimating steady velocities and diffusivities from measurements of steady tracer concentrations. Now consider the situation where the velocities and diffusivities are known (they may be unsteady), and tracer data have been collected in recent times, but estimates of previous tracer concentrations are required. There may be some previous data for the tracers; if, for example, the tracer is bomb tritium ^3He, then it may be safely assumed that the concentration was negligible prior to 1950. Previous tracer data may also be available at open boundaries of the oceanic region under consideration. The estimation problem for the previous tracer distribution closely resembles the problems already encountered in the preceding sections; the similarities will now be exploited.

To make the hindcast data requirements more specific, suppose that the tracer concentration $C = C(\mathbf{x}, t)$ obeys the time-dependent advection-diffusion equation

$$C_t + \mathbf{u} \cdot \nabla C = G + \nabla \cdot (\mathbf{K}_H \nabla C), \qquad (8.11.1)$$

where $\mathbf{u} = \mathbf{u}(\mathbf{x}, t)$ is the known velocity field, and \mathbf{K}_H is a known diffusivity tensor, while $G = G(\mathbf{x}, t)$ is a prior estimate of the tracer sources, if any. According to the convention used here, the notation indicates a two-dimensional problem, but extension to three dimensions is straightforward. The diffusion equation (8.11.1) has a unique solution in a region R for $0 \leq t \leq T$ for a given source strength G, provided an initial condition is given in R:

$$C(\mathbf{x}, 0) = C_I(\mathbf{x}), \qquad (8.11.2)$$

and a boundary condition is given for $0 \leq t \leq T$, such as

$$C(\mathbf{x}, t) = C_B \Big(S(\mathbf{x}), t \Big) \qquad (8.11.3)$$

for each point \mathbf{x} on the boundary B. The path parameter s on B has the value $s = S(\mathbf{x})$ at the point \mathbf{x}. The data set, which is always finite in practice, may be expressed in the form

$$\mathbf{d} = \mathcal{L}\,[C] + \boldsymbol{\epsilon}, \qquad (8.11.4)$$

where \mathbf{d} is the finite-dimensional data vector, \mathcal{L} is the vector of linear measurement functionals and $\boldsymbol{\epsilon}$ is the vector of unknown measurement errors.

Let C_G denote the solution of (8.11.1)–(8.11.3). In general, the prior estimates G, C_I and C_B are poor, hence the prior misfit $\mathbf{d} - \mathcal{L}\,[C_G]$ will greatly exceed prior estimates of the measurement error $\boldsymbol{\epsilon}$. It would be concluded that C_G is a poor estimate of the true concentration C. Consequently, the prior errors $g = g\,(\mathbf{x},\,t)$, $\phi = \phi(\mathbf{x})$ and $\theta = \theta(s, t)$ should be admitted in, respectively, (8.11.1), (8.11.2) and (8.11.3). The errors may be estimated using weighted least-squares, that is, by minimizing a quadratic penalty functional $\mathcal{J} = \mathcal{J}[C]$ closely resembling (8.7.8).

Exercise 8.11.1 Derive the Euler-Lagrange equations for the inverse estimates \hat{g}, $\hat{\phi}$, $\hat{\theta}$, $\hat{\boldsymbol{\epsilon}}$ and \hat{C}. Explain why these yield the global minimum of \mathcal{J}. □

Exercise 8.11.2 Express the inverse estimate \hat{C} in the preceding exercise as the sum of the prior estimate C_G and a linear combination of representer fields. Derive advection-diffusion equations for the representers and their adjoints. Show that the necessary integrations of these equations are well posed. □

Exercise 8.11.3 Derive posterior error covariances for the inverse estimates in the preceding exercises. □

Exercise 8.11.4 Under what circumstances might the preceding inverse calculation be ill-conditioned, and at which stage in the calculation would this ill-conditioning appear? □

For a highly simplified illustration of the preceding analysis, see Wunsch (1988a). A one-dimensional Kalman filter was developed by Budgell (1982) for transient tracers in an open channel.

9

PRIMITIVE-EQUATION MODELS

9.1 Introduction

The quasi-geostrophic circulation models considered in Chapters 3–8 contain mechanisms of fundamental importance, such as barotropic instability, baroclinic instability (Pedlosky, 1987) and frontogenesis (Stone, 1966). Nevertheless, only the more complex primitive-equation models (Lorenz, 1967) can reasonably be expected to have a close resemblance to major ocean current systems. The many assumptions underlying the simpler models do not hold in the real ocean. Consider, for example, the Antarctic Circumpolar Current system (ACC). A meridional section of the density field is shown in Fig. 9.1.1 (Patterson & Whitworth, 1990). Zonal geostrophic velocities of several tens of centimeters per second may be inferred, implying a Rossby number as small as 10^{-2}. However, the bathymetry has changes of order unity in the meridional direction alone, as do the depths of the isopycnal surfaces. Moreover, in some of the eddies and rings associated with the ACC, the Rossby number may approach a value of 10^{-1} (Bryden, 1983). Thus for varying reasons it may be concluded that the horizontal field of velocity in the ACC is significantly divergent, and therefore poorly represented in the quasi-geostrophic approximation. Similar remarks may be made about the Gulf Stream (Stommel, 1960) and about the Kuroshio (Stommel & Yoshida, 1972). Equatorial ocean dynamics are fundamentally non-geostrophic, even on seasonal or even interannual time scales, owing to the significance of Kelvin waves (Philander, 1990).

Figure 9.1.1. Distribution of the density parameter $\sigma = (\rho - 1) \times 10^3$, where ρ is the potential density, along 135°W. Units are milligrams per cubic centimeter (after Patterson & Whitworth, 1990).

Another major difference between quasi-geostrophic (QG) and primitive-equation (PE) models is the active nature of the thermo-dynamics in the latter. That is, a mean static profile of density is not imposed upon the system. Rather, the total stratification evolves in accordance with the conservation laws. This provides essential information for ocean acoustics at the relatively small scale, and for ocean climate at the large scale. The first numerical integration of a PE model, by Bryan & Cox (1968a,b) was, accordingly, a major achievement.

Several integrations of world ocean circulation models have been performed recently (for example, Semtner & Chervin, 1988), but most integrations have been regional for reasons of computational economy or lack of extensive data coverage (for example, Hurlburt & Thompson, 1980; Spall & Holland, 1991). The great complexity of these PE models, together with their sensitive dependence upon the many forcing fields, boundary data and initial data, makes the case *a fortiori* for combining the models with all available observations of the circulation, in order to achieve a more realistic solution. Thus far, observations have been combined with regional PE models using the simple sequential techniques for data assimilation described in §2.2, §3.2 and §4.3. With the important exception of coarse-resolution, large-scale models of ocean circulation, there appears to be little prospect of applying the optimal sequential techniques described in §3.3, owing to the huge proportions of the error covariance matrices. In any event the "forward problem" for open-ocean PE models is, as will be seen, ill-posed owing to difficulties with boundary conditions. An inverse formulation resolves the ill-posedness. The need for an inverse formulation goes beyond merely dealing with problematical boundary conditions, but the issue of ill-posedness provides the focus for this chapter.

9.2 Boundary conditions: shallow-water theory

Both the number and nature of boundary conditions needed for an open-ocean QG model were determined by Charney, Fjörtoft & von Neumann (1950). Vorticity values must be provided on the inflow boundary in order to integrate the vorticity equation, which is of first-order hyperbolic type if the advecting velocity field is known (Courant & Hilbert, 1962). Given the interior distribution of vorticity, the interior streamfunction is determined by solving an elliptic equation

subject to Dirichlet boundary conditions. That is, streamfunction values are required on the entire boundary at all times. These values also serve to identify the inflow boundary: see §8.2. The CFvN problem is ill-posed whenever there are isolated points on the boundary where the flow is tangential to the boundary. Even though it causes serious problems, this is a relatively subtle difficulty compared to that found in open-ocean PE models, where there is a difficulty even in choosing the *number* of boundary conditions.

The discussion in this section and subsequent sections is based on arguments due to Oliger & Sundström (1978). These in turn are based on the analysis in Courant & Hilbert (1962), which is expressed in a powerful matrix notation bearing no resemblance to the standard formulation of PE models. The concepts of hyperbolicity and symmetric hyperbolicity are developed by Courant & Hilbert at length; eventually the development is motivated by investigations of well-posedness, including existence of smooth solutions. Oliger & Sundström (1978) move back and forth between abstract and conventional notation, with a tacit assumption that the reader needs only to be reminded of the classical analysis. The presentation here will be expressed entirely in conventional notation, and will not assume familiarity with the classical analysis.

The discussion begins with the linear shallow-water equations (Pedlosky, 1987, Chapter 3):

$$\mathbf{u}_t = -g\nabla h + \mathbf{F}\,, \qquad (9.2.1)$$

and

$$h_t + H\nabla \cdot \mathbf{u} = 0\,, \qquad (9.2.2)$$

where $\mathbf{u} = \mathbf{u}(\mathbf{x}, t)$ is the planar velocity field, $h = h(\mathbf{x}, t)$ is the elevation of the free surface, g is the gravitational acceleration, H is the mean water depth and \mathbf{F} is the force per unit mass. The issue is: what is the smallest set of initial and boundary conditions which ensures a unique solution for (9.2.1) and (9.2.2)? Since the dynamics are linear, and since only linear initial and boundary conditions will be considered, it suffices to determine the smallest set of homogeneous initial and boundary conditions which ensure that there is only a trivial solution for the homogeneous forms of (9.2.1) and (9.2.2). Suppose, therefore, that $\mathbf{F} \equiv 0$. The following energy integral is

readily derived:

$$K(T) - K(0) = -gH \int_0^T dt \int_B ds\, \mathbf{u} \cdot \mathbf{n}h, \qquad (9.2.3)$$

where $K = K(t)$ is the total mechanical energy:

$$K(t) = \tfrac{1}{2} \int_R d\mathbf{x}\big(H|\mathbf{u}|^2 + gh^2\big), \qquad (9.2.4)$$

while R is the open-ocean region with boundary B, and \mathbf{n} is the outward normal on B. It is clearly the case that if

$$\mathbf{u} = 0, \quad h = 0, \qquad \text{at } t = 0, \qquad (9.2.5)$$

and, if either

$$h = 0, \qquad\qquad \text{on } B, \qquad (9.2.6)$$

or

$$\mathbf{u} \cdot \mathbf{n} = \gamma\,(g/H)^{1/2}h, \qquad \text{on } B, \qquad (9.2.7)$$

where $\gamma \geq 0$, then

$$K(T) \leq 0. \qquad (9.2.8)$$

Since K is a positive-definite quadratic functional in \mathbf{u} and h, it follows from (9.2.8) that

$$\mathbf{u}(\mathbf{x}, T) \equiv 0, \qquad h(\mathbf{x}, T) \equiv 0 \qquad (9.2.9)$$

for any $T \geq 0$. That is, the homogeneous problem has only the trivial solution, and so the solution of the inhomogeneous problem is unique. As might be expected, the argument requires that the state of the flow must be specified initially: see (9.2.5). In addition, one condition must be specified on the entire boundary at all times. The choice is (9.2.6) or (9.2.7); different choices may be made at different places or at different times, at the risk of loss of continuity of the solution. In particular, (9.2.7) is the rigid boundary condition when $\gamma = 0$. If $\gamma > 0$ then $K(T) < K(0)$: the region is losing energy as the interior fluid does work against the hydrostatic pressure field

at the boundary. The alternative boundary conditions (9.2.6) and (9.2.7) are homogeneous, but the uniqueness theorem holds even if the chosen condition is inhomogeneous: the difference between any two solutions having the same boundary inhomogeneity satisfies the homogeneous form of the condition.

Exercise 9.2.1 Express the shallow-water equations in the matrix form

$$\mathbf{A}^{(t)}\mathbf{q}_t + \mathbf{A}^{(x)}\mathbf{q}_x + \mathbf{A}^{(y)}\mathbf{q}_y = \mathbf{G}\,, \qquad (9.2.10)$$

where $\mathbf{q}^* = (\mathbf{u}^*, h)$ and $\mathbf{G}^* = (\mathbf{F}^*, 0)$. Subscripts denote partial derivatives. Construct three singular matrices by taking linear combinations of the coefficient matrices. That is, establish that the system (9.2.1)–(9.2.2) is *hyperbolic* (Courant & Hilbert, 1962). □

Exercise 9.2.2 Verify that all of the coefficient matrices in (9.2.10) are symmetric, or else may be simultaneously symmetrized by a non-singular linear transformation (rescaling) of the state vector \mathbf{q}. Also, show that at least one linear combination of the coefficient matrices is definite. That is, establish that (9.2.1)–(9.2.2) is *symmetric-hyperbolic*, and has a *time-like direction* (Courant & Hilbert, 1962).
 □

Exercise 9.2.3 Prove that a symmetric-hyperbolic system, such as the symmetric form of (9.2.1)–(9.2.2), must be hyperbolic. □

Exercise 9.2.4 Derive the energy integral (9.2.3) in matrix notation. Hint: assume that (9.2.10) is in symmetric form, and left-multiply (9.2.10) with the state vector \mathbf{q}. □

Consider next the linear shallow-water equations, with advection by a uniform mean flow \mathbf{U}:

$$\mathbf{u}_t + \mathbf{U} \cdot \nabla \mathbf{u} = -g\, \nabla h + \mathbf{F}\,, \qquad (9.2.11)$$

and

$$h_t + \mathbf{U} \cdot \nabla h + H \nabla \cdot \mathbf{u} = 0\,. \qquad (9.2.12)$$

This linearization is consistent in bounded regions if the rigid parts of the boundary are tangential to \mathbf{U}, or conversely if all parts to which \mathbf{U} is incident are open. The energy integral for unforced flow ($\mathbf{F} \equiv \mathbf{0}$)

is

$$K(T) - K(0) = - \int\limits_0^T dt \int\limits_B ds \left\{ \tfrac{1}{2} \mathbf{U} \cdot \mathbf{n} (H|\mathbf{u}|^2 + gh^2) + gH\mathbf{u} \cdot \mathbf{n}h \right\} ,$$

(9.2.13)

where K is again given by (9.2.4). Uniqueness of solutions to (9.2.11) and (9.2.12) follows as before, once boundary conditions are chosen such that the right-hand side (rhs) of (9.2.13) is negative. That is, linear combinations of the normal velocity $\mathbf{u} \cdot \mathbf{n}$, the tangential velocity $\mathbf{u} \cdot \mathbf{s}$ and the scaled interface height $(g/H)^{1/2}h$ are sought such that the restriction of the quadratic form

$$\mathbf{Q} = \begin{pmatrix} \mathbf{U} \cdot \mathbf{n} & 0 & c \\ 0 & \mathbf{U} \cdot \mathbf{n} & 0 \\ c & 0 & \mathbf{U} \cdot \mathbf{n} \end{pmatrix}$$

(9.2.14)

to the corresponding subspace of \Re^3 is positive. Then the vector $\left(\mathbf{u} \cdot \mathbf{n}, \mathbf{u} \cdot \mathbf{s}, (g/H)^{1/2}h \right)$ has *null projections* in the directions defined by the eigenvectors associated with the negative eigenvalues of \mathbf{Q}. In (9.2.14), c is the phase speed $(gH)^{1/2}$. The eigenvalues of \mathbf{Q} are $\mathbf{U} \cdot \mathbf{n}, \mathbf{U} \cdot \mathbf{n} - c$ and $\mathbf{U} \cdot \mathbf{n} + c$: accordingly there are four cases to consider.

(i) $0 < \mathbf{U} \cdot \mathbf{n} < c$ (subcritical outflow). In this case \mathbf{Q} has only one negative eigenvalue: $\mathbf{U} \cdot \mathbf{n} - c$. The associated eigenvector is $(1, 0, -1)$, and so the one boundary condition

$$\mathbf{u} \cdot \mathbf{n} = (g/H)^{1/2}h$$

(9.2.15)

ensures that the rhs of (9.2.13) is negative. In fact, (9.2.13) is negative provided

$$\mathbf{u} \cdot \mathbf{n} = \gamma \, (g/H)^{1/2}h$$

(9.2.16)

and

$$\gamma + c/U > \left[(c/U)^2 - 1 \right]^{1/2} ,$$

(9.2.17)

where $U = \mathbf{U} \cdot \mathbf{n}$.

(ii) $0 < c < \mathbf{U} \cdot \mathbf{n}$ (supercritical outflow). In this case \mathbf{Q} has no negative eigenvalues, and so no boundary conditions are needed in order to ensure that the rhs of (9.2.13) is negative.

(iii) $-c < \mathbf{U} \cdot \mathbf{n} < 0$ (subcritical inflow). In this case \mathbf{Q} has two negative eigenvalues: $\mathbf{U} \cdot \mathbf{n}$ and $\mathbf{U} \cdot \mathbf{n} - c$, with associated eigenvectors $(0, 1, 0)$ and $(1, 0, -1)$, respectively. Consequently two boundary conditions are needed, to ensure that the rhs of (9.2.13) is negative:

$$\mathbf{u} \cdot \mathbf{s} = 0 \qquad (9.2.18)$$

and

$$\mathbf{u} \cdot \mathbf{n} = -(g/H)^{1/2} h\,. \qquad (9.2.19)$$

The latter may be replaced with

$$\mathbf{u} \cdot \mathbf{n} = -\gamma\, (g/H)^{1/2} h \qquad (9.2.20)$$

provided

$$|\gamma + (c/U)| < \left[(c/U)^2 - 1\right]^{1/2}\,, \qquad (9.2.21)$$

where $U = |\mathbf{U} \cdot \mathbf{n}|$.

(iv) $\mathbf{U} \cdot \mathbf{n} < -c$ (supercritical inflow). In this final case all three eigenvalues of \mathbf{Q} are negative, and so it is necessary to specify the three boundary conditions

$$\mathbf{u} \cdot \mathbf{s} = 0\,, \qquad (9.2.22)$$

$$\mathbf{u} \cdot \mathbf{n} = 0 \qquad (9.2.23)$$

and

$$h = 0\,, \qquad (9.2.24)$$

in order to ensure that the rhs of (9.2.13) is negative.

The four cases (i)–(iv) are summarized in Fig. 9.2.1.

Exercise 9.2.5 Repeat Exercises 9.2.1–9.2.4, with application to (9.2.11), (9.2.12). □

Exercise 9.2.6 Include the Coriolis acceleration $f\,\mathbf{e}_3 \times \mathbf{u}$ in (9.2.11). Verify that the preceding energy arguments and boundary conditions are unaltered. Identify the eigenvalues of \mathbf{Q} as a single mode which conserves potential vorticity, plus a pair of gravity waves. □

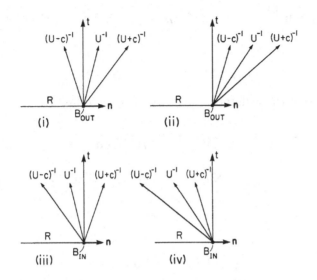

Figure 9.2.1. Four cases to be considered when choosing boundary conditions for the shallow-water equations, linearized about a mean advecting velocity **U**. The normal component of velocity at the boundary B is $U = \mathbf{U} \cdot \mathbf{n}$. The slopes of the rays are inversely proportional to the eigenvalues $U - c$, U and $U + c$ of the quadratic form **Q**: see (9.2.13) and (9.2.14). Case (i), subcritical outflow, $0 < U < c$; case (ii), supercritical outflow, $c < U$; case (iii), subcritical inflow, $-c < U < 0$; and case (iv), supercritical inflow, $U < -c$. The correct number of boundary conditions in each case is 1, 0, 2 and 3, respectively.

Oliger & Sundström (1978) point out that the preceding method for devising open boundary conditions, using a uniqueness theorem, may be extended to the nonlinear shallow-water equations:

$$\mathbf{u}_t + \mathbf{u} \cdot \boldsymbol{\nabla} \mathbf{u} = -g \boldsymbol{\nabla} h + \mathbf{F} \qquad (9.2.25)$$

and

$$h_t + \mathbf{u} \cdot \boldsymbol{\nabla} h + h \boldsymbol{\nabla} \cdot \mathbf{u} = 0, \qquad (9.2.26)$$

subject to the initial conditions

$$\mathbf{u}(\mathbf{x}, 0) = \mathbf{u}_I(\mathbf{x}), \qquad h(\mathbf{x}, 0) = h_I(\mathbf{x}). \qquad (9.2.27)$$

Indeed, let $\mathbf{u}^{(1)}$, $h^{(1)}$ and $\mathbf{u}^{(2)}$, $h^{(2)}$ be two solutions of (9.2.25)–(9.2.27). Then the difference fields $\boldsymbol{\mu} = \mathbf{u}^{(1)} - \mathbf{u}^{(2)}$ and $\chi = h^{(1)} - h^{(2)}$ satisfy

$$\boldsymbol{\mu}_t + \mathbf{u}^{(1)} \cdot \boldsymbol{\nabla} \boldsymbol{\mu} = -g \boldsymbol{\nabla} \chi - \boldsymbol{\mu} \cdot \boldsymbol{\nabla} \mathbf{u}^{(2)} \qquad (9.2.28)$$

and

$$\chi_t + \mathbf{u}^{(1)} \cdot \nabla\chi + h^{(1)}\nabla \cdot \boldsymbol{\mu} = -\boldsymbol{\mu} \cdot \nabla h^{(2)} - \chi\nabla \cdot \mathbf{u}^{(2)} \quad (9.2.29)$$

subject to

$$\boldsymbol{\mu}(\mathbf{x}, 0) = \mathbf{0}, \qquad \chi(\mathbf{x}, 0) = 0. \quad (9.2.30)$$

The energy of the difference fields evolves according to

$$\frac{d\mathcal{K}}{dt} + \mathcal{F} = \int_R \Big\{ \tfrac{1}{2} g\chi^2 \nabla \cdot \mathbf{u}^{(1)} - g\chi^2 \nabla \cdot \mathbf{u}^{(2)} + g\chi\,\boldsymbol{\mu} \cdot \nabla h^{(1)}$$

$$- g\chi\,\boldsymbol{\mu} \cdot \nabla h^{(2)} - h^{(1)}\boldsymbol{\mu} \cdot (\boldsymbol{\mu} \cdot \nabla)\mathbf{u}^{(2)} \Big\} \, d\mathbf{x}, \quad (9.2.31)$$

where

$$\mathcal{K} = \tfrac{1}{2} \int_R \Big(h^{(1)}|\boldsymbol{\mu}|^2 + g\chi^2 \Big) \, d\mathbf{x} \quad (9.2.32)$$

and

$$\mathcal{F} = \int_B \Big\{ \tfrac{1}{2}\mathbf{u}^{(1)} \cdot \mathbf{n}\big(h^{(1)}|\boldsymbol{\mu}|^2 + g\chi^2\big)$$

$$+ gh^{(1)}\boldsymbol{\mu} \cdot \mathbf{n}\chi \Big\} \, ds. \quad (9.2.33)$$

Hence

$$\frac{d\mathcal{K}}{dt} + \mathcal{F} \leq A\mathcal{K} \quad (9.2.34)$$

where

$$A = \left(\tfrac{11}{2}\right) \max \Big\{ |\nabla \cdot \mathbf{u}^{(i)}|, \; \big(g/h^{(i)}\big)^{1/2}|\nabla h^{(i)}| \Big\}. \quad (9.2.35)$$

The maximum in (9.2.35) is taken for $i = 1$ and 2, for all \mathbf{x} in R, and for $0 \leq t \leq T$. Integrating (9.2.34) yields

$$\mathcal{K}(T) - \mathcal{K}(0) \leq - \int_0^T e^{A(T-t)} \mathcal{F}(t) \, dt. \quad (9.2.36)$$

The cases (i)–(iv) may now be revisited, with $\mathbf{u}^{(1)}$ and $h^{(1)}$ replacing \mathbf{U} and H respectively. The nonlinear analysis is valid only if $h^{(1)}$ and $h^{(2)}$ remain strictly positive everywhere in R, and for $0 \leq t \leq T$.

Some insight into the behavior of the water depth may be gained from the continuity equation (9.2.26), which may be integrated to yield

$$h(\mathbf{x}, t) = h_I(\mathbf{x}) \exp\left\{ - \int_0^t \delta(\mathbf{X}(\mathbf{x}, t|r), r)\, dr \right\}, \qquad (9.2.37)$$

where $\delta(\mathbf{x}, t) = \nabla \cdot \mathbf{u}(\mathbf{x}, t)$ and $\mathbf{X}(\mathbf{x}, t|r)$ is the position, at time r, of the particle passing through \mathbf{x} at time t. That is, the water depth does not vanish, provided the integrated Lagrangian history of the flow divergence remains finite.

<u>Exercise 9.2.7</u> What may be inferred from (9.2.36) concerning the well-posedness of the shallow-water equations? □

Integrations of the nonlinear shallow-water equations on regions of the sphere (Robert & Yakimiw, 1986; Yakimiw & Robert, 1990) indicate that boundary conditions are major sources of error in short-range atmospheric prediction on synoptic scales. The errors in the region of interest can be satisfactorily reduced by expanding the region three-fold in each direction. Errors are then limited to the vast "buffer-region" so created. This extreme measure is found necessary even if the boundary data (all three variables: $\mathbf{u} \cdot \mathbf{n}$, $\mathbf{u} \cdot \mathbf{s}$ and h) are introduced as gently as possible using a nudging scheme (Davies, 1976). A massive expansion of the region of integration is not attractive in the ocean, where initial data are collected with great difficulty and expense.

9.3 Boundary conditions: stratified flow

Stable density stratification supports internal gravity waves, which have much slower phase speeds than those of external gravity waves. Classifying the flow as subcritical or supercritical, and so determining the correct number of open boundary conditions, becomes involved. The analysis of Oliger & Sundström (1978) uses depth as a coordinate, but, as they indicate, other choices lead to the same conclusions. Specific volume α will be used here. The equations of motion are, in homogeneous form,

$$\mathbf{u}_t + \mathbf{u} \cdot \nabla \mathbf{u} = -\nabla m, \qquad (9.3.1)$$

$$m_\alpha = p \qquad (9.3.2)$$

and

$$p_{\alpha t} + \nabla \cdot (\mathbf{u} p_\alpha) = 0, \tag{9.3.3}$$

where the gradient operation ∇ is defined on surfaces of constant α. In (9.3.1) and (9.3.2), m is the Montgomery potential:

$$m = \alpha p + gz \tag{9.3.4}$$

(see, for example, Bleck & Smith, 1990). This last relation yields the depth z, once m and p have been found as functions of \mathbf{x}, α and t. The free-surface condition expresses continuity of pressure:

$$p = p^{(a)}, \qquad \text{at } \alpha = \alpha^{(a)}, \tag{9.3.5}$$

where $p^{(a)}$ is the air pressure and the ocean-atmosphere interface is located at

$$\alpha = \alpha^{(a)}(\mathbf{x}, t). \tag{9.3.6}$$

The ocean bottom is located at

$$z = z^{(b)}(\mathbf{x}, t), \tag{9.3.7}$$

where

$$\alpha = \alpha^{(b)}(\mathbf{x}, t) \tag{9.3.8}$$

and

$$m^{(b)} = \alpha^{(b)} p^{(b)} + gz^{(b)}. \tag{9.3.9}$$

The isopycnal PE model (9.3.1)–(9.3.3) makes the hydrostatic approximation and assumes adiabatic motion with no change in internal energy (Pedlosky, 1987, Chapter 1), hence

$$\varpi \equiv \alpha_t + \mathbf{u} \cdot \nabla \alpha + w \alpha_z \equiv 0, \tag{9.3.10}$$

and so there is no diapycnal advection in either (9.3.1) or (9.3.3). The Boussinesq approximation is *not* made.

Linearizing the system about the uniform velocity \mathbf{U}, and about the static profiles $M = M(\alpha)$, $P = P(\alpha) = dM(\alpha)/d\alpha$ yields

$$\mathbf{u}_t + \mathbf{U} \cdot \nabla \mathbf{u} = -\nabla m, \tag{9.3.11}$$

$$m_\alpha = p \qquad (9.3.12)$$

and

$$p_{\alpha t} + \mathbf{U} \cdot \nabla p_\alpha + \frac{dP}{d\alpha} \nabla \cdot \mathbf{u} = 0, \qquad (9.3.13)$$

where \mathbf{u}, m and p are now additive perturbations.

<u>Exercise 9.3.1</u> Show that the energy budget for (9.3.11)–(9.3.13), with boundary conditions (9.3.5)–(9.3.9) at $\alpha = \alpha^{(a)}$, $\alpha^{(b)}$, respectively, in a region R with boundary B, is

$$\frac{d}{dt} \frac{1}{2} \int_{\alpha^{(b)}}^{\alpha^{(a)}} d\alpha \int_R d\mathbf{x} \left\{ -\frac{dP}{d\alpha} |\mathbf{u}|^2 + p^2 \right\}$$

$$= - \int_B ds \int_{\alpha^{(b)}}^{\alpha^{(a)}} d\alpha \left\{ \tfrac{1}{2} \mathbf{U} \cdot \mathbf{n} \left(-\frac{dP}{d\alpha} |\mathbf{u}|^2 + p^2 \right) \right.$$

$$\left. - \frac{dP}{d\alpha} \mathbf{u} \cdot \mathbf{n} \int_{\alpha^{(b)}}^{\alpha} p \, d\alpha' \right\}. \qquad (9.3.14)$$

Recall that P is a monotonically decreasing function of α. □

The budget (9.3.14) strongly resembles the time derivative of (9.2.13), where K is given by (9.2.4). However, the resemblance is incomplete, since the energy loss at the boundary B in (9.3.14) is not a quadratic functional in the circulation variables \mathbf{u} and p. It is therefore not obvious how to choose a minimal set of linear boundary conditions which ensures a loss of energy.

The situation at the boundary is revealed by separating variables according to

$$\mathbf{u}' (\mathbf{x}, \alpha, t) = \mathbf{u}' (\mathbf{x}, t) A(\alpha), \qquad (9.3.15)$$

$$m(\mathbf{x}, \alpha, t) = m' (\mathbf{x}, t) A(\alpha) \qquad (9.3.16)$$

and

$$p(\mathbf{x}, \alpha, t) = p' (\mathbf{x}, t) \frac{dA}{d\alpha} (\alpha), \qquad (9.3.17)$$

yielding

$$\mathbf{u}'_t + \mathbf{U} \cdot \nabla \mathbf{u}' = -\nabla m' \qquad (9.3.18)$$

and

$$m'_t + \mathbf{U} \cdot \nabla m' + c^2 \, \nabla \cdot \mathbf{u}' = 0 , \qquad (9.3.19)$$

where the separation constant c also appears in the equation for the "vertical" structure:

$$\frac{d^2 A}{d\alpha^2} - c^{-2} \frac{dP}{d\alpha} A = 0 . \qquad (9.3.20)$$

The perturbation boundary conditions for (9.3.20) are

$$\frac{dA}{d\alpha} = 0 , \qquad \text{at } \alpha = \alpha^{(a)} , \qquad (9.3.21)$$

and

$$A = \alpha \frac{dA}{d\alpha} , \qquad \text{at } \alpha = \alpha^{(b)} , \qquad (9.3.22)$$

where $\alpha^{(a)}$ and $\alpha^{(b)}$ are approximated by constants. That is, the "horizontal" structure is determined by the linearized shallow-water equations (9.3.18) and (9.3.19), with a phase speed c determined by the "vertical" Sturm-Liouville problem (9.3.20)–(9.3.22).

Some estimates for c may be obtained by simple manipulations of the Sturm-Liouville problem. First, consider the identity

$$\left(\alpha^{(b)}\right)^{-1} \left(A^{(b)}\right)^2 + \int_{\alpha^{(b)}}^{\alpha^{(a)}} \left\{ \left(\frac{dA}{d\alpha}\right)^2 + c^{-2} \frac{dP}{d\alpha} A^2 \right\} d\alpha = 0 . \qquad (9.3.23)$$

The external mode of A is approximately independent of α, and so (9.3.23) becomes

$$c_0^2 \approx \alpha^{(b)} \left(P^{(b)} - P^{(a)}\right) , \qquad (9.3.24)$$

or

$$c_0^2 \approx gH , \qquad (9.3.25)$$

where H is the mean depth. The internal modes of A are denumerable, they are orthogonal to $(dP/d\alpha)$ in $\alpha^{(b)} < \alpha < \alpha^{(a)}$, and they have

vertical scales $(\alpha^{(a)} - \alpha^{(b)})/j_n$, where $j_n = O(n)$ as $n \to \infty$, so their phase speeds are given by

$$c_n^2 \approx \alpha^{(b)} \left| \overline{\frac{dP}{d\alpha}} \right| (\alpha^{(a)} - \alpha^{(b)})/j_n^2. \qquad (9.3.26)$$

The overbar in (9.3.26) denotes a vertical mean. The ratios of the internal phase speeds c_n to the external speed c_0 are roughly given by

$$c_n^2 \approx c_0^2 (\Delta\alpha/\bar{\alpha})/j_n^2, \qquad (9.3.27)$$

where $\Delta\alpha = \alpha^{(a)} - \alpha^{(b)}$ and $\bar{\alpha} = \frac{1}{2}(\alpha^{(a)} + \alpha^{(b)})$. In particular, $c_n/c_0 = O(n^{-1})$ as $n \to \infty$. For the Southern Ocean density field shown in Fig. 9.1.1, the first few phase speeds south of the circumpolar front are: $c_0 \approx 221$ m s^{-1}, $c_1 \approx 0.97$ m s^{-1}, $c_2 \approx 0.54$ m s^{-1}, $c_3 \approx 0.30$ m s^{-1}, and $c_4 \approx 0.21$ m s^{-1}. Thus a flow speed of 0.6 m s^{-1} is subcritical with respect to the external mode and the gravest internal mode. Two conditions would be needed at inflow boundaries for each of these two modes, while one would be required for each at outflow. However, the same fluid speed (0.6 m s^{-1}) is supercritical with respect to all higher internal modes. Hence three inflow boundary conditions, but no outflow boundary conditions, would be required for each of the higher modes.

It may be concluded that there is no difficulty in choosing open boundary conditions for the linearized isopycnal model (9.3.11)–(9.3.13), provided the model is integrated by projecting onto the vertical modes. If it is integrated otherwise, say using finite differences in α (i.e., using layers of constant density), and if the same open boundary conditions are chosen for each layer, then some vertical modes must be either overspecified or underspecified. A linear, two-layer, one-dimensional model serves as an illustration. At most two boundary conditions are needed. The mean advecting velocity in each layer is $U = 0.2$ m s^{-1}. The external phase speed in the following calculation is $c_0 \cong 220$ m s^{-1} (in which case the mean flow is subcritical), while the single internal phase speed c_1 is either 0.5 m s^{-1} (subcritical mean flow) or else 0.1 m s^{-1} (supercritical mean flow). Initial and boundary values are provided by exact analytical solutions in the form of a simple *internal* wave. Numerical and analytical solutions for the upper-layer Montgomery potential $m_1(\mathbf{x}, t)$ are compared in Fig. 9.3.1. It is consistent to impose one boundary condition (9.2.16) at each end of the domain when $c_1 = 0.5$ m s^{-1} (the mean flow is

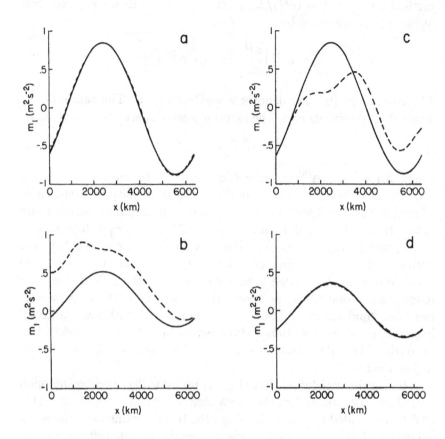

Figure 9.3.1. Numerical solutions (dashed lines) and exact analytical solutions (solid lines) for the upper-layer Montgomery potential $m_1(x, t)$ in a linear, two-layer, one-dimensional isopycnal model. Initial and boundary values for the numerical integrations are provided by the analytical solution, which is a simple *internal* wave. The mean advecting velocity U is 0.2 m s^{-1}. The external phase speed is $c_0 \cong 220$ m s^{-1}; the internal phase speed c_1 is either 0.5 m s^{-1} (panels (a) and (c)) or else is 0.1 m s^{-1} (panels (b) and (d)). The one boundary condition (9.2.16) is specified at each end of the domain in panels (a) and (b), while two ((9.2.18)–(9.2.19)) are specified upstream in panels (c) and (d).

subcritical with respect to both modes): see panel (a); it is inconsistent to do so if $c_1 = 0.1$ m s^{-1} (the mean flow is supercritical with respect to the internal mode): see panel (b). Specification of two upstream boundary conditions (9.2.18)–(9.2.19) is inconsistent for either value of c_1: see panels (c) and (d); however, a good result

is obtained for $c_1 = 0.1$ m s^{-1}, since the external mode is excited only by truncation error: see panel (d). Computational boundary conditions, in addition to the physical boundary conditions mentioned above, were needed in these calculations owing to the use of centered difference schemes in the horizontal direction.

Exercise 9.3.2 Derive an energy budget for the nonlinear PE model (9.3.1)–(9.3.3). □

Exercise 9.3.3 Derive the inequality

$$\int_{\alpha^{(b)}}^{\alpha^{(a)}} \left(m(\alpha) - m(\alpha^{(b)}) \right)^2 d\alpha \leq \left(\alpha^{(a)} - \alpha^{(b)} \right)^2 \int_{\alpha^{(b)}}^{\alpha^{(a)}} \left(\frac{\partial m}{\partial \alpha} \right)^2 d\alpha \quad (9.3.28)$$

for any function m such that the rhs of (9.3.28) is finite. Hint: use Fourier series. Recall also the triangle inequality:

$$\left(\int_{\alpha^{(b)}}^{\alpha^{(a)}} \mathbf{u} \cdot \mathbf{n} \left(m - m^{(b)} \right) d\alpha \right)^2 \leq \int_{\alpha^{(b)}}^{\alpha^{(a)}} (\mathbf{u} \cdot \mathbf{n})^2 d\alpha \int_{\alpha^{(b)}}^{\alpha^{(a)}} \left(m - m^{(b)} \right)^2 d\alpha .$$

$$(9.3.29)$$

Now consider (9.3.14). Allow linear boundary conditions at each level, such as $\mathbf{u} \cdot \mathbf{s} = 0$, $p = 0$ and $\mathbf{u} \cdot \mathbf{n} \propto p$. Apply the above inequalities to (9.3.14), leading to bounds on the energy indicating that the "shallow-water cases" (i)–(iv) in §9.2 hold here also, implying that at least the linear stratified open-ocean PE model is well posed. Explain why this conclusion contradicts Oliger & Sundström (1978), and explain why this argument is false. □

Exercise 9.3.4 Consider the following PE model, which includes changes in internal energy subject to the shallow Boussinesq approximation (Gill, 1982), and which has a vertical coordinate proportional to the logarithm of the pressure (Eliassen, 1949):

$$\mathbf{u}_t + (\mathbf{u} \cdot \nabla)\mathbf{u} + w\mathbf{u}_z = -\nabla \phi, \quad (9.3.30)$$

$$g \frac{\theta}{\theta_0} = \phi_z, \quad (9.3.31)$$

$$\theta_t + \mathbf{u} \cdot \nabla \theta + w\,\theta_z = 0\,, \tag{9.3.32}$$

and

$$\nabla \cdot \mathbf{u} + w_z = 0\,. \tag{9.3.33}$$

The "depth" is given by $z = -H\,ln(p/p_0)$, while $w = (dz/dt)$ is the "vertical velocity"; ϕ is the geopotential, and θ is potential temperature. The reference temperature θ_0 and pressure p_0 are constants. A linearized form of this PE model is

$$\mathbf{u}_t + \mathbf{U} \cdot \nabla \mathbf{u} = -\nabla \phi\,, \tag{9.3.34}$$

$$g\frac{\theta}{\theta_0} = \phi_z\,, \tag{9.3.35}$$

$$\theta_t + \mathbf{U} \cdot \nabla \theta + w\,\frac{d\Theta}{dz} = 0\,, \tag{9.3.36}$$

and

$$\nabla \cdot \mathbf{u} + w_z = 0\,, \tag{9.3.37}$$

where \mathbf{U} is a uniform flow and $\Theta = \Theta(z)$ is the stable, mean profile of potential temperature.

Express the linearized PE model (9.3.34)–(9.3.37) in matrix form. Verify that the coefficient matrices are symmetric. Show that the system is hyperbolic. That is, show that there are five not-necessarily-distinct, singular linear combinations of the coefficient matrices (Norbury & Cullen, 1985). On the other hand show that the system is not symmetric-hyperbolic, that is, no linear combination of the coefficient matrices is definite. Discuss the implications (or lack thereof) for the uniqueness of solutions of the system, for the continuous dependence of solutions upon initial and boundary data, and for the existence of solutions (see Courant & Hilbert, 1962, Chapter VI, §10; Reid, 1976, p. 116). Would the structure of the system (9.3.34)–(9.3.37) be significantly altered if it included advection by a vertical mean velocity W, contributing the terms $W\mathbf{u}_z$ and $W\theta_z$ to the left-hand sides of (9.3.34) and (9.3.36), respectively?. □

Exercise 9.3.5 Separate the independent variables in (9.3.34)–(9.3.37), leading to the linear shallow-water equations for the "horizontal" dependence and to a Sturm-Liouville problem for the "verti-

cal" dependence. Discuss appropriate choices for boundary conditions (Oliger & Sundström, 1978). □

9.4 Inverse formulation

The inevitability of ill-posed boundary conditions for open-ocean PE models prompts the search for an inverse formulation. The goals would be fields of velocity, pressure, density, and so forth, that would most nearly fit the dynamics, the initial conditions, the boundary conditions and any other available data. Since the problem of posing boundary conditions arises even with linear models, it is appropriate to examine linear inverse formulations first. Again, the discussion begins with the shallow-water equations linearized about a uniform mean flow \mathbf{U}:

$$\mathbf{u}_t + \mathbf{U} \cdot \nabla \mathbf{u} = -c\nabla q + \mathbf{F} + \boldsymbol{\mu} \qquad (9.4.1)$$

and

$$q_t + \mathbf{U} \cdot \nabla q + c\nabla \cdot \mathbf{u} = \chi. \qquad (9.4.2)$$

The field $q = (g/H)^{1/2}h$ in (9.4.1) and (9.4.2) has been scaled to have the dimensions of velocity, thereby symmetrizing the matrix-operator form of the equations. The fields μ and χ are errors in the prior estimates \mathbf{F} and 0 for the respective source terms. As always, these errors may also include unresolved eddy fluxes of momentum and mass, respectively. Prior estimates for initial and boundary values of *all* components of (\mathbf{u}, q) are given by (\mathbf{u}_I, q_I) and (\mathbf{u}_B, q_B), respectively. It suffices here to admit just one additional measurement (\mathbf{u}_d, q_d) for (\mathbf{u}, q), at the isolated space-time point (\mathbf{X}_1, T_1). (For simplicity, no distinction will be made between row and column vectors in this section.) Then a suitable quadratic penalty functional may be defined by

$$\mathcal{J}[\mathbf{u}, q] = \left(r^2 l^2 T\right)^{-1} \int_0^T dt \int_R d\mathbf{x} \left(|\boldsymbol{\mu}|^2 + \chi^2\right)$$

$$+ (lT)^{-1} \int_0^T dt \int_B ds \left(|\mathbf{u} - \mathbf{u}_B|^2 + \left(q - q_B\right)^2\right)$$

$$+ l^{-2} \int_R d\mathbf{x} \left(|\mathbf{u} - \mathbf{u}_I|^2 + (q - q_I)^2\right)$$

$$+ |\mathbf{u}_d - \mathbf{u}_1|^2 + (q_d - q_1)^2 , \qquad (9.4.3)$$

where $\mathbf{u}_1 \equiv \mathbf{u}(\mathbf{X}_1, T_1)$, $q_1 \equiv q(\mathbf{X}_1, T_1)$, r is the inverse time scale $(g/H)^{-1/2}$ and l is a length scale for the region R and boundary B. Scales aside, the residuals in (9.4.3) have been given uniform and independent weights, for simplicity alone. The Euler-Lagrange system, for local minima $(\hat{\mathbf{u}}, \hat{q})$ of the quadratic functional \mathcal{J}, is

$$-\hat{\boldsymbol{\mu}}_t - \mathbf{U} \cdot \boldsymbol{\nabla}\, \hat{\boldsymbol{\mu}} - c\boldsymbol{\nabla}\, \hat{\chi} = (rl)^2 T(\mathbf{u}_d - \hat{\mathbf{u}}_1)\delta\delta , \qquad (9.4.4)$$

$$-\hat{\chi}_t - \mathbf{U} \cdot \boldsymbol{\nabla}\, \hat{\chi} - c\boldsymbol{\nabla}\, \hat{\boldsymbol{\mu}} = (rl)^2 T(q_d - \hat{q}_1)\delta\delta , \qquad (9.4.5)$$

$$\hat{\mathbf{u}}_t + \mathbf{U} \cdot \boldsymbol{\nabla}\, \hat{\mathbf{u}} + c\boldsymbol{\nabla}\, \hat{q} = \mathbf{F} + \hat{\boldsymbol{\mu}} , \qquad (9.4.6)$$

$$\hat{q}_t + \mathbf{U} \cdot \boldsymbol{\nabla}\, \hat{q} + c\boldsymbol{\nabla} \cdot \hat{\mathbf{u}} = \hat{\chi} , \qquad (9.4.7)$$

subject to

$$\hat{\boldsymbol{\mu}} = \mathbf{0} , \qquad \hat{\chi} = 0 , \qquad (9.4.8)$$
$$\text{at } t = T ,$$

$$\hat{\mathbf{u}} = \mathbf{u}_I + \left(r^2 T\right)^{-1} \hat{\boldsymbol{\mu}} , \qquad \hat{q} = q_I + \left(r^2 T\right)^{-1} \hat{\chi} , \qquad (9.4.9)$$
$$\text{at } t = 0 ,$$

and

$$\mathbf{U} \cdot \mathbf{n}\, \hat{\mu} + cn\hat{\chi} + \left(r^2 l\right)\left(\hat{\mathbf{u}} - \mathbf{u}_B\right) = 0, \qquad (9.4.10)$$

$$c\hat{\mu} \cdot \mathbf{n} + \mathbf{U} \cdot \mathbf{n}\, \hat{\chi} + \left(r^2 l\right)\left(\hat{q} - q_B\right) = 0, \qquad (9.4.11)$$
$$\text{on } B.$$

In (9.4.4) and (9.4.5), $\delta\delta = \delta(\mathbf{x} - \mathbf{X}_1)\,\delta(T - T_1)$. Note that the values of the model residuals corresponding to $(\hat{\mathbf{u}}, \hat{q})$ are denoted by $(\hat{\mu}, \hat{\chi})$. The use of carets on adjoint variables was avoided in previous chapters, for simplicity. However, a more precise notation seems essential here.

The above inverse formulation may be regarded as having resolved the ill-posedness of the open-ocean modeling problem for the linearized shallow-water equations, if (9.4.8)–(9.4.11) provide just enough initial and boundary conditions to ensure that the system (9.4.4)–(9.4.7) has a unique solution. It suffices to demonstrate that the homogeneous problem $\left(\mathbf{u}_d = 0,\ q_d = 0,\ \mathbf{F} = 0,\ \mathbf{u}_I = 0\ q_I = 0,\right.$ $\mathbf{u}_B = 0,\ q_B = 0\big)$ has only the trivial solution.

The following identity may be obtained from the homogeneous forms of (9.4.4) and (9.4.5):

$$\int_0^T dt \int_R d\mathbf{x}\left(|\hat{\mu}|^2 + \hat{\chi}^2\right) - \int_0^T dt \int_B ds \Big\{ \hat{\mathbf{u}} \cdot \left(\mathbf{U} \cdot \mathbf{n}\hat{\mu} + cn\,\hat{\chi}\right)$$

$$+ \hat{q}\left(c\hat{\mu} \cdot \mathbf{n} + \mathbf{U} \cdot \mathbf{n}\,\hat{\chi}\right) \Big\} - \int_R d\mathbf{x}\left[\hat{\mathbf{u}} \cdot \hat{\mu} + \hat{q}\,\hat{\chi}\right]_{t=0}^{t=T}$$

$$+ \left(r^2 l^2 T\right)^{-1}\left(|\hat{\mathbf{u}}_1|^2 + \hat{q}_1^2\right) = 0. \qquad (9.4.12)$$

Substituting the homogeneous forms of (9.4.8)–(9.4.11) into the left-hand side of (9.4.12) leads to a positive-definite quadratic functional. It may then be inferred that

$$\hat{\mu} \equiv \mathbf{0}, \qquad \hat{\chi} \equiv 0, \qquad (9.4.13)$$
$$\text{in } R \text{ for } 0 \le t \le T,$$

$$\hat{\mathbf{u}} \equiv \mathbf{0}, \qquad \hat{q} \equiv 0, \qquad (9.4.14)$$
$$\text{on } B \text{ for } 0 \le t \le T,$$

$$\hat{\mathbf{u}} \equiv 0, \qquad \hat{q} \equiv 0, \qquad\qquad (9.4.15)$$

$$\text{in } R \text{ at } t = 0$$

and

$$\hat{\mathbf{u}}(\mathbf{X}_1, T_1) = 0, \qquad \hat{q}(\mathbf{X}_1, T_1) = 0. \qquad (9.4.16)$$

Indeed, an analysis of the 6×6 quadratic form in the boundary integral in (9.4.12) shows that the form has three positive eigenvalues, thus three linear boundary conditions are needed for the 6-vector $(\hat{\mu}, \hat{\chi}, \hat{\mathbf{u}}, \hat{q})$ in order to ensure that the boundary integral is negative. The conditions are defined by the eigenvectors associated with the positive eigenvalues. The conditions which are provided, namely (9.4.10) and (9.4.11), are precisely those required. It remains to demonstrate that $(\hat{\mathbf{u}}, \hat{q})$ is the null vector. As usual the following identity may be derived from the homogeneous form of (9.4.6) and (9.4.7), making use of (9.4.13):

$$\frac{d}{dt} \tfrac{1}{2} \int_R d\mathbf{x}(|\hat{\mathbf{u}}|^2 + \hat{q}^2) = -l \int_B ds \left\{ \mathbf{U} \cdot \mathbf{n} \, \tfrac{1}{2}(|\hat{\mathbf{u}}|^2 + \hat{q}^2) \right.$$

$$\left. + c\,\hat{\mathbf{u}} \cdot \mathbf{n}\,\hat{q} \right\}. \qquad (9.4.17)$$

Now (9.4.14) and (9.4.15) ensure that

$$\hat{\mathbf{u}} \equiv 0, \qquad \hat{q} \equiv 0, \qquad\qquad (9.4.18)$$

$$\text{in } R \text{ for } 0 \le t \le T,$$

which finally establishes the uniqueness of extrema of the penalty functional \mathcal{J}. Since \mathcal{J} is quadratic and positive-definite in \mathbf{u} and q, it follows that the extremum of \mathcal{J} is the minimum.

The preceding argument deserves careful review. The objective is to prove that the system of equations (9.4.4)–(9.4.7) and terminal conditions (9.4.8) and (9.4.9) define a well-posed problem when supplemented with the three boundary conditions (9.4.10), (9.4.11). The assertion is proved once it is shown that the three boundary conditions are just sufficient to define a unique solution, that is, they are just sufficient to establish that the homogeneous problem has only the trivial solution. The first part of the above proof shows that they are indeed just sufficient to establish (9.4.13)–(9.4.16). The second

part merely confirms that these null conditions ensure that $(\hat{\mathbf{u}}, \hat{q})$ $\equiv (\mathbf{0}, 0)$, thereby completing the proof. The null conditions are in fact more than sufficient for this purpose, as previously discussed (that is why the inhomogeneous forward problem is ill-posed), but the null conditions have been *derived from* the Euler-Lagrange equations, rather than being *imposed upon* them.

Exercise 9.4.1 Derive the nonlinear Euler-Lagrange equations for an inverse formulation of the nonlinear shallow-water equations (9.2.25)–(9.2.26), the inverse being defined by a penalty functional analogous to (9.4.3). Attempt to prove the well-posedness of the nonlinear Euler-Lagrange equations, by combining the methods of §9.2 and §9.4. □

Exercise 9.4.2 Devise an inverse formulation of the linearized isopycnal model (9.3.11)–(9.3.13). Derive the Euler-Lagrange system for local minima of the quadratic penalty functional. Use energy arguments, as above, to establish that the system has a unique solution, provided that the mean stratification is stable. □

Thus far, a linearized PE model with open boundaries has been formulated as an inverse problem. It has been established that the Euler-Lagrange equations have precisely the correct set of initial and boundary conditions which ensure uniqueness of solutions. However, the ill-posedness of the original, linearized PE model cannot be regarded as resolved until a well-posed solution algorithm has been found for the Euler-Lagrange system. Computing costs aside, the Gelfand & Fomin sweep algorithm and the representer method have both been found suitable in the context of linear QG models, having either periodic or open boundary conditions. These methods are also suitable for linear QG and PE models having entirely rigid boundaries, but are in general unsuitable for stratified PE models with open boundaries. Both methods reduce the Euler-Lagrange equations to a system of initial-value problems. No matter how the Euler-Lagrange boundary conditions are partitioned amongst the model and adjoint equations, there will, as discussed in §9.3, be either an underdetermining or an overdetermining of the initial-value problems unless a modal decomposition is performed in the vertical. This last option is not always attractive for linear models, much less the nonlinear models of interest, if the advecting velocity varies with depth or if the total depth varies around the open boundary.

There is a further difficulty in applying the sweep algorithm or representer method to the Euler-Lagrange equations related to nonlinear PE models. The nonlinear Euler-Lagrange system must be iterated, in order to obtain a sequence of linear Euler-Lagrange systems amenable to such solution methods. Since the sequence of iterated solutions is not descending the gradient of the nonlinear penalty functional, a proof of convergence of the sequence would not be trivial. Proving that the sequence is bounded in a suitable class of functions would ensure the existence of convergent subsequences. As a first step it would be essential to prove that solutions of the original nonlinear PE model are bounded. Energy arguments readily establish that the flow variables are square-integrable in space and time, but the gradients of the flow variables need not be bounded. Indeed, even the nonlinear shallow-water equations (9.2.25) and (9.2.26) include the dynamics of the textbook "piston problems," which lead to the formulation of infinite gradients in a finite time (see, for example, Goldstein, 1960). Further, the divergence of the nonlinear momentum equation (9.2.25) is

$$\delta_t + \mathbf{u} \cdot \nabla \delta + \delta^2 + 2(v_x\, u_y - u_x\, v_y) = -g\nabla^2 h + \nabla \cdot \mathbf{F}\,, \qquad (9.4.19)$$

where $\delta = \nabla \cdot \mathbf{u}$. Establishing bounds on δ using (9.4.19) is difficult. The essence of the difficulty may be found in the ordinary differential equation

$$\frac{d\delta}{dt} + \delta^2 = 0\,. \qquad (9.4.20)$$

The solution of this Riccati equation is

$$\delta(t) = \left(t + \delta(0)^{-1}\right)^{-1}\,. \qquad (9.4.21)$$

If $\delta(0)$ is negative, then $\delta(t)$ is unbounded as $t \to -\delta(0)^{-1}$. Nevertheless, PE models are routinely integrated for relatively short but useful times, during which the divergence field remains acceptable. It may be remarked that singular behavior in the divergence field would coincide with singular behavior of the mass field: see (9.2.37). On the other hand such behavior may reflect physical reality, as opposed to singular behavior due to inconsistent conditions at open boundaries. It may finally be remarked that, on account of these difficulties, there appears to be no prospect of proving the well-posedness of an inverse

formulation of a nonlinear PE model, other than for short smoothing intervals $0 \leq t \leq T$ which preclude the development of genuine flow singularities. *The well-posedness of the formulation for linearized models must, therefore, suffice as a guide.*

The method of gradient descent is another candidate for finding local extrema of the penalty functional, for an inverse formulation of an open-ocean PE model. As mentioned in §5.9, considerable experience with this method has been obtained for QG and PE models, subject to assumptions of perfect dynamics and perfectly estimated forcing, etc. These assumptions reduce the dimension of the gradient vector to the number of computational degrees of freedom in space, which itself may be considerable. The assumptions also lead to efficient algorithms for calculating the gradient vector by integrating the adjoint model equations backward in time. This efficiency is not available for open-ocean PE models, owing to the inevitable ill-posedness of such integrations, as discussed in §9.3. When perfectly known dynamics and forcing are not assumed, calculation of the now very long gradient vector may be expedited by *substituting* the best estimate for the flow fields *into* the adjoint equations: see §5.9. Covariances of posterior errors can only be obtained by calculating, and then inverting, the Hessian form associated with the penalty functional. This massive task has only been accomplished so far for steady models, low-resolution models, or models with perfectly known dynamics. It has already been pointed out in §5.9 that the calculations of both the gradient and the Hessian form are conventionally carried out in the space of all admissible flow fields, whereas they only need be carried out in the finite-dimensional space of observable fields. The latter are defined for nonlinear problems by linearization about the best estimate, and are no more numerous than the observations.

In summary, a minimization algorithm for finding the inverse of an open-ocean PE model must

i) avoid the integration of initial-value problems, either backward or forward in time.

Ideally, it would also

ii) yield the global minimum of the penalty functional,

iii) yield posterior error statistics with little additional effort,

iv) be equally easily applied to linear and nonlinear models,

v) lead to a criterion for antenna design, and

vi) be easily implemented on computers having massively parallel architectures.

As a final requirement, it would of course

vii) resolve the ill-posedness of the CFvN problem for open-ocean QG models, which problem is also embedded in open-ocean PE models.

Exercise 9.4.3 Prove the assertion, in requirement (*vii*), about open-ocean PE models. What other difficulties arise in PE models, when there are isolated points on the boundaries where the flow is tangential to the boundary? □

Exercise 9.4.4 Which of the preceding requirements is met by the method of simulated annealing described in §6.3? □

10

OUTSTANDING PROBLEMS

10.1 The general circulation

i) Introduction

The general circulation of the oceans is an essential component of the thermodynamic system which determines global climate. The contributions of the oceans to the poleward fluxes of heat and water, for example, are clearly significant if not yet reliably known (Lorenz, 1967). It is widely accepted that modeling has improved our understanding of the general ocean circulation, but the objectives of ocean modeling are evolving along with the models themselves. Goal 1 of the World Ocean Circulation Experiment (WOCE) restates the grand objective of physical oceanography:

> To develop models useful for predicting climate
> change and to collect the data necessary to test
> them

(WOCE, 1988). So far models have been developed by exploring the consequences of adding ever more physics and ever more detail. The earliest developments showed that westward intensification in subtropical gyres could be attributed to a combination of the β-effect with flow at high Reynolds number (Stommel, 1948; Munk, 1950). The nonlinear effect of vorticity advection was shown by Bryan (1963) to lead to time-dependent flow on the β-plane at high Reynolds number, even if the flow is steadily forced. An imposed mean density stratification allows the Lorenz cycle of energy exchange between mean and eddy forms of available potential energy and kinetic energy

283

(Holland & Lin, 1975a,b). Stratification, determined internally by thermodynamics, is found to develop plausible thermohaline structure in response to reasonably representative mean surface fluxes of heat, salt and momentum (Bryan & Cox, 1968a,b; Bryan & Lewis, 1979). Finally, highly detailed global models, just capable of resolving meso-scale eddies in the subtropical gyres (Robinson, 1983), are found to be at least as complex as our knowledge of the general circulation, and to possess striking similarities to most major features of the general circulation (Semtner & Chervin, 1988). While many difficulties remain, especially in the parameterization of unresolved mixing processes, modeling the general circulation of the oceans has gained a significant level of credibility. However, it must be concluded that WOCE Goal 1 remains far off, owing to the inadequacy of existing sets of initial data and surface data, together with the demonstrable unpredictability of the circulation in the existing models.

Model refinement continues. For example, Bleck & Smith (1990) exchanged isobaric coordinates for isopycnal coordinates. The appeal of the latter coordinate system is that motion is predominantly isopycnal; in particular the mixing of passive tracers at least is predominantly isopycnal. The problem of parameterizing convective adjustment in isobaric coordinates (see also §10.3) is exchanged for the problem of representing the surface ventilation of isopycnal layers. Other refinements include surface mixed layers (for example: Huang & Bryan, 1987) and sea ice (Oberhuber, 1993a,b).

First steps are being taken toward combining general ocean circulating models with observations. Ideally this would involve time-dependent, eddy-resolving primitive-equation models, together with data of comparable resolution. Global inverse calculations of this kind are not yet feasible with available computers. It is likely that computer size and speed will be adequate within a few years (see §10.4); then mathematical issues will have to be addressed, such as the conditioning or convergence of algorithms for inverse problems having very large numbers of degrees of freedom ($>10^7$). However, prospects for obtaining subsurface eddy-resolving data are bleak. Should a technical breakthrough be made, such as a "smart" drifter capable of continuous dead-reckoning of its position using correlation sonar, it will then be necessary to decide if time-dependent, high-resolution inverse calculations are really worthwhile. Accurate forecasting of individual eddies such as in the East Australian Current, or boundary current meanders such as in the Kuroshio, may be necessary for the

prediction of local climatic variability, but it is improbable that such detail impacts significantly upon global climate. Rather, it is the mean eddy fluxes which are almost certainly significant. But is it necessary to simulate the actual oceanic eddies in order to estimate the mean fluxes accurately? If so, then inversions using eddy-resolving data will be required. Otherwise, it may suffice to use only low-resolution data which, while guiding the large-scale circulation, do not directly influence the circulation on smaller scales. It would then be consistent to use only coarse-resolution models. Non-uniform resolution may be necessary in order to represent narrow mean boundary currents, for example. Mean fluxes due to unresolved eddies would be the natural candidates for model residuals. Indeed, our global knowledge of eddy statistics is barely adequate to allow reasonable estimates of prior covariances, much less the calibration of closure schemes. Inevitably, coarse-resolution models will include closure schemes, even if only in the form of eddy diffusivities, and at this stage it must be assumed that the error level in such schemes must be about 100%. That is, the full level of eddy variability should be used when estimating prior covariances of model residuals.

These introductory remarks may be concluded with the observation that generalized inversion of coarse-resolution models of the general circulation of the oceans is feasible with extant computer technology.

ii) Steady, quasi-geostrophic circulation models

As might be expected, the first tests of inversion methods for general oceanic circulation models have involved steady data and models. This is, for obvious reasons, a sensible start even if the real oceanic circulation has no long-term mean and even if the steady solutions of the circulation models are unstable.

Schröter & Wunsch (1986) experimented with single-layer and two-layer steady, nonlinear quasi-geostrophic models in a rectangular basin. Rather than seek a weighted least-squares best-fit to the prior information, they chose instead to explore the range of circulations corresponding to forcing fields and data both subject to lower and upper bounds. In particular they sought admissible solutions which minimize or maximize various properties of the circulation, such as the transport in the western boundary current, the total potential energy or the total kinetic energy. After replacing the model with a finite-difference approximation, Schröter & Wunsch expressed the

problem as follows:

$$\text{Minimize } F(\mathbf{\Psi}) \text{ subject to } \mathbf{L} \leq \mathbf{M}(\mathbf{\Psi}) \leq \mathbf{U}, \qquad (10.1.1)$$

where $\mathbf{\Psi}$ is a finite-dimensional vector, namely the grid-point values of the streamfunction, \mathbf{M} represents the finite set of nonlinear model difference operators, while \mathbf{L} and \mathbf{U} are vectors of lower and upper bounds on the forcing field (the curl of the wind stress). The scalar F is exemplified by discrete analogues of

a) $F = F_1 = \psi_{max}$: the transport around the closed basin;

b) $F = F_2 = \int \psi^2 dx$: the total potential energy in the basin; and

c) $F = F_3 = \int |\nabla \psi|^2 dx$: the total kinetic energy in the basin.

The model constraints (10.1.1) were augmented with additional constraints to the effect that the differences between the measurements of admissible circulations and measurements of the "fully posed" circulation (satisfying $\mathbf{M}(\mathbf{\Psi}) = \mathbf{0}$) should be bounded above and below. Altimetric (§4.3) and tomographic (§6.3) measurements were considered. The augmented problem retained the general form (10.1.1). After augmenting the circulation vector $\mathbf{\Psi}$ to include streamfunction values, measurements and forcing values, the problem may be expressed in the form:

$$\text{Minimize } F(\mathbf{X}) \text{ subject to } \mathbf{C}(\mathbf{X}) = \mathbf{0} \text{ and } \mathbf{L} \leq \mathbf{X} \leq \mathbf{U}. \quad (10.1.2)$$

Note that some components of \mathbf{L} have the value $-\infty$, while some components of \mathbf{U} are $+\infty$; that is, some components (the streamfunction values) of the augmented state vector are unconstrained. Following Luenberger (1984), an augmented penalty function may be defined by

$$\mathcal{F}(\mathbf{X}, \boldsymbol{\lambda}) = F(\mathbf{X}) + \boldsymbol{\lambda}^* \mathbf{C}(\mathbf{X}) + W\mathbf{C}(\mathbf{X})^* \mathbf{C}(\mathbf{X}). \qquad (10.1.3)$$

The Lagrange multiplier $\boldsymbol{\lambda}$ appends the strong constraint $\mathbf{C}(\mathbf{X}) = \mathbf{0}$ on the minimum of $F(\mathbf{X})$; if the positive weight W is sufficiently large and if \mathcal{F} is a convex function of \mathbf{X}, then F is also convex. Thus the minimization of $\mathcal{F}(\mathbf{X}, \boldsymbol{\lambda})$, subject to $\mathbf{L} \leq \mathbf{X} \leq \mathbf{U}$, is an example of the convex programming problem considered in §1.9. Convexity of F and \mathcal{F} ensures that F has a unique minimum. See Schröter & Wunsch (1986) for details of their solution method, including convergence criteria and scalings.

Figure 10.1.1. (a) Contours of the normalized Lagrange multiplier $\lambda(\mathbf{x})$ for the boundedness constraints in the augmented penalty functional (10.1.3). The vector $\boldsymbol{\lambda}$ denotes the values of λ on a rectangular grid. (b) The corresponding solution $\psi(\mathbf{x})$ for the streamfunction; it maximizes the kinetic energy $F_3 = \int |\nabla\psi|^2\, d\mathbf{x}$. The maximum value is 32.27 units (the minimum value is 8.22 units) (after Schröter & Wunsch, 1986).

Exercise 10.1.1 Express the minimization problem (10.1.1) in the form (10.1.2). □

Exercise 10.1.2 Let X_j be one of the bounded state variables: $-\infty < L_j \leq X_j \leq U_j < +\infty$. Assume that, at the minimum of F, X_j attains one of its bounds: either $X_j = L_j$ or $X_j = U_j$. Prove that either

$$\lambda_j = -\frac{\partial F}{\partial L_j} \quad \text{or} \quad \lambda_j = -\frac{\partial F}{\partial U_j}, \tag{10.1.4}$$

while

$$\lambda_j = 0 \tag{10.1.5}$$

if $L_j < X_j < U_j$ at the minimum of F. □

Schröter & Wunsch concluded from (10.1.4) and (10.1.5) that "by making maps of the Lagrange multipliers of our problem we can thus easily and conveniently display those constraints (observational or dynamical) to which our particular answer is the most sensitive and so determine which are most important." An example of their calculations is shown in Fig. 10.1.1. The objective is to maximize

the total kinetic energy (that is, to minimize $F = -F_3$) subject to spatially uniform upper and lower bounds on the uncertainty in the wind stress curl. The corresponding range for the transport is ± 12 Sv, given a prior estimate ("fully posed" solution) of 50 Sv. Other model parameters are: Rossby number $\varepsilon = 0.01$, linear drag coefficient $D_0 = 0.05$. Normalized values of the Lagrange multiplier λ are mapped in Fig. 10.1.1a. The positive sign of the values indicates that the curl is at its upper bound. The kinetic energy is clearly the most sensitive to the upper bound in the curl over the northwest corner of the basin. The circulation with maximum kinetic energy is shown in Fig. 10.1.1b. The maximum value of F_3 is 32.27 in dimensionless units. The curl which minimizes F_3 attains its lower bound everywhere, and the value of F_3 is sensitive to that bound over a wide region in the middle of the basin. The minimal value of F_3 is 8.22 units.

Other experiments described by Schröter & Wunsch involved sensitivity to bounds on parameter values, and bounds on data misfits. They remarked on the cubic dependence of the number of computations upon the number of variables (their calculations were based on a 13×13 spatial grid). McIntosh (1987b) pointed out that if the relative vorticity $\nabla^2 \psi$ is merely known to be bounded, then the divergence component $u_x = -\psi_{xy}$ may even have a logarithmic singularity. Such behavior would be incompatible with the use of finite-difference methods. Thus the penalty functional \mathcal{F} should be augmented with a term penalizing the streamfunction gradient, or perhaps higher derivatives. In any event, Schröter & Wunsch have formulated some appealing objectives for general circulation modeling, in terms of sensitivity to bounds on prior estimates of forcing fields, measurements and model parameters.

The least-squares generalized inverse of a steady, single-layer, nonlinear quasi-geostrophic model and synthetic data set was constructed by Tziperman & Thacker (1989) as a prelude to considering a more realistic general circulation model. They used a novel technique for constructing the steady inverse. Formally, a time-dependent problem was solved, in which the smoothing interval $0 \le t \le T$ was very small compared to the time scales of the motion. The penalty functional included the mean square of the first time derivative of the streamfunction.

Exercise 10.1.3 Derive the Euler-Lagrange equations for extrema

of the penalty functional

$$\mathcal{K}[\psi] = \mathcal{J}[\psi] + W_0 \int_0^T dt \int d\mathbf{x} \, \psi_t(\mathbf{x}, t)^2 \,, \qquad (10.1.6)$$

where \mathcal{J} is defined by (5.3.1). Note that q appearing in (5.3.1) is the error in the prior forcing estimate F for the linear quasi-geostrophic model (5.2.2). Using the methods of §5.3, prove that the extremum is unique provided that each prior error is given a positive weight. □

Exercise 10.1.4 Assume that the prior forcing estimate F appearing in the penalty functional \mathcal{K} (see (10.1.6), (5.2.2) and (5.3.1)) is steady: $F \equiv F(\mathbf{x})$. Show that, in general, there is no steady solution to the Euler-Lagrange equations if all the weights are positive. □

Exercise 10.1.5 Consider again the inverse problem in the preceding exercise. Suppose that the prior estimate of the initial streamfunction is given no weight: $V(\mathbf{x}_1, \mathbf{x}_2) \equiv 0$. In the limit of a vanishing weight also for the prior forcing estimate: $W(\mathbf{x}_1, t_1, \mathbf{x}_2, t_2) \to 0$, show that

a) there are many steady solutions to the Euler-Lagrange equations;

b) these inverse solutions interpolate the data; and

c) the dynamics of the model have no influence upon the solutions.
 □

Exercise 10.1.6 Replace $\mathcal{K}[\psi]$, defined in (10.1.6), with the penalty functional

$$\mathcal{M}[\psi, \lambda] = \mathcal{J}[\psi] + \int_0^T dt \int d\mathbf{x} \, \lambda(\psi_t - L\psi - F - q) \,, \qquad (10.1.7)$$

where $\lambda(\mathbf{x}, t)$ is a Lagrange multiplier for the model (5.2.2), and \mathcal{J} is again defined in (5.3.1). Show that the limit $V, W \to 0$ yields the results (a), (b) and (c) of the preceding exercise. □

Tziperman & Thacker (1989) found minima of their penalty functional using a gradient-descent method. The gradient was effectively calculated by backward integration of the adjoint equation (see §5.9). The latter governs the Lagrange multiplier for the forcing error. The

Hessian matrix of second derivatives of the penalty functional was
calculated by simple finite-differencing of the gradient

$$\frac{\partial^2 \mathcal{K}}{\partial \alpha_i \partial \alpha_j} \cong \left(\left. \frac{\partial \mathcal{K}}{\partial \alpha_i} \right|_{\alpha_j + \delta\alpha_j} - \left. \frac{\partial \mathcal{K}}{\partial \alpha_i} \right|_{\alpha_j} \right) \frac{1}{\delta\alpha_j}, \qquad (10.1.8)$$

where α_i and α_j are scalar arguments of \mathcal{K} (streamfunction values,
forcing values or parameter values). If \mathcal{K} has N such arguments, then
evaluation of (10.1.8) for all i and j requires $2N$ model integrations.
In the experiments of Tziperman & Thacker, $N = 450$. Nevertheless
the computing task was relatively modest since each integration over
the short interval $0 \leq t \leq T$ involved only one time step. Examination
of the eigenvalues of the Hessian showed poor conditioning, consistent
with slow convergence of the gradient-descent calculations. That is,
taking steps away from a minimum, in the direction of minor axes of
the Hessian, would not increase the penalty functional significantly. A
linear transformation of the arguments of \mathcal{K} may improve condition-
ing of the Hessian, and hence accelerate convergence; the search for
preconditioning transformations would be an essential but expensive
task if the circulation model has many degrees of freedom.

iii) Fully stratified circulation models

Several teams of investigators are presently attempting to con-
struct inverses of fully stratified general oceanic circulation models.
Conservation laws for both salinity and temperature are included,
together with approximate equations of state for seawater. Data sets
include the Levitus atlas of mean hydrographic fields (Levitus, 1982)
and global fields of wind stress curl (Hellerman & Rosenstein, 1983).
A recent addition to the data sets is a comprehensive hydrographic
atlas of the North Atlantic compiled by Fukumori, Martel & Wunsch
(1991).

It appears, and is not entirely unexpected, that mesoscale vari-
ability related to momentum advection (that is, barotropic and baro-
clinic instability) poses serious technical problems. Most efforts, there-
fore, essentially allow only geostrophic momentum balances. Eddy
diffusion of momentum may also be included, in order to allow closed
circulation in closed basins.

Thus the dynamics of present inverse models of the general oceanic
circulation closely resemble the dynamics of the β-spiral and other
models examined in Chapter 1; enough ideas have been introduced,

and it is time to get detailed results for the global ocean as more data become available during WOCE.

10.2 Parameter estimation

Theories of ocean circulation are derived from the laws of hydrodynamics. The only empirical constants appearing in these laws are molecular diffusivities, specific heat capacities and other microscopic thermodynamic or chemical parameters. All have been determined with sufficient precision by controlled experiments in the laboratory. An economical description of the circulation, on the other hand, necessitates averaging the nonlinear hydrodynamic laws over the smallest resolvable scales in space and time. Such averaging leads to the well-known turbulence closure problem. It is customary for ocean circulation models to be closed by the use of turbulent or eddy diffusivities for mechanical, thermodynamic and chemical fields (Phillips, 1977). Specification of averaged transfer rates at the ocean surface involves additional parameters, such as the drag coefficients in expressions for turbulent wind stress. In some particularly simple models, such as reduced-gravity quasi-geostrophic models, the effects of vortex stretching due to density stratification are represented in the streamfunction-vorticity relationship by a coefficient dependent upon the internal Rossby radius of deformation (Pedlosky, 1987, Chapter 3).

Climatological parameters such as eddy diffusivities, drag coefficients and deformation radii cannot be determined in the laboratory; they possess large space and time scales of variability. Prior estimates of the values of such parameters, let alone the parameterizations themselves, must be suspect, and so it makes sense to vary the parameters in order to obtain improved hindcasts. Such calculations may be viewed as the use of ocean circulation data in order to improve parameter estimates for use in subsequent forecasts.

Parameter estimation has long been of interest to engineers. The variety of parameterizations which may be found in practical models is naturally vast, and the opportunities for adapting estimation schemes to specific problems are equally great. Ljung & Söderström (1987) remark that the area has been described as a "fiddler's paradise." Their text develops the subject of sequential parameter estimation in an orderly fashion, with due regard to theoretical issues of stability and convergence, as well as to practical aspects of computation. The

treatment is limited to engineering models described by systems of ordinary differential equations.

Much practical information and experience may be found in the engineering literature, but the principles of least-squares estimation are clear enough, and the nature of ocean circulation models is so specialized that it is best to approach oceanic parameter estimation directly. The computational constraints alone are idiosyncratic, while the mathematical structures of the models and parameterizations are sufficiently unique that any substantial theoretical progress can only be obtained on a model-by-model basis. Perhaps the most relevant engineering discipline is groundwater hydrology (Carrera & Neuman, 1986a,b,c; Yeh, 1986).

Several oceanic examples of parameter estimation are described below.

i) Drag coefficients and eddy viscosities

Yu & O'Brien (1991) estimated the wind stress drag coefficient and the oceanic eddy viscosity profile using data from the *Long-Term Upper Ocean Study–3* (LOTUS–3; Tarbell, Pennington & Briscoe, 1984). Their model for the oceanic surface Ekman layer may be expressed as

$$\mathbf{u}_t + f\mathbf{e_3} \times \mathbf{u} = (A\mathbf{u}_z)_z \, , \qquad (10.2.1)$$

where $\mathbf{u} = \mathbf{u}(z,t)$ is the horizontal velocity and $A = A(z)$ is the oceanic eddy viscosity. Note that the horizontal pressure gradient does not contribute significantly to the boundary-layer accelerations (Pedlosky, 1987, Chapter 4). The condition of imposed wind stress τ on the flat ocean surface is

$$\rho_w \, A \frac{\partial \mathbf{u}}{\partial z} = \tau = \rho_a \, C \, |\mathbf{u}_a| \, \mathbf{u}_a \, , \qquad (10.2.2)$$

where ρ_w is the water density, ρ_a is the air density, \mathbf{u}_a is the wind velocity, and C is the drag coefficient which is to be estimated along with $A(z)$. A condition of no diffusive flux of momentum at some depth H is

$$A \frac{\partial \mathbf{u}}{\partial z} = 0 \, , \qquad \text{at } z = -H \, . \qquad (10.2.3)$$

An initial condition is given by

$$\mathbf{u}(z,0) = \mathbf{u}_I(z) \, . \qquad (10.2.4)$$

Observations of $\mathbf{u}(z,t)$ were collected during LOTUS–3 using current meters, set at many depths and recording virtually continuously in time. Wind velocity was measured from a tower. A weighted least-squares best-fit to the observations was sought by minimizing the penalty functional

$$J = W_A \int_{-H}^{0} (A - A_0)^2 dz + W_C (C - C_0)^2$$

$$+ w \int_0^T dt \sum_{m=1}^{M} |\mathbf{u}_m - \mathbf{d}_m|^2 \qquad (10.2.5)$$

subject to the strong constraints (10.2.1), (10.2.2) and (10.2.3). The initial condition (10.2.4) was ignored. Formally, it may be regarded as having been included as a weak constraint with zero weight. In (10.2.5), W_A, W_C and w are positive weights, while C_0 and $A_0 = A_0(z)$ are prior estimates of the parameters C and $A = A(z)$. The velocity data are the vectors $\mathbf{d}_m = \mathbf{d}_m(t)$, while $\mathbf{u}_m = \mathbf{u}_m(t)$ denotes $\mathbf{u}(z_m, t)$, where z_m is the mth measurement depth ($1 \le m \le M$). Let $\mathcal{K} = \mathcal{K}[\mathbf{u}, A, C, \boldsymbol{\lambda}, \boldsymbol{\theta}, \boldsymbol{\phi}]$ denote the penalty functional J with the strong constraints appended; $\boldsymbol{\lambda}(z,t)$, $\boldsymbol{\theta}(t)$ and $\boldsymbol{\phi}(t)$ are the Lagrange multipliers for (10.2.1), (10.2.2) and (10.2.3), respectively. Then extrema of \mathcal{K} satisfy, in particular,

$$\frac{\partial \mathcal{K}}{\partial A(z)} = 0, \qquad -H \le z \le 0, \qquad (10.2.6)$$

and

$$\frac{\partial \mathcal{K}}{\partial C} = 0. \qquad (10.2.7)$$

Exercise 10.2.1 Show that the extremal conditions (10.2.6), (10.2.7) are, respectively,

$$W_A \left(\hat{A} - A_0 \right) + \int_0^T \hat{\mathbf{u}}_z \cdot \boldsymbol{\lambda}_z \, dt = 0 \qquad (10.2.8)$$

and

$$W_C \left(\hat{C} - C_0 \right) - \rho_a \int_0^T |\mathbf{u}_a| \, \mathbf{u}_a \cdot \boldsymbol{\theta} \, dt = 0, \qquad (10.2.9)$$

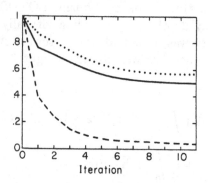

Figure 10.2.1. Successive values of the penalty functional \mathcal{K} (solid line), the magnitude g of its gradient (dashed line), and the mean square of the posterior data misfit (dotted line) in the parameter estimation problem of Yu & O'Brien (1991).

where $\hat{A} = \hat{A}(z)$ and \hat{C} are values of A and C at extrema of \mathcal{K}. Show that $\boldsymbol{\theta}(t) = \boldsymbol{\lambda}(0, t)$. \qquad □

Exercise 10.2.2 Derive the adjoint equations, boundary conditions, initial condition and final condition for $\boldsymbol{\lambda}$. Recall that (10.2.4) has been given no weight. \qquad □

It is of fundamental importance to observe that even though the original model (10.2.1)–(10.2.4) is linear in the oceanic velocity \mathbf{u}, the parameter estimation problem is nonlinear. This is clear from (10.2.8) and (10.2.9), since $\hat{\mathbf{u}}$ and $\boldsymbol{\lambda}$ have nonlinear dependences upon \hat{A} and \hat{C}.

Rather than solve the extremal conditions directly, Yu & O'Brien (1991) minimized \mathcal{K} using a descent algorithm, as in §5.9. They used LOTUS-3 data taken at depths from 5 m to 100 m, and taken every 15 minutes for 10 days. The prior estimates for the parameters were: $A_0(z) = 10^{-5}$ m^2 s^{-1} at all depths, and $C_0 = 1.34 \times 10^{-3}$. Successive values of the penalty functional \mathcal{K}, the magnitude g of its gradient with respect to A and C, and the sum of the squared data misfits $|\mathbf{d}_m - \hat{\mathbf{u}}_m|^2$ are shown in Fig. 10.2.1. All have been scaled with their prior values. The sequences of viscosity profiles $A(z)$ and drag coefficients C are shown in Figs. 10.2.2 and 10.2.3. It is striking that while the viscosity profile changes substantially, the drag coefficient changes only by about 5% as the gradient is descended. However, it is not clear that there is any meaningful distinction between varying

Figure 10.2.2. Successive estimates of the eddy viscosity profile $A(z)$ (after Yu & O'Brien, 1991).

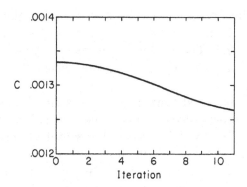

Figure 10.2.3. Successive estimates of the drag coefficient C (after Yu & O'Brien, 1991).

$A(0)$ and varying C in the surface stress condition (10.2.2). Note that no smoothness criterion has been explicitly imposed upon $A(z)$; it might have a discontinuity at $z = 0$, for example. Yu & O'Brien argued that the estimated eddy viscosity, which decreases markedly with depth, is consistent with the suppression of turbulent mixing by strong stratification in the thermocline. Not shown here are time series of observed and estimated velocities at selected depths. These

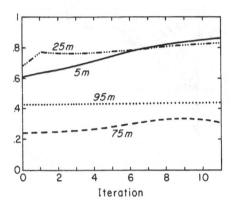

Figure 10.2.4. Successive correlation coefficients between observed time series of current and estimates, at selected depths (after Yu & O'Brien, 1991).

are dominated by inertial oscillations; there is reasonable agreement at the inertial time scale. Correlation coefficients at selected depths are shown in Fig. 10.2.4. The agreement between observed and estimated velocities is clearly best at 5 m and 25 m, where the simple inertial oscillations dominate.

It may be remarked that imposing the simple model (10.2.1)–(10.2.3) as a strong constraint upon the estimation of A and C is unreasonable, in light of the neglect of horizontal advection of momentum and in light of the errors in the turbulence closure schemes. It may also be remarked that a possible solution is $\hat{A}(z) \equiv A_0$, $\hat{C} = C_0$, and all Lagrange multipliers vanishing identically, provided an initial velocity estimate $\hat{\mathbf{u}}(z, 0)$ exists such that the data are interpolated: $\hat{\mathbf{u}}(z_m, t) = \mathbf{d}_m(t) \, (1 \leq m \leq M, \quad 0 \leq t \leq T)$. The dominance of the inertial oscillations in the data indicates that there are only one or two degrees of freedom in the current meter time series at each depth. On the other hand, not every depth in the numerical model is metered. Thus, the number of degrees of freedom in the data does not greatly exceed the number of degrees of freedom in the initial velocity for the model. It is only these excess degrees of freedom in the data which lead to revisions of the prior parameters. There is, accordingly, a danger that the posterior estimates may be sensitive to

the noise in the data. This sensitivity could be reduced by penalizing the roughness of the initial velocity and the eddy viscosity profile.

For estimates of bottom drag coefficients in a tidal channel, see Panchang & O'Brien (1988); phase speeds in reduced-gravity models of equatorial waves are estimated in Smedstad & O'Brien (1991).

ii) *Deformation radii*

The limited-area, reduced-gravity quasi-geostrophic model described in §8.10 has as a parameter the internal deformation radius ρ. This climatological parameter appears in the streamfunction-vorticity relation (8.10.2); the term proportional to ρ^{-2} is the contribution of the stratification to the potential vorticity. Since ρ only appears in the expression ρ^{-2}, it is convenient to introduce the derived parameter $\sigma = \rho^{-2}$, with the range $0 \leq \sigma \leq \infty$. Varying about a prior estimate σ_0 may yield a better fit to data, and presumably a more realistic choice for σ. Accordingly, the penalty functional \mathcal{J}, defined by (8.10.12), is replaced by an augmented functional $\mathcal{K} = \mathcal{K}[\psi, \sigma]$ given by

$$\mathcal{K} = \mathcal{J} + W_{\sigma\sigma}(\sigma - \sigma_0)^2 + 2\lambda \bullet (\nabla^2 \psi - \sigma\psi + f(y) - \zeta). \quad (10.2.10)$$

Note the use of the Lagrange multiplier $\lambda(\mathbf{x}, t)$ in order to append (8.10.2) as a strong constraint. The conditions for an extremum of \mathcal{K} now include

$$\frac{\partial \mathcal{K}}{\partial \sigma} = 0. \quad (10.2.11)$$

Exercise 10.2.3 Show that (10.2.11) yields

$$\hat{\sigma} = \sigma_0 + W_{\sigma\sigma}^{-1} \lambda \bullet \hat{\psi}, \quad (10.2.12)$$

where $\hat{\sigma}$ is the value of σ at an extremum of \mathcal{K}. □

The condition (10.2.12) is easily solved by inclusion in the iteration scheme for solving the Euler-Lagrange equations (8.10.15)–(8.10.26): see also Exercises 8.10.1–8.10.3. In particular, (10.2.12) is replaced with

$$\hat{\sigma}^{(n+1)} = \sigma_0 + W_{\sigma\sigma}^{-1} \lambda^{(n)} \bullet \hat{\psi}^{(n)}, \quad (10.2.13)$$

where $\lambda^{(n)}$ and $\hat{\psi}^{(n)}$ depend upon the nth parameter estimate $\hat{\sigma}^{(n)}$: see (8.10.15), (8.10.17) and (8.10.24).

When the Laplacian operator ∇^2 is subjected to Dirichlet conditions on the boundary B, the eigenvalues of the operator are strictly negative. Let κ be the eigenvalue having the smallest magnitude $|\kappa|$. Then the eigenvalues of the elliptic operator E are also strictly negative; the eigenvalue of smallest magnitude is $\kappa - \sigma$. Since $|\kappa - \sigma|^{-1} \leq |\kappa|^{-1}$, the inverse of E is uniformly bounded for all $\sigma > 0$. As a consequence, $\lambda^{(n)}$ and $\hat{\psi}^{(n)}$ are bounded uniformly with respect to σ. By (10.2.13), if $\lambda^{(n)}$ and $\hat{\psi}^{(n)}$ are uniformly bounded with respect to n, then so is $\hat{\sigma}^{(n)}$. In such circumstances the sequence $\hat{\sigma}^{(n)}$ must have at least one limit point.

Statistical linearization yields an estimate of the error variance for $\hat{\sigma}$, that is, the posterior error variance for σ. As always, the prior error variance $Q_{\sigma\sigma} \equiv \overline{(\sigma - \sigma_0)^2}$ is the natural choice for $W_{\sigma\sigma}^{-1}$. As in §5.6, the estimates will actually be for the extended Kalman smoother, rather than for the generalized inverse. That is, terms arising from the functional variation of the advecting velocity will be ignored. Hence,

$$\lambda = \hat{\mathbf{b}}^* \, \boldsymbol{\alpha} \,, \tag{10.2.14}$$

where $\boldsymbol{\alpha}$ is a vector of adjoint representer fields, while $\hat{\mathbf{b}}$ is the vector of representer coefficients. The iteration subscripts will be dropped; this analysis applies to perturbations about the posterior fields and parameter values. The condition (10.2.13) yields

$$\sigma - \hat{\sigma} = \sigma - \sigma_0 - Q_{\sigma\sigma} \, \hat{\psi} \bullet \boldsymbol{\alpha}^* \hat{\mathbf{b}} \,. \tag{10.2.15}$$

The vector of representer coefficients is $\hat{\mathbf{b}} = \mathbf{P}^{-1}\mathbf{h}$, where $\mathbf{P} = \mathbf{R} + \mathbf{w}^{-1}$ is the stabilized representer matrix and $\mathbf{h} = \mathbf{d} - \mathcal{L}[\psi_F]$ is the prior data misfit. Fluctuations in σ contribute the term $(\sigma - \sigma_0)\boldsymbol{\alpha} \bullet \hat{\psi}$ to \mathbf{h}; some algebra leads eventually to

$$C_{\hat{\sigma}\hat{\sigma}} = \overline{(\sigma - \hat{\sigma})^2} = Q_{\sigma\sigma} - Q_{\sigma\sigma}^2 \, \mathbf{v}^* \mathbf{P}^{-1} \mathbf{v} \,, \tag{10.2.16}$$

where $\mathbf{v} = \boldsymbol{\alpha} \bullet \hat{\psi}$.

10.3 Discontinuous fields and discontinuous dynamics

The preceding chapters have made repeated use of a standard result in the calculus of variations: the extrema of penalty functionals

satisfy the associated Euler-Lagrange equations. That is, if $\mathcal{J} = \mathcal{J}[\psi]$ is a penalty functional, and if variations $\delta\psi$ in ψ lead to variations $\delta\mathcal{J}$ in \mathcal{J} expressible in the form

$$\delta\mathcal{J} = \int d\mathbf{x} \int_0^T dt \left\{ \delta\psi(\mathbf{x}, t)\, \mathcal{E}\psi(\mathbf{x}, t) + O(\delta\psi)^2 \right\}, \qquad (10.3.1)$$

where \mathcal{E} is an operator acting on ψ, then $\hat{\psi}$ is an extremum for \mathcal{J} provided

$$\mathcal{E}\hat{\psi}(\mathbf{x}, t) = 0 \qquad (10.3.2)$$

for all \mathbf{x} and t. It is to be understood that (10.3.1) may include initial, final or boundary contributions, in which case (10.3.2) will include corresponding conditions. The conclusion (10.3.2) is the fundamental lemma of the calculus of variations (Courant & Hilbert, 1953, p. 185). It is valid if $\mathcal{E}\hat{\psi}$ is continuous in \mathbf{x} and t, and if extrema are sought amongst twice-continuously-differentiable functions $\psi(\mathbf{x}, t)$ which are consistent with any initial or boundary conditions. Thus far it has been assumed that the circulation fields, the model dynamics and the measurement functionals (both of the latter are contained in \mathcal{E}) meet these conditions.

i) Discontinuous fields

The continuity conditions for the fundamental lemma can be violated in several ways. As discussed in Chapters 8 and 9, discontinuous or non-smooth solutions may develop in the "forward" problem owing to ill-posed boundary conditions. Such behavior may also arise in well-posed problems owing to nonlinearity: the existence of smooth solutions of the quasi-geostrophic equations in large time intervals has only been proved in the case of isothermal conditions on horizontal boundaries (Bennett & Kloeden, 1980; Bennett, 1984), while no such "classical" theorem "in the large" has been proved for the primitive-equation.

Discontinuity may be essential to the description of fields on certain times scales. For example, Ghil, Mullhaupt & Pestiaux (1987) modeled the advance and retreat of Quaternary glaciation using Boolean delay equations. In their approach, the volume V of the Northern Hemisphere ice sheet, the global temperature T and the intensity C of the deep thermohaline circulation take only the values 0 and 1 ("low" and "high"). They are related through the following

Figure 10.3.1. Asymptotically periodic solutions of Boolean delay equations which model Quaternary glaciation. Time is measured in millennia (ka). (a) Weyl scenario: deep-water turnover time $\tau_{dw} = 1$ ka; (b) Worthington scenario: $\tau_{dw} = 10$ ka (after Ghil *et al.*, 1987).

time-delay equations:

$$V(t) = T(t - \tau_{acc}) \qquad (10.3.3)$$

(higher temperature leads to a stronger hydrological cycle and hence greater snow accumulation, with a growth time scale τ_{acc});

$$T(t) = \{not V(t - \tau_{if})\} \, and \, \{not C(t - \tau_{dw})\} \qquad (10.3.4)$$

(more ice, which spreads with time scale τ_{if}, increases albedo yielding lower temperatures, while stronger thermohaline circulation, which turns over in time τ_{dw}, also lowers temperature);

$$not C(t) = \{not V(t)\} \, and \, V(t - \tau_{dw}) \qquad (10.3.5)$$

(melt water does not sink, while surface freezing raises surface salinity leading to greater sinking). Ghil *et al.* (1987) demonstrated that solutions of the coupled system (10.3.3)–(10.3.5) tend to become periodic after a finite time with period $2\tau_{dw}$. The shapes of the periodic solutions (Fig. 10.3.1) depend upon the values of the time delays (which are measured in units of 1 ka $= 10^3$ years). In particular, each cycle contains significant intervals of glaciation ($V = 1$) provided $\tau_{dw} > \tau_{if} + \tau_{acc}$. This behavior is representative of records of Quaternary glaciation. It is perhaps naive to consider smoothing such

Figure 10.3.2. Monthly averages of daily production rates from St. Margaret's Bay, Nova Scotia. Mean and 95% confidence limits. The spring bloom March-May is evident, but its short time scales are not resolved (after Bennett & Denman, 1989).

records using simple systems of Boolean delay equations like (10.3.3)–(10.3.5), or to consider estimating the time delays from the records using inverse methods, but the idea is certainly appealing. However, the theory of estimation from digital signals is not fully developed; it appears that only discrete values for the independent variable have been considered.

Boolean delay equations might also be used to model plankton blooms in the ocean. Springtime concentrations grow from negligible levels to the carrying capacity of the environment (Fig. 10.3.2) within days (Mooney et al., 1990).

One thing is certain: least-squares estimation is very inefficient for highly intermittent processes, such as fronts and blooms, having probability distributions not well characterized by means and variances.

ii) Discontinuous dynamics

Convective overturning is precluded from quasi- geostrophic models, which impose stable profiles of mean density upon the circulation. Primitive-equation models, on the other hand, allow the entire thermohaline structure to evolve in accordance with the appropriate conservation laws, and so convective overturning is admitted. Such

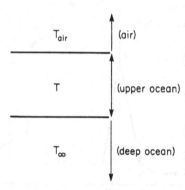

Figure 10.3.3. A trivial model of the thermodynamics of the upper ocean. The air temperature T_{air}, the upper ocean temperature T and deep ocean temperature T_∞ are functions of time only. The upper ocean exchanges heat with the air according to Newton's law of cooling (10.3.6), while deep convective adjustment is modeled by the simple constraint $T \geq T_\infty$.

motions have time scales which are too short to be resolved at present, so it is current practice to parameterize convection in one of several simple ways. The simplest imposes linear constraints on vertical differences in density. The solution stays on the constraint boundary during convective overturning, according to this model. The approach may be illustrated as follows. Let $T = T(t)$ be the time-dependent temperature of the upper ocean (Fig. 10.3.3), which exchanges heat with the atmosphere in accordance with Newton's law of cooling:

$$\dot{T} = -\gamma(T - T_{\text{air}}), \qquad (10.3.6)$$

subject to the constraint that the surface temperature T does not drop below that of the deep ocean:

$$T \geq T_\infty. \qquad (10.3.7)$$

The choice $T_{\text{air}}(t) = A + a\sin\omega t$ represents a simple sinusoidal oscillation about a mean state; the parameters T_∞, A, a, ω and the cooling coefficient γ are all constants. The case of interest is $A - a < T_\infty < A$. A feasible initial condition is provided by

$$T(0) = T_0, \qquad (10.3.8)$$

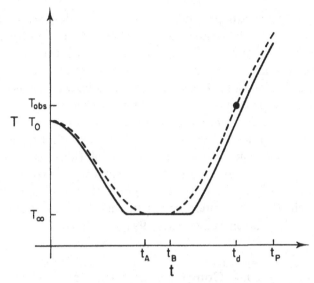

Figure 10.3.4. Generalized inversion of a simple model of convective overturning. Solid line: prior estimate of temperature $T = T(t)$, subject to the lower bound $T \geq T_\infty$. Solid circle: data value at $t = t_\mathrm{d}$. Broken line: posterior estimate, subject to $T \geq T_\infty$. The penalty functional $\mathcal{J} = \mathcal{J}[T, t_A, t_B]$ depends upon the temperature, and also upon the unknown times t_A and t_B for the onset and the end of convective overturning, respectively. Note that the estimate is smooth at t_A and at t_B (after Miller, Zaron & Bennett, 1994).

provided $T_0 \geq T_\infty$. An example of a solution of (10.3.6) and (10.3.8), subject to (10.3.7), is shown in Fig. 10.3.4. Now suppose an observation T_obs is obtained at time t_d:

$$T_\mathrm{obs} = T(t_\mathrm{d}) + \theta, \qquad (10.3.9)$$

where θ is the observational error. In order to smooth the observation using (10.3.6) and (10.3.8), subject to the constraint (10.3.7), it is necessary to speculate that, for example, T is at the deep temperature T_∞ only for $t_A < t < t_B$, and that $t_B < t_P = 2\pi/\omega$. Otherwise t_A and t_B are unknown. A suitable penalty functional is provided by

$$\mathcal{J}[T, t_A, t_B] = W \int_0^{t_A} \left\{ \dot{T} - \gamma(T - T_\mathrm{air}) \right\}^2 dt$$

$$+ w\gamma^2 \int_{t_A}^{t_B} (T_\infty - T_\mathrm{air})^2 dt + w \int_{t_B}^{t_P} \left\{ \dot{T} - \gamma(T - T_\mathrm{air}) \right\}^2 dt$$

$$+ V\big(T(0) - T_0\big)^2 + w\big(T_\mathrm{obs} - T(t_\mathrm{d})\big)^2, \qquad (10.3.10)$$

where W, V and w are positive weights. The global minimum of
(10.3.10) may be sought subject to (10.3.7) and to the condition that T
be continuous at t_A and t_B. The heating rate, which is proportional to
\dot{T}, need not be continuous, since a very rapid convective adjustment is
being modeled. The Euler-Lagrange equations for variations of \mathcal{J} may
be derived in piecewise fashion; in this trivial problem they are easily
solved in terms of elementary functions. The continuity conditions
at the unknown times t_A and t_B yield transcendental equations for
t_A and t_B. These equations have multiple solutions; not all lead to
smooth temperatures at t_A and t_B. A smooth inverse estimate is
shown in Fig. 10.3.4. The prior *assumption* of only a single interval,
during which $T = T_\infty$, underscores further the non-uniqueness of
solutions (Miller, Zaron & Bennett, 1994).

10.4 Computer architecture

i) SIMD and MIMD architecture

The development of inverse methods has benefitted enormously
from the widespread availability of powerful personal computers and
workstations. The latter in particular can have as much memory
as a supercomputer, such as those in the CRAY series, and are fast
enough to give comparable "wall clock" times. That is, a dedicated
workstation may well complete a job, while the same job waits for its
turn to execute on a shared supercomputer. The latter are usually
only available at remote sites, making code development tedious and
file transfers slow. Nevertheless, it is the incipient availability of su-
percomputers possessing quite incomprehensible levels of performance
(one teraflops, or 10^{12} floating-point operations per second) which is
stimulating so much interest in ocean modeling. The examination
of rigorous and general inverse methods in the preceding chapters
anticipates the availability of these computers. There will be com-
petition for access to the "teraflops" machines, so it still behooves
oceanographers to seek the most efficient algorithms for solving their
inverse problems. It may be recalled that for iterative methods,
efficiency and conditioning are virtually synonymous.

There is a consensus in the computer industry that highly parallel
computers will be used to achieve the teraflops level of performance
by about 1995 (Denning & Tichy, 1990). Such computers are already
reaching speeds of 10 gigaflops or 10^{10} flops (Bromley *et al.*, 1991).

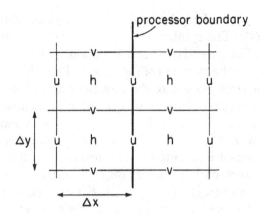

Figure 10.4.1. Distributions of variables on the Arakawa C-grid. In SIMD architecture, each processor is assigned a single (u, v, h) triplet. In MIMD architecture, each processor is assigned a set of triplets. The solid line marks a boundary between two MIMD processors, interior to the flow region. Both processors store the boundary values of u. In order to update u on the boundary, the processors must exchange values of h adjacent to the boundary. This is achieved by passing messages through buffers.

The two most promising architectures are SIMD (single-instruction stream, multiple-data) and MIMD (multiple-instruction stream, multiple-data) (Tanenbaum, 1990). Well-known examples are the Connection Machine (Hillis, 1989) and the Intel iPSC, respectively. Both may be programmed in high-level languages; the Connection Machine (CM) uses FORTRAN–90 (Metcalf & Reid, 1990), while the iPSC accepts FORTRAN–77 with message-passing extensions.

The SIMD architecture of the CM is most easily explained to ocean modelers in terms of an ideal application: integrating the linear shallow-water equations (9.2.1), (9.2.2) in a square (spatial) domain, with doubly-periodic boundary conditions, using centered differences on the C-grid (Arakawa, 1966) and explicit time-stepping (see Fig. 10.4.1). The FORTRAN–90 instructions for updating the fields are

$$u = u - c1 * \big(h - cshift(h, -1, 1)\big) + dt * F, \qquad (10.4.1a)$$

$$v = v - c2 * \big(h - cshift(h, -1, 2)\big) + dt * G, \qquad (10.4.1b)$$

$$h = h - c3 * \big(cshift(u, 1, 1) - u\big)$$
$$- c4 * \big(cshift(v, 1, 2) - v\big), \qquad (10.4.1c)$$

where $\Delta t, \Delta x, \Delta y$ are the grid spacings, $dt = \Delta t$, $c1 = g\Delta t/\Delta x$,

$c2 = g\Delta t/\Delta y$, $c3 = H\Delta t/\Delta x$, and $c4 = H\Delta t/\Delta y$, while $\mathbf{u} = (u, v)$ and $\mathbf{F} = (F, G)$. The symbols u, v, h, , F and G represent arrays, while $c1$–$c4$ and dt are scalars; the instructions (10.4.1) are performed simultaneously on the elements of the arrays. Thus the time to execute (10.4.1a) is the time to execute the subtractions, two multiplications, one addition and one circular shift operation (which replaces $h(i, j)$ with $h(i - 1, j)$ for $2 \le i \le I$, $1 \le j \le J$ and replaces $h(1, j)$ with $h(I, j)$ for $1 \le j \le J$, where $I \times J$ is the array dimension). These instructions are executed simultaneously on all array elements by $I \times J$ processors, each having its own memory containing values of the single elements $u(i, j)$, $v(i, j)$, $h(i, j)$, $F(i, j)$, $G(i, j)$, together with the constants $c1$–$c4$ and dt. The processors are connected to form multidimensional hypercubes, but the natural toroidally wrapped NEWS (North-East-West-South) grid for this application may be selected using proprietary compiler directives. The largest CMs yet built have $2^{16} = 65,536$ processors. These machines would be most efficient for 256×256 grids, or $64 \times 1,024$ grids, for example, or even 16 sets of shallow-water equations approximated on 32×128 grids. Very much larger grids can be accommodated by partitioning local memory. Each memory module then holds array values from many parts of the grid. The availability of 128–kilobyte memory modules allows very large such "virtual processor ratios" (vpr's). The programming of virtual processors is transparent to the user, who merely dimensions as big an array as is needed or will fit in the system. With a vpr of 32, and counting each of $+$, $-$, $*$, $/$, and $cshift$ as one instruction, the code fragment (10.4.1) executes at 8 gigaflops on 2^{16} ($=$ "64k") processors.

It is straightforward to expand the above FORTRAN–90 fragment into code for the free-surface primitive equations (9.3.1)–(9.3.9), which happen to be expressed in isopycnal coordinates. It seems to be advantageous to direct the compiler to assign a water column to each processor, and to treat the vertical index explicitly (P. Schopf, personal communication). If the vpr exceeds unity, then each processor will compute several water columns. The Montgomery potential m is obtained from the pressure p by summing (9.3.2), (9.3.4) using a standard FORTRAN do-loop construction. The code fragment is

$$
\left.
\begin{array}{l}
m(:,:,KK) = a(KK) * p(:,:,KK + 1) + g * zb \\
do \quad 10 \quad k = KK - 1, 1, -1 \\
m(:,:,k) = m(:,:,k + 1) + da(k + 1) * p(:,:,k + 1) \\
10 \quad continue
\end{array}
\right\} \quad (10.4.2)
$$

where KK is the number of isopycnal layers, $a(k) = \alpha_k$ is the specific volume and $da(k) = \Delta\alpha_k$ is the thickness of the kth layer. Note that m and p are three-dimensional arrays, zb is a two-dimensional array of ocean depths, while g is the scalar gravitational acceleration. The colons in strings such as $(:,:,k)$ indicate that the instructions are to be performed simultaneously in the first two coordinates. An eight-layer code of this kind executes at a rate in excess of 4 gigaflops on a 64k CM.

The phase speed of gravity waves on the ocean surface is typically 200 m s^{-1}, while that of the gravest internal mode is about 2 m s^{-1}. Thus the computational stability criterion for a free-surface, primitive-equation model with explicit time-stepping is dominated by the surface mode. This may be ameliorated by operator-splitting (Bleck & Smith, 1990), so that the external or surface mode may be stepped sufficiently rapidly while the internal modes are stepped much more slowly. Larger time steps for the external mode would be possible if an unconditionally stable, explicit scheme were used (Richardson, 1991), but these schemes require far more algebra than do conditionally stable schemes. Another alternative would be to make the rigid-lid approximation; by making the external phase speed infinite, the symmetric-hyperbolic system for the displacement of the free surface is replaced with an elliptic equation for the pressure just beneath the rigid surface. Exact, direct elliptic solvers are not well suited to the world's ocean basins, which have such irregular coastlines. Iterative methods, such as successive over-relaxation, are relatively easily implemented in FORTRAN–90 using logical masks to impose conditions on irregular boundaries (Boghosian, 1990). Similar remarks apply to the inversion of matrices involved in unconditionally stable implicit schemes for time-stepping the external mode (or all modes). The ideal combination of ocean physics and numerics has yet to be determined for SIMD computers such as the CM, and for programming languages such as FORTRAN–90.

The other promising computer architecture (MIMD) has fewer processors than are found in SIMD machines, typically between 2^7 and 2^{10} in number, but MIMD processors are at least one or two orders of magnitude more powerful than those in a SIMD machine. Each MIMD processor may be programmed independently of the others, but the processors may share data by passing "messages" through buffers. Thus each processor may execute a different instruction stream; synchronization is effected only by the availability of needed

messages. One processor may be reserved in the role of "manager," assigning fresh tasks to the "worker" processors as previous tasks are computed. Processors may be given staggered starts in order to avoid input/output (I/O) bottlenecks as tasks are completed. Sophisticated scheduling of this kind becomes essential if the overall job has been decomposed into a large number of widely disparate tasks with heavy I/O requirements.

For the purposes of ocean modeling, it may well be sufficient to use MIMD in a relatively unsophisticated way. The ocean basins or regions of interest may be decomposed into as many subregions as there are processors. Time-dependent circulation within each subregion may be updated by conventional time-stepping, coded in FORTRAN–77, for example, but updating the circulation fields on boundaries requires special treatment. Consider again the linear shallow-water equations (9.2.1), (9.2.2), approximated with centered differences on the C-grid as in Fig. 10.4.1. Values of the eastward velocity u, on a meridional boundary, must be stored in the local memories of the two processors computing the circulation in the subregions separated by the boundary. The values of u are used to evaluate the zonal divergence Hu_x when updating the surface displacement h. Each processor must update the boundary value of u; thus they must share values of h adjacent to the boundary in order to evaluate the gradient gh_x. This is effected by message-passing, that is, by calling proprietary subroutines which send or receive data to or from buffers. Clearly, a processor cannot continue to compute the flow in a subregion until it has received the necessary messages. Efficient routing is therefore of prime importance; early examples of MIMD machines employed hypercube connections between processors, but non-periodic NEWS grids (two-dimensional meshes) now seem to be favored. These are naturally suited to ocean modeling. Detailed analyses of the efficiency of a computational fluid-dynamics code on a MIMD machine (the NCUBE/ten) are reported by Gustafson, Montry & Benner (1988). The ocean modeler intending to use MIMD computers is confronted by the same choices which confront the SIMD user: explicit or implicit physics, and explicit or implicit numerics. While code for SIMD machines must be entirely rewritten in a language with array-handling capabilities such as FORTRAN–90, code for MIMD machines may be written, message-passing aside, in a conventional language such as FORTRAN–77. However, the addition of message-passing can expand the code by factors of two or three.

ii) "Parallelizing" inverse methods

The preceding remarks on highly parallel computers have addressed the possibility of accelerating a single model integration by fine (SIMD) or coarse (MIMD) domain decomposition. Most of the algorithms for solving inverse problems involve many model integrations, both backward and forward, and so highly parallel computers will make possible the acceleration of these algorithms. In the case of gradient descent it should be possible to accelerate stepping down the gradient, in addition to accelerating the calculation of the gradient vector using the adjoint equations. An entirely different approach to accelerating the representer algorithm is possible on a MIMD machine, or on a network of workstations. A single processor code in FORTRAN–77 may be ported to a MIMD machine with the addition of a few proprietary instructions. In terms of the quasi-geostrophic inverse model of Chapter 5, the "parallelized" representer algorithm is as follows (Bennett & Baugh, 1992):

a) Each processor calculates the prior streamfunction estimate ψ_F, by integrating the model (5.4.2) over the *entire* (spatial) domain, subject to the prior initial condition (5.4.3). This presupposes that each processor has sufficient local memory, and is sufficiently fast, to complete such a calculation in a reasonable time.

b) Each processor calculates the prior measurement misfit

$$\mathbf{h} = \mathbf{d} - \boldsymbol{\mathcal{L}}[\psi_F], \qquad (10.4.3)$$

where \mathbf{d} is the data vector and $\boldsymbol{\mathcal{L}}$ the vector of measurement functionals (both vectors have M components).

c) The mth processor calculates the mth representer $r_m(\mathbf{x}, t)$ ($1 \leq m \leq M$) by a backward integration of the mth component of the adjoint representer equation (5.4.6) and final condition (5.4.7) over the entire domain, followed by a forward integration of the mth component of the forward representer equation (5.4.4) subject to the initial condition (5.4.5).

d) The mth processor calculates the mth column of the representer matrix \mathbf{R} by taking all M measurements of r_m: see (5.4.10).

e) The mth processor ($1 \leq m \leq M$) broadcasts the mth column of \mathbf{R} to all the other $M - 1$ processors.

f) Each processor receives the other $M - 1$ columns of \mathbf{R} from the other $M - 1$ processors.

g) Each processor assembles the complete representer matrix **R** from the M columns.

h) Each processor calculates the weighted posterior measurement misfit

$$\mathbf{w}(\mathbf{d} - \mathcal{L}[\hat{\psi}]) = \mathbf{P}^{-1}\mathbf{h}, \qquad (10.4.4)$$

where **w** is the data weight.

i) Each processor integrates the Euler-Lagrange equation (5.3.4) backward from the final condition (5.3.5) (note that the measurement term which couples (5.3.4) to (5.3.2) is now explicitly determined), and then integrates the model equation (5.3.2) from the initial condition (5.3.3) to obtain the posterior streamfunction estimate $\hat{\psi}$.

As already mentioned, the parallel code may be developed from a serial code, in FORTRAN-77, for example, with the addition of a few proprietary subroutine calls. For the Intel iPSC computer these calls are:

$$
\left.
\begin{array}{l}
\vdots \\[4pt]
P = numnode() \\
j = mynode() \\[2pt]
\vdots \\[4pt]
call\ csend\big(j, R(1,j),\ 4*M, -1, 0\big) \\[4pt]
\vdots \\[4pt]
do\ 10 \quad k = 1, M-1 \\
call\ crecv(-1, COL,\ 4*M) \\
l = infotype() \\[4pt]
\vdots \\[4pt]
do\ 20 \quad j = 1, M \\
20 \quad R(j, l) = COL(j) \\
10 \quad continue
\end{array}
\right\} . \qquad (10.4.5)
$$

The call to *numnode* informs the program of the number of available processors or "nodes." The call to *mynode* identifies the processor to the program, which then calculates the corresponding representer. Note that a copy of the program resides on each processor. The call to *csend* sends a message. The first argument is information about the message, here the identity of the processor originating the

message. The argument $R(1, j)$ is the first word in the message, here the first element in the jth column of the $M \times M$ representer matrix \mathbf{R}; the argument $4 * M$ is the single-precision message length in bytes; the argument -1 indicates that the message is to be sent globally, that is, broadcast to all other processors; the argument 0 is an error code. The call to *crecv* receives messages from another processor, storing the message in the one-dimensional array COL. The call to *infotype* obtains the first argument in the call to *csend* (made on another processor) which originated the message, so that COL can be inserted in the appropriate column of \mathbf{R}.

If there are more representers than there are available processors ($M > P$), then each processor must calculate at least M/P representers. This entails additional "bookkeeping," which may be written in FORTRAN–77. The efficiency of the algorithm is discussed in detail in Bennett & Baugh (1992). Briefly, if F, R and E are the times taken to calculate the prior estimate, a representer and the posterior estimate, respectively, then the execution time T_P using P processors is

$$T_P = F + (M/P)R + E, \qquad (10.4.6)$$

provided M/P is an integer. The efficiency of the parallel algorithm, as measured by T_1/T_P, approaches P from below as $M/P \to \infty$. The estimate (10.4.6) is in close agreement with experimental values obtained during an iteration of the nonlinear quasi-geostrophic inverse problem described in §7.6: see Fig. 10.4.2.

Minimization of a penalty functional by simulated annealing with the Metropolis algorithm, as discussed in §6.3, is also amenable to highly parallel programming. A parallel Metropolis algorithm for quantum chromodynamics is described by Brickner & Baillie (1989). That algorithm, written in the parallel language *LISP for the Connection Machine, is readily translated to FORTRAN–90. It is also simply but tediously adapted to the problem of minimizing a penalty functional \mathcal{J}, such as in (9.4.3), for an inverse shallow-water model. The flow variables u, v and q are simultaneously perturbed everywhere; the resulting local perturbations to \mathcal{J} are evaluated simultaneously, and simultaneous decisions are made to accept or reject the flow perturbations. A representative code fragment in FORTRAN–90

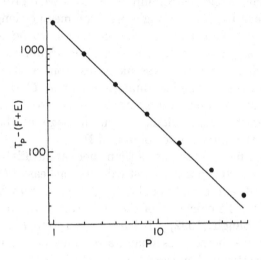

Figure 10.4.2. Execution time T_P versus the number of processors P, for a linear generalized inverse, using the representer algorithm and a MIMD computer (the Intel iPSC860). The times for calculating the prior and posterior estimates are F and E respectively: see (10.4.6). Time is measured in seconds (after Bennett & Baugh, 1992).

is as follows:

$$
\left.
\begin{aligned}
&\vdots \\
&Jpert = Jnew - Jold \\
&ratio = exp(-Jpert/theta) \\
&call\ cmf_random(rand, 0.0) \\
&where\ (ratio.gt.rand) \\
&\qquad Jold = Jnew \\
&endwhere \\
&\vdots
\end{aligned}
\right\} \qquad (10.4.7)
$$

The variables *Jpert, Jnew, Jold, ratio* and *rand* are all arrays of the same shape and size (conformable arrays). There is one array component for each summand in the numerical approximation to the penalty functional \mathcal{J}. The annealing parameter *theta* is a scalar. The proprietary FORTRAN–90 subroutine *cmf_random* returns an array

of independent pseudo-random numbers, uniformly distributed in the
unit interval. The *where–endwhere* block replaces *Jold* with *Jnew*,
thereby accepting the perturbation to the local penalty functional,
wherever *ratio* exceeds *rand.*

Given a large enough SIMD computer, the simulated annealing
algorithm could be implemented in parallel with respect to space and
time, but it is conceptually comforting to treat the time axis serially.
A complication with the penalty functional \mathcal{J} is that perturbing,
say, the velocity component u at one point in space-time alters the
residuals μ, ν and χ at a number of points, and so perturbing the
entire grid must be performed in several sweeps in order to preserve
the independence of the decisions. The random sequence of trials
generated by simulated annealing may be used to estimate moments
of posterior errors (Metropolis *et al.*, 1953).

An example of generalized inversion by simulated annealing is
shown in Fig. 10.4.3. The dynamics are those of nonlinear shallow-
water waves: (9.2.25), (9.2.26), restricted here to motion along one
space coordinate x. Reference solutions are obtained by integrating
the shallow-water equations, for about one period in time, in a channel
subject to periodic boundary conditions at $x = 0$ and at $x = L$.
The reference solution provides initial values and boundary values for
a "nested" integration in the channel subsection $L/4 \leq x \leq 3L/4$.
Denoting mean values of the water velocity u and depth h by U and
H, respectively, the flow is characterized by $U(gH)^{-\frac{1}{2}} \approx 2.5$. That is,
the flow is supercritical. Two physical boundary conditions should be
specified at $x = L/4$, which is always upstream in this integration. No
physical boundary conditions should be specified at $x = 3L/4$, which
is always downstream here. Fig. 10.4.3a shows the result, after about
a period, of specifying one physical boundary condition at $x = L/4$
and one at $x = 3L/4$. A computational boundary condition is needed
at each end of the subsection; simple extrapolation formulae are used
here. The inaccuracy of the integration is seeded by the computational
conditions, but is fundamentally the result of ill-posedness. (The
same combination of physical and computational conditions yields
accurate results for subcritical flow, that is, for well-posed integra-
tions.) A generalized inverse estimate after about a period is shown in
Fig.10.4.3b. Both u and h are specified at each end of the subsection.
Depth data are also provided at a number of interior points, at every
time interval. A weighted least-squares best fit to the system, for the
subsection and for about one period in time, is obtained by simulated

annealing. The supercritical reference solution is accurately recovered. Since annealing involves no integration, the time interval for explicit finite-difference approximations to the dynamics is not restricted by a computational stability criterion (Bennett & Chua, 1992).

10.5 Prior estimates

The conventional definition of an ocean model includes dynamics, forcing for the dynamics, initial conditions and boundary conditions. In recognition of errors in the prior estimates for the forcing, initial values and boundary values, the definition is extended in Chapter 5 to include estimates of the means and covariances of the various errors. These covariances are the natural choice for the (reciprocals of the) weights, when combining the model and a data set into a generalized inverse problem. The example shown in Chapter 7 (compare Figs. 7.6.3 and 7.6.9) shows a poor inverse estimate, which is the typical result if the prior covariances are badly estimated. As in Gauss-Markov interpolation, the essential features of the covariances which must be reliably estimated are the variances and the decorrelation scales. If the covariances are correctly estimated, then the minimal residual is a χ^2-variable with as many degrees of freedom as there are independent data. Significance tests may therefore be used to accept or reject the covariance estimates

While plausible estimates are usually available for the covariances of wind errors, initial errors and perhaps even boundary errors, there has been little opportunity for oceanographers to estimate the statistics of dynamical errors. Hua, McWilliams, & Owens (1986) produced maps of individual terms and of the total residual in potential vorticity balances during the POLYMODE Local Dynamics Experiment. This intensive survey was conducted in a small region of the western North Atlantic (200 km × 200 km, approximately). Quoting from Hua *et al.* (1986), "... the uncertainties in our estimates of the terms in the potential vorticity balance are larger than the expected departures from quasi-geostrophic dynamics. Therefore, we shall not be very concerned with the smallness of the residual to the expected balance, although, of course, this residual contains important ... information about the accuracy of our estimates." In general terms, Hua *et al.* found that the residuals had "somewhat smaller amplitudes and scales" than those of the local rates of change of potential vorticity, "indicating a modest degree of success in demonstrating overall

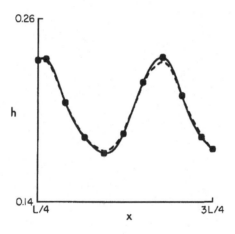

Figure 10.4.3. (a) Nested integration of the nonlinear shallow-water equations, in one space coordinate x. Solid line: reference solution for water depth h after about one period in time. Broken line: ill-posed integration. (b) Generalized inverse estimate of h, with interior data for h, obtained by simulated annealing in the nesting subsection and over about one period in time. Solid line: reference solution. Broken line: annealing solution. Solid circles: depth data.

balance." It may be immediately inferred that inverse estimates of circulation in the region of the Local Dynamics Experiment could only impose the quasi-geostrophic balance of potential vorticity as a very weak constraint; the variance of the dynamical error would have to be around 50% of the variance of the local rate of change of potential vorticity, while the internal deformation radius would provide a sensible estimate for the decorrelation length scale. The subsurface float data collected by Rossby, Price & Webb (1986) during the same survey indicate decorrelation time scales of 7–8 days.

The dynamical error variance of basin-scale or global primitive-equation ocean models resides in the marginally resolved scales. At present, these are the highly active mesoscales such as the internal deformation radius. Regional estimates of dynamical errors are consequently essential to the inversion of oceanic general circulation models. These statistics should be conditional upon the climatological state of the ocean circulation.

BIBLIOGRAPHY

Adams, R. A. (1975). *Sobolev Spaces*. New York: Academic Press.

Amodei, L. (1995). Solution approchée pour un problème d'assimilation de données meteorologiques avec prise en compte de l'erreur de modèle. Comptes Rendus de l'Academie des Sciences, 321, série II a, 1087–1094.

Anderson, D. L. T., Sheinbaum, J. & Haines, K. (1996). Data assimilation in ocean models. *Reports on Progress in Physics*, 59, 1209–1266.

Anderson, D. L. T. & Willebrand, J. (eds.) (1989). *Oceanic Circulation Models: Combining Data and Dynamics*. Dordrecht: Kluwer Academic Publishers.

Arakawa, A. (1966). Computational design for long-term numerical integrations of the equations of atmospheric motion. *Journal of Computational Physics*, 1, 119–143.

Aziz, A. K., Wingate, J. W. & Balas, M. J. (eds.) (1977). *Control Theory of Systems Governed by Partial Differential Equations*. New York: Academic Press.

Backus, G. (1970a). Inference from inadequate and inaccurate data, I. *Proceedings of the National Academy of Sciences*, 65, 1–7.

Backus, G. (1970b). Inference from inadequate and inaccurate data, II. *Proceedings of the National Academy of Sciences*,

65, 281–287.

Backus, G. (1971). Inference from inadequate and inaccurate data. In *Mathematical Problems in the Geophysical Sciences: Lectures in Applied Mathematics*, No. 14. Providence: American Mathematical Society.

Barbieri, R. & Schopf, P. W. (1982). *Oceanographic Applications of the Kalman Filter.* NASA Technical Memorandum, No. TM83993. Greenbelt: National Aeronautics and Space Administration, Goddard Space Flight Center.

Barth, N. & Wunsch, C. (1990). Oceanographic experiment design by simulated annealing. *Journal of Physical Oceanography*, 20, 1249–1263.

Batchelor, G. K. (1960). *The Theory of Homogeneous Turbulence.* Cambridge University Press.

Behringer, D., Birdsall, T., Brown, M., Cornuelle, B., Heinmiller, R., Knox, R., Metzger, K., Munk, W., Spiesberger, J., Spindel, R., Webb, D., Worcester, P. & Wunsch, C. (1982). A demonstration of ocean acoustic tomography. *Nature*, 299, 121–125.

Bennett, A. F. (1978). Poleward heat fluxes in Southern Hemisphere oceans. *Journal of Physical Oceanography*, 8, 785–798.

Bennett, A. F. (1984). Predictability and frontogenesis. In *Predictability of Fluid Motions*, ed. G. Holloway & B. J. West, pp. 295–302. New York: American Institute of Physics.

Bennett, A. F. (1985). Array design by inverse methods. *Progress in Oceanography*, 15, 129–156.

Bennett, A. F. (1990). Inverse methods for assessing ship-of-opportunity networks and estimating circulation and winds from tropical expendable bathythermograph data. *Journal of Geophysical Research*, 95, 16,111–16,148.

Bennett, A. F. & Baugh, J. R. (1992). A parallel algorithm for variational assimilation in oceanography and meteorology. *Journal of Atmospheric and Oceanic Technology*, 9, 426–433.

Bennett, A. F. & Budgell, W. P. (1987). Ocean data assimilation and the Kalman filter: spatial regularity. *Journal of Physical Oceanography*, 17, 1583–1601.

Bennett, A. F. & Budgell, W. P. (1989). The Kalman smoother for a linear quasi-geostrophic model of ocean circulation. *Dynamics of Atmospheres and Oceans*, 13, 219–267.

Bennett, A. F. & Chua, B. S. (1992). Open-ocean modeling as an inverse problem: the primitive equations. *Monthly Weather Review*, 122, 1326–1336.

Bennett, A. F., Chua, B. S., Harrison, D. E. & McPhaden, M. J. (1998) Generalized inversion of Tropical Atmosphere-Ocean (TAO) data and a coupled model of the tropical Pacific. *Journal of Climate*, 11, 1768–1792.

Bennett, A. F., Chua, B. S. & Leslie, L. M. (1996). Generalized inversion of a global numerical weather prediction model. *Meteorology and Atmospheric Physics*, 60, 165–178.

Bennett, A. F., Chua, B. S. & Leslie, L. M. (1997). Generalized inversion of a global numerical weather prediction model, II: analysis and implementation. *Meteorology and Atmospheric Physics*, 62, 129–140.

Bennett, A. F. & Denman, K. L. (1989). Large-scale patchiness due to an annual plankton cycle. *Journal of Geophysical Research*, 94, 823–829.

Bennett, A. F. & Kloeden, P. E. (1978). Boundary conditions for limited area forecasts. *Journal of Atmospheric Science*, 35, 990–996.

Bennett, A. F. & Kloeden, P. E. (1980). The simplified quasigeostrophic equations: existence and uniqueness of strong solutions. *Mathematika*, 27, 287–311.

Bennett, A. F. & Kloeden, P. E. (1981a). The quasigeostrophic equations: approximation, predictability and equilibrium spectra of solutions. *Quarterly Journal of the Royal Meteorological Society*, 107, 121–136.

Bennett, A. F. & Kloeden, P. E. (1981b). The ill-posedness of open
ocean models. *Journal of Physical Oceanography*, 11, 1027–1029.

Bennett, A. F. & Kloeden, P. E. (1982). The periodic quasigeo-
strophic equations: existence and uniqueness of strong solutions.
Proceedings of the Royal Society of Edinburgh, A91, 185–203.

Bennett, A. F., Leslie, L. M., Hagelberg, C. R. & Powers, P. E. (1992).
Tropical cyclone prediction using a barotropic model initialized
by a generalized inverse method. *Monthly Weather Review*, 121,
1714–1729.

Bennett, A. F. & McIntosh, P. C. (1982). Open ocean modeling as
an inverse problem: tidal theory. *Journal of Physical Oceanog-
raphy*, 12, 1004–1018.

Bennett, A. F. & Miller, R. N. (1991). Weighting initial conditions
in variational assimilation schemes. *Monthly Weather Review*,
119, 1098–1102.

Bennett, A. F. & Thorburn, M. A. (1992). The generalized inverse of
a nonlinear quasigeostrophic ocean circulation model. *Journal
of Physical Oceanography*, 22, 213–230.

Bergthorsson, P. & Döös, B. R. (1955). Numerical weather map
analysis. *Tellus*, 7, 329–340.

Binder, K. & Heermann, D. W. (1988). *Monte Carlo Simulation in
Statistical Physics*. Berlin: Springer-Verlag.

Bleck, R. & Smith, L. T. (1990). A wind-driven isopycnic coordinate
model of the North and Equatorial Atlantic Ocean. 1. Model
development and supporting experiments. *Journal of Geophys-
ical Research*, 95, 3273–3285.

Boghosian, B. M. (1990). Computational physics on the Connection
Machine. *Computers in Physics*, 4, 14–33.

Bretherton, F. P., Davis, R. E. & Fandry, C. B. (1976). A technique
for objective analysis and design of oceanographic experiments
applied to MODE-73. *Deep-Sea Research*, 23, 559–582.

Brickner, R. G. & Baillie, C. F. (1989). Pure gauge QCD on the Connection Machine. *International Journal of High Speed Computing*, 1, 303–320.

Broecker, W. S. & Peng, T.-H. (1982). *Tracers in the Sea*. Palisades: Lamont-Doherty Geological Observatory, Columbia University.

Bromley, M., Heller, S., McNerney, J. T. & Steele, G. L., Jr. (1991). Fortran at ten gigaflops: the Connection Machine convolution compiler. In *Proceedings of the ACM SIGPLAN '91 Conference on Programming Language Design and Implementation*, pp. 145–156. Toronto: Association for Computing Machinery.

Bryan, K. (1963). A numerical investigation of a nonlinear model of a wind-driven ocean. *Journal of the Atmospheric Sciences*, 20, 594–606.

Bryan, K. & Cox, M. D. (1968a). A nonlinear model of an ocean driven by wind and differential heating: Part I. Description of the three-dimensional velocity and density fields. *Journal of the Atmospheric Sciences*, 25, 945–967.

Bryan, K. & Cox, M. D. (1968b). A nonlinear model of an ocean driven by wind and differential heating: Part II. An analysis of the heat, vorticity and energy balance. *Journal of the Atmospheric Sciences*, 25, 968–978.

Bryan, K. & Lewis, L. J. (1979). A water mass model of the world ocean. *Journal of Geophysical Research*, 84, 2503–2517.

Bryden, H. L. (1983). The Southern Ocean. In *Eddies in Marine Science*, ed. A. R. Robinson, pp. 265–277. New York: Springer-Verlag.

Bryson, A. E. & Ho, Y.-C. (1975). *Applied Optimal Control*. New York: Hemisphere.

Budgell, W. P. (1982). A dynamic-stochastic approach for modelling advection-dispersion processes in open channels. In *Time Series Methods in Hydrosciences*, ed. A. H. El-Shaarawi & S. R. Esterby, pp. 244–263. Amsterdam: Elsevier Scientific Publishing Company.

Budgell, W. P. (1987). Stochastic filtering of linear shallow water wave processes. *SIAM Journal of Science and Statistical Computing*, 8, 152–170.

Cane, M. A. & Sarachik, E. S. (1981). The response of a linear baroclinic equatorial ocean to periodic forcing. *Journal of Marine Research*, 39, 651–693.

Carrera, J. & Neuman, S. P. (1986a). Estimation of aquifer parameters under transient and steady state conditions: 1. Maximum likelihood method incorporating prior information. *Water Resources Research*, 22, 199–210.

Carrera, J. & Neuman, S. P. (1986b). Estimation of aquifer parameters under transient and steady state conditions: 2. Uniqueness, stability, and solution algorithms. *Water Resources Research*, 22, 211–227.

Carrera, J. & Neuman, S. P. (1986c). Estimation of aquifer parameters under transient and steady state conditions: 3. Application to synthetic and field data. *Water Resources Research*, 22, 228–242.

Charney, J. G., Fjörtoft, R. & von Neumann, J. (1950). Numerical integration of the barotropic vorticity equation. *Tellus*, 2, 237–254.

Chelton, D. B., Freilich, M. H. & Haines, B. J. (1996). Satellite radar remote sensing of the ocean. *(to appear)*.

Cohn, S. E. & Parrish, D. F. (1991). The behavior of forecast error covariances for a Kalman filter in two dimensions. *Monthly Weather Review*, 119, 1757–1785.

Coleman, T. F. & Van Loan, C. (1988). *Handbook for Matrix Computations*. Philadelphia: Society for Industrial and Applied Mathematics.

Courant, R. & Hilbert, D. (1953). *Methods of Mathematical Physics*, vol. 1. New York: Wiley Interscience.

Courant, R. & Hilbert, D. (1962). *Methods of Mathematical Physics*, vol. 2. New York: Wiley Interscience.

Courtier, P., Derber, J., Errico, R., Louis, J-F. & Vukicevic, T. (1993). Important literature on the use of adjoint, variational methods and the Kalman filter in meteorology. *Tellus*, 45A, 342–357.

Courtier, P. & Talagrand, O. (1987). Variational assimilation of meteorological observations with the adjoint vorticity equation. II: Numerical results. *Quarterly Journal of the Royal Meteorological Society*, 113, 1329–1347.

Cressman, G. P. (1958). Barotropic divergence and very long atmospheric waves. *Monthly Weather Review*, 86, 293–297.

Curtain, R. F. & Pritchard, A. J. (1978). *Infinite Dimensional Linear Systems Theory*. Lecture Notes in Control Theory and Information Sciences, vol. 8. Berlin: Springer-Verlag.

Daley, R. A. (1991). *Atmospheric Data Analysis*. Cambridge University Press.

Davies, H. C. (1976). A lateral boundary formulation for multi-level prediction models. *Quarterly Journal of the Royal Meteorological Society*, 102, 405–418.

Davis, R. E. (1978). On estimating velocity from hydrographic data. *Journal of Geophysical Research*, 83, 5507–5509.

Denman, K. L. & Freeland, H. J. (1985). Correlation scales, objective mapping and a statistical test of geostrophy over the continental shelf. *Journal of Marine Research*, 43, 517–539.

Denning, P. J. & Tichy, W. F. (1990). Highly parallel computation. *Science*, 250, 1217–1222.

Dongarra, J. J., Moler, C. B., Bunch, J. R. & Stewart, G. W. (1979). *Linpack Users' Guide*. Philadelphia: Society for Industrial and Applied Mathematics.

Egbert, G. D. (1998). Tidal data inversion: interpolation and inference.. *Progress in Oceanography*, 40, 53–80.

Egbert, G. D. & Bennett, A. F. (1996). Data assimilation methods for ocean tides. In *Modern Approaches to Data Assimilation in Ocean Modelling*, ed. P. Malanotte-Rizzoli, pp. 147–179. Amsterdam: Elsevier Press.

Egbert, G. D., Bennett, A. F. & Foreman, M. G. G. (1994). TOPEX/POSEIDON tides estimated using a global inverse model. *Journal of Geophysical Research*, 99, 24821–24852.

Eliassen, A. (1949). The quasi-static equations of motion with pressure as independent variable. *Geofysiske Publikasjoner*, 17, 1–44.

Eliassen, A. & Kleinschmidt, E. (1957). Dynamic meteorology. In *Handbuch der Physik*, ed. J. Bartels, pp. 1–154. Berlin: Springer-Verlag.

Elsberry, R. L. (1990). International experiments to study tropical cyclones in the western North Pacific. *Bulletin of the American Meteorological Society*, 71, 1305–1316.

Fiadeiro, M. E. & Veronis, G. (1984). Obtaining velocities from tracer distributions. *Journal of Physical Oceanography*, 14, 1734–1746.

Fix, G. J. (1975). Finite element models for ocean circulation problems. *SIAM Journal of Applied Mathematics*, 29, 371–387.

Freeland, H. & McIntosh, P. (1989). The vorticity balance on the southern British Columbia continental shelf. *Atmosphere-Ocean*, 27, 643–657.

Fu, L.-L. (1981). The general circulation and meridional heat transport of the subtropical South Atlantic determined by inverse methods. *Journal of Physical Oceanography*, 11, 1171–1193.

Fu, L.-L., Vazquez, J. & Perigaud, C. (1991). Fitting dynamical models to the Geosat sea level observations in the tropical Pacific Ocean. Part I: A free wave model. *Journal of Physical Oceanography*, 21, 798–809.

Fukumori, I., Martel, F. & Wunsch, C. (1991). The hydrography of the North Atlantic in the early 1980s. An atlas. *Progress in Oceanography*, 27, 1–110.

Gandin, L. S. (1965). *Objective Analysis of Meteorological Fields.* Jerusalem: Israel Program for Scientific Translations.

Gaspar, P. & Wunsch, C. (1989). Estimates from altimeter data of barotropic Rossby waves in the northwestern Atlantic Ocean. *Journal of Physical Oceanography*, 19, 1821–1844.

Gelb, A. (ed.) (1974). *Applied Optimal Estimation.* Cambridge: MIT Press.

Gelfand, I. M. & Fomin, S. V. (1963). *Calculus of Variations.* Englewood Cliffs: Prentice-Hall.

Ghil, M., Ide, K., Bennett, A., Courtier, P., Kimoto, M., Nagata, M., Saiki, M. & Sato, N., (eds) (1997). *Data Assimilation in Meteorology and Oceanography, Theory and Practice.* Tokyo: The Meteorological Society of Japan.

Ghil, M., Mullhaupt, A. & Pestiaux, P. (1987). Deep water formation and Quaternary glaciations. *Climate Dynamics*, 2, 1–10.

Gill, A. E. (1982). *Atmosphere-Ocean Dynamics.* San Diego: Academic Press.

Goldenberg, S. B. & O'Brien, J. J. (1981). Time and space variability of tropical Pacific wind stress. *Monthly Weather Review*, 109, 1190–1207.

Goldstein, S. (1960). *Lectures on Fluid Mechanics.* London: Interscience.

Gustafson, J. L., Montry, G. R. & Benner, R. E. (1988). Development of parallel methods for a 1024-processor hypercube. *SIAM Journal on Scientific and Statistical Computing*, 9, 609–638.

Hagelberg, C. R. (1992). *Existence of a Solution to a Variational Data Assimilation Method in Two-dimensional Hydrodynamics.*

Ph.D. thesis, Department of Mathematics, Oregon State University, Corvallis.

Haidvogel, D. B., Robinson, A. R. & Schulman, E. E. (1980). The accuracy, efficiency and stability of three numerical methods with application to open ocean problems. *Journal of Computational Physics*, 34, 1–53.

Haidvogel, D. B. & Robinson, A. R. (eds.) (1989). Special issue on data assimilation. *Dynamics of Atmospheres and Oceans*, 13, 171–518.

Haltiner, G. J. & Williams, R. T. (1980). *Numerical Weather Prediction*, 2nd ed. New York: John Wiley & Sons.

Hartman, P. (1973). *Ordinary Differential Equations*. Baltimore: Hartman.

Haskel, K. H. & Hanson, R. J. (1981). An algorithm for linear least squares problems with equality and nonnegativity constraints. *Mathematical Programming*, 21, 98–118.

Hastenrath, S. (1980). Heat budget of tropical ocean and atmosphere. *Journal of Physical Oceanography*, 10, 159–170.

Hellerman, S. & Rosenstein, M. (1983). Normal monthly wind stress over the world ocean with error estimates. *Journal of Physical Oceanography*, 13, 1093–1104.

Hillis, W. D. (1989). *The Connection Machine*. Cambridge: MIT Press.

Holland, G. J., Leslie, L. M., Ritchie, E. A., Dietachmayer, G. S., Powers, P. F. & Klink, M. (1991). An interactive analysis and forecast system for tropical cyclone motion. *Weather and Forecasting*, 6, 415–424.

Holland, W. R. & Lin, L. B. (1975a). On the generation of mesoscale eddies and their contribution to the oceanic general circulation. I. A preliminary numerical experiment. *Journal of Physical Oceanography*, 5, 642–657.

Holland, W. R. & Lin, L. B. (1975b). On the generation of mesoscale eddies and their contribution to the oceanic general circulation. II. A parameter study. *Journal of Physical Oceanography*, 5, 658–669.

Holland, W. R. & Malanotte-Rizzoli, P. (1989). Assimilation of altimeter data into an ocean circulation model: space versus time resolution studies. *Journal of Physical Oceanography*, 19, 1507–1534.

Hua, B. L., McWilliams, J. C. & Owens, W. B. (1986). An objective analysis of the POLYMODE Local Dynamics Experiment. Part II: Streamfunction and potential vorticity fields during the intensive period. *Journal of Physical Oceanography*, 16, 506–522.

Huang, R. X. & Bryan, K. (1987). A multilayer model of the thermohaline and wind-driven ocean circulation. *Journal of Physical Oceanography*, 17, 1909–1924.

Hurlburt, H. E. & Thompson, J. D. (1980). A numerical study of Loop Current intrusions and eddy shedding. *Journal of Physical Oceanography*, 10, 1611–1651.

Jackson, J. D. (1962). *Classical Electrodynamics*. New York: John Wiley & Sons.

Joyce, T. M., Wunsch, C. & Pierce, S. D. (1986). Synoptic Gulf Stream velocity profiles through simultaneous inversion of hydrographic and acoustic doppler data. *Journal of Geophysical Research*, 91, 7573–7585.

Judovich, V. I. (1966). A two-dimensional problem of unsteady flow of an ideal incompressible fluid across a given domain. *American Mathematical Society Translations*, 57, 277–304.

Kalman, R. E. (1960). A new approach to linear filter and prediction problems. *Journal of Basic Engineering*, 82, 35–45.

Kalman, R. E. & Bucy, R. S. (1961). New results of linear filtering and prediction theory. *Journal of Basic Engineering*, 83, 95–108.

Killworth, P. D. (1979). A note on velocity determination from hydrographic data. *Journal of Geophysical Research*, 84, 5093–5094.

Kindle, J. C. (1986). Sampling strategies and model assimilation of altimetric data for ocean monitoring and prediction. *Journal of Geophysical Research*, 91, 2418–2432.

Kirkpatrick, S., Gelatt, C. & Vecchi, M. (1983). Optimization by simulated annealing. *Science*, 220, 671–680.

Kraichnan, R. H. (1965). Lagrangian-history closure for turbulence. *Physics of Fluids*, 8, 575–598.

Ladyzhenskaya, O. A. (1969). *The Mathematical Theory of Viscous Incompressible Flow*, 2nd English ed. New York: Gordon & Breach.

Lanczos, C. (1961). *Linear Differential Operators*. New York: Van Nostrand.

Lawson, C. L. & Hanson, R. J. (1974). *Solving Least Squares Problems*. Englewood Cliffs: Prentice-Hall.

Levitus, S. (1982). *Climatological Atlas of the World Ocean*. Professional Papers, vol. 13. Rockville: National Oceanic and Atmospheric Administration.

Lighthill, M. J. (1958). *Fourier Analysis and Generalised Functions*. Cambridge University Press.

Ljung, L. & Söderström, T. (1987). *Theory and Practice of Recursive Identification*. Cambridge: MIT Press.

Lorenc, A. C. (1986). Analysis methods for numerical weather prediction. *Quarterly Journal of the Royal Meteorological Society*, 112, 1177–1194.

Lorenz, E. N. (1967). *The Nature and Theory of the General Circulation of the Atmosphere*. Geneva: World Meteorological Organization.

Luenberger, D. C. (1984). *Linear and Nonlinear Programming*. Reading: Addison-Wesley.

Mackas, D. L., Denman, K. L. & Bennett, A. F. (1987). Least squares multiple tracer analysis of water mass composition. *Journal of Geophysical Research*, 92, 2907–2918.

Malanotte-Rizzoli, P. (ed) (1996). *Modern Approaches to Data Assimilation in Ocean Modeling*. Amsterdam: Elsevier Press.

Malanotte-Rizzoli, P. & Holland, W. R. (1986). Data constraints applied to models of the ocean general circulation. Part I: The steady case. *Journal of Physical Oceanography*, 16, 1665–1682.

Malanotte-Rizzoli, P. & Holland, W. R. (1988). Data constraints applied to models of the ocean general circulation. Part II: The transient, eddy-resolving case. *Journal of Physical Oceanography*, 18, 1093–1107.

Mariano, A. J. (1990). Contour analysis: a new approach for melding geophysical fields. *Journal of Atmospheric and Oceanic Technology*, 7, 285–295.

Marshall, J. C. (1985). Determining the ocean circulation and improving the geoid from satellite altimetry. *Journal of Physical Oceanography*, 15, 330–349.

McIntosh, P. C. (1984). *Tidal Modelling as an Inverse Problem*. Ph.D thesis, Department of Mathematics, Monash University, Melbourne.

McIntosh, P. C. (1987a). Systematic design of observational arrays. *Journal of Physical Oceanography*, 17, 885–902.

McIntosh, P. C. (1987b). Comment on "Solution of nonlinear finite-difference ocean models by optimization methods with sensitivity and observational strategy analysis". *Journal of Physical Oceanography*, 17, 839–840.

McIntosh, P. C. (1990). Oceanographic data interpolation: objective analysis and splines. *Journal of Geophysical Research*, 95, 13,529–13,541.

McIntosh, P. C. & Bennett, A. F. (1984). Open ocean modeling as an inverse problem: M_2 tides in Bass Strait. *Journal of Physical Oceanography*, 14, 601–614.

McPhaden, M. J. & Taft, B. A. (1984). *Proceedings of the First International TOGA Workshop on Thermal Sampling*. US-TOGA Publication, No. 3. Boulder: Universities Corporation for Atmospheric Research.

McWilliams, J. C. (1976). Maps from the Mid-Ocean Dynamics Experiment: Part I. Geostrophic streamfunction. *Journal of Physical Oceanography*, 6, 810–827.

Meditch, J. S. (1970). On state estimation for distributed parameter systems. *Journal of the Franklin Institute*, 290, 49–59.

Mellor, G. L. & Ezer, T. (1991). A Gulf Stream model and an altimetry assimilation scheme. *Journal of Geophysical Research*, 96, 8779–8795.

Menke, W. (1984). *Geophysical Data Analysis: Discrete Inverse Theory*. Orlando: Academic Press.

Mercier, H. (1989). A study of the time-averaged circulation in the western North Atlantic by simultaneous nonlinear inversion of hydrographic and current meter data. *Deep-Sea Research*, 36, 297–313.

Metcalf, M. & Reid, J. (1990). *Fortran 90 Explained*. Oxford: Clarendon Press.

Metropolis, N., Rosenbluth, A. W., Rosenbluth, M. N., Teller, A. H. & Teller, E. (1953). Equation of state calculations by fast computing machines. *The Journal of Chemical Physics*, 21, 1087–1092.

Miller, R. N. (1986). Toward the application of the Kalman filter to regional open ocean modeling. *Journal of Physical Oceanography*, 16, 72–86.

Miller, R. N. (1987). *Theory and Practice of Data Assimilation for Oceanography*. Reports in Meteorology and Oceanography, No. 26. Cambridge: Harvard University.

Miller, R. N. (1989). Direct assimilation of altimetric differences using the Kalman filter. *Dynamics of Atmospheres and Oceans*, 13, 317–333.

Miller, R. N. & Bennett, A. F. (1988). Numerical simulation of flows with locally characteristic boundaries. *Tellus*, 40A, 303–323.

Miller, R. N. & Cane, M. A. (1989). A Kalman filter analysis of sea level height in the tropical Pacific. *Journal of Physical Oceanography*, 19, 773–790.

Miller, R. N., Robinson, A. R. & Haidvogel, D. B. (1983). A baroclinic quasi-geostrophic open ocean model. *Journal of Computational Physics*, 50, 38–70.

Miller, R. N., Zaron, E. D. & Bennett, A. F. (1994). Data assimilation in models with convective adjustment. *Monthly Weather Review*, 122, 2607–2613.

Mooers, C. N. K. & Robinson, A. R. (1984). Turbulent jets and eddies in the California current and inferred cross-shore transports. *Science*, 223, 51–53.

Mooney, H., Bretherton, F. P., Baker, D. J., Jr., Burke, K. C., Clark, W. C., Davis, M. B., Dickinson, R. E., Imbrie, J., Kates, R. W., Malone, T. F., McElroy, M. B., Moore, B. S., III, Mosley-Thompson, E. S., Risser, P. G. & Sellers, P. J. (1990). *Research Strategies for the U.S. Global Change Research Program*. Washington, DC: National Academy Press.

Moore, A. M. (1991). Data assimilation in a quasi-geostrophic open-ocean model of the Gulf Stream region using the adjoint method. *Journal of Physical Oceanography*, 21, 398–427.

Moore, A. M., Cooper, N. S. & Anderson, D. L. T. (1987). Initialization and data assimilation in models of the Indian Ocean. *Journal of Physical Oceanography*, 17, 1965–1977.

Munk, W. H. (1950). On the wind-driven ocean circulation. *Journal of Meteorology*, 7, 79–93.

Noble, B. & Daniel, J. W. (1988). *Applied Linear Algebra*, 3rd ed. Englewood Cliffs: Prentice-Hall.

Norbury, J. & Cullen, M. J. P. (1985). A note on the properties of the primitive hydrostatic equations of motion. *Quarterly Journal of the Royal Meteorological Society*, 111, 1135–1137.

Oberhuber, J. M. (1993a). Simulation of the Atlantic circulation with a coupled sea ice-mixed layer-isopycnal general circulation model. Part I: model description. *Journal of Physical Oceanography*, 23, 808-829.

Oberhuber, J. M. (1993b). Simulation of the Atlantic circulation with a coupled sea ice-mixed layer-isopycnal general circulation model. Part II: model experiment. *Journal of Physical Oceanography*, 23, 830–845.

Olbers, D. J., Wenzel, M. & Willebrand, J. (1985). The inference of North Atlantic circulation patterns from climatological hydrographic data. *Reviews of Geophysics*, 23, 313–356.

Oliger, J. & Sundström, A. (1978). Theoretical and practical aspects of some initial boundary value problems in fluid dynamics. *SIAM Journal of Applied Mathematics*, 35, 419–446.

Panchang, V. G. & O'Brien, J. J. (1988). On the determination of hydraulic model parameters using the adjoint state formulation. In *Modeling Marine Systems*, ed. A. M. Davies, pp. 5–18. Boca Raton: CRC Press.

Parker, R. L. (1994). *Geophysical Inverse Theory*. Princeton: Princeton University Press.

Patterson, S. L. & Whitworth, T. (1990). Physical oceanography. In *Antarctic Sector of the Pacific*, ed. G. P. Glasby, pp. 55–93. Amsterdam: Elsevier.

Pedlosky, J. (1987). *Geophysical Fluid Dynamics*, 2nd ed. New York: Springer-Verlag.

Peressini, A. L., Sullivan, F. E. & Uhl, J. J., Jr. (1988). *The Mathematics of Nonlinear Programming.* New York: Springer-Verlag.

Philander, S. G. (1990). *El Niño, La Niña, and the Southern Oscillation.* San Diego: Academic Press.

Phillips, N. A. (1986). The spatial statistics of random geostrophic modes and first-guess errors. *Tellus,* 38A, 314–332.

Phillips, O. M. (1977). *Dynamics of the Upper Ocean,* 2nd ed. Cambridge University Press.

Pickard, G. L. & Emery, W. J. (1990). *Descriptive Physical Oceanography,* 5th enlarged ed. Oxford: Pergamon Press.

Pierce, S. D. & Joyce, T. M. (1988). Gulf Stream velocity structure through inversion of hydrographic and acoustic Doppler data. *Journal of Geophysical Research,* 93, 2227–2236.

Pinardi, N. & Robinson, A. R. (1987). Dynamics of deep thermocline jets in the POLYMODE region. *Journal of Physical Oceanography,* 17, 1163–1188.

Press, W. H., Flannery, B. P., Teukolsky, S. A. & Vetterling, W. T. (1986). *Numerical Recipes: The Art of Scientific Computing.* Cambridge University Press.

Provost, C. & Salmon, R. (1986). A variational method for inverting hydrographic data. *Journal of Marine Research,* 44, 1–34.

Rasmussen, E. M. & Carpenter, T. H. (1983). Variations in tropical sea surface temperature and surface wind fields associated with the Southern Oscillation/El Niño. *Monthly Weather Review,* 110, 354–384.

Reid, C. (1976). *Courant in Göttingen and New York. The Story of an Improbable Mathematician.* New York: Springer-Verlag.

Reid, W. T. (1968). Generalized inverses of differential and integral operators. In *Theory and Applications of Generalized Inverse of*

Matrices, ed. T. L. Boullion & P. L. Odell, pp. 1–25. Lubbock: Texas Technical College.

Richardson, J. L. (1991). Visualizing quantum scattering on the CM-2 supercomputer. *Computer Physics Communications*, 63, 84–94.

Rienecker, M. M. & Miller, R. N. (1991). Ocean data assimilation using optimal interpolation with a quasi-geostrophic model. *Journal of Geophysical Research*, 96, 15,093–15,103.

Rienecker, M. M., Mooers, C. N. K. & Robinson, A. R. (1987). The evolution of mesoscale features off northern California: dynamical interpolation and forecast experiments. *Journal of Physical Oceanography*, 17, 1189–1213.

Robert, A. & Yakimiw, E. (1986). Identification and elimination of an inflow boundary computational solution in limited area model integrations. *Atmosphere-Ocean*, 24, 369–385.

Robinson, A. R. (ed.) (1983). *Eddies in Marine Science*. Berlin: Springer-Verlag.

Robinson, A. R. (1987). Predicting ocean currents, fronts, and eddies. In *Three-Dimensional Models of Marine and Estuarine Dynamics*, ed. J. C. J. Nihoul & B. M. Jamart, pp. 89–112. Amsterdam: Elsevier Science.

Robinson, A. R., Carton, J. A., Mooers, C. N. K., Carter, E. F., Rienecker,M. M., Smith, J. A. & Leslie, W. G. (1984). A real-time dynamical forecast of ocean synoptic/mesoscale eddies. *Nature*, 309, 782–783.

Robinson, A. R., Carton, J. A., Pinardi, N. & Mooers, C. N. K. (1986). Dynamical forecasting and dynamical interpolation: an experiment in the California current. *Journal of Physical Oceanography*, 16, 1561–1579.

Robinson, A. R. & Haidvogel, D. B. (1980). Dynamical forecast experiments with a barotropic open ocean model. *Journal of Physical Oceanography*, 10, 1909–1920.

Robinson, A. R. & Leslie, W. G. (1985). Estimation and prediction of oceanic eddy fields. *Progress in Oceanography*, 14, 485–510.

Robinson, A. R., Spall, M. A. & Pinardi, N. (1988). Gulf stream simulations and the dynamics of ring and meander processes. *Journal of Physical Oceanography*, 18, 1811–1853.

Roemmich, D. (1980). Estimation of meridional heat flux in the North Atlantic by inverse methods. *Journal of Physical Oceanography*, 10, 1972–1983.

Rossby, T., Price, J. & Webb, D. (1986). The spatial and temporal evolution of a cluster of SOFAR floats in the POLYMODE Local Dynamics Experiment (LDE). *Journal of Physical Oceanography*, 16, 428–442.

Sasaki, Y. (1955). A variational study of the numerical prediction based on the variational principle. *Journal of the Meteorological Society of Japan*, 33, 262–275.

Schlitzer, R. (1987). Renewal rates of East Atlantic deep water estimated by inversion of ^{14}C data. *Journal of Geophysical Research*, 92, 2953–2969.

Schröter, J. (1989). Driving of non-linear time-dependent ocean models by observation of transient tracers–a problem of constrained optimisation. In *Oceanic Circulation Models: Combining Data and Dynamics*, ed. D. L. T. Anderson & J. Willebrand, pp. 257–285. Dordrecht: Kluwer Academic Publishers.

Schröter, J. & Wunsch, C. (1986). Solution of nonlinear finite-difference ocean models by optimization methods with sensitivity and observational strategy analysis. *Journal of Physical Oceanography*, 16, 1855–1874.

Semtner, A. J., Jr. & Chervin, R. M. (1988). A simulation of the global ocean circulation with resolved eddies. *Journal of Geophysical Research*, 93, 15,502–15,522.

Shapiro, R. (1970). Smoothing, filtering, and boundary effects. *Reviews in Geophysics and Space Physics*, 8, 359–387.

Sheinbaum, J. & Anderson, D. L. T. (1990a). Variational assimilation of XBT data. Part I. *Journal of Physical Oceanography*, 20, 672–688.

Sheinbaum, J. & Anderson, D. L. T. (1990b). Variational assimilation of XBT data. Part II: Sensitivity studies and use of smoothing constraints. *Journal of Physical Oceanography*, 20, 689–704.

Shen, C. Y., McWilliams, J. C., Taft, B. A., Ebbesmeyer, C. C. & Lindstrom, E. J. (1986). The mesoscale spatial structure and evolution of dynamical and scalar properties observed in the northwestern Atlantic Ocean during the POLYMODE local dynamics experiment. *Journal of Physical Oceanography*, 16, 454–482.

Smedstad, O. M. & O'Brien, J. J. (1991). Variational data assimilation and parameter estimation in an equatorial Pacific Ocean model. *Progress in Oceanography*, 26, 179–241.

Smith, J. A., Mooers, C. N. K. & Robinson, A. R. (1985). Estimation of baroclinic quasi-geostrophic modal amplitudes from XBT/CDT survey data. *Journal of Atmospheric and Oceanic Technology*, 2, 491–507.

Sommerfeld, A. (1952). *Electrodynamics*. New York: Academic.

Spall, M. A. & Holland, W. R. (1991). A nested primitive equation model for oceanic applications. *Journal of Physical Oceanography*, 21, 205–220.

Spall, M. A. & Robinson, A. R. (1990). Regional primitive equation studies of the Gulf Stream meander and ring formation region. *Journal of Physical Oceanography*, 20, 985–1016.

Stewart, R. H. (1985). *Methods of Satellite Oceanography*. Berkeley: University of California Press.

Stommel, H. (1948). The westward intensification of wind-driven ocean currents. *Transactions, American Geophysical Union*, 29, 202–206.

Stommel, H. (1960). *The Gulf Stream.* Berkeley: University of California Press.

Stommel, H. (1987). *A View of the Sea.* Princeton: Princeton University Press.

Stommel, H. (1989). Why we are oceanographers. *Oceanography,* 2, 48–54.

Stommel, H. & Schott, F. (1977). The beta spiral and the determination of the absolute velocity field from hydrographic station data. *Deep-Sea Research,* 24, 325–329.

Stommel, H. & Yoshida, K. (1972). *Kuroshio: Its Physical Properties.* Tokyo: University of Tokyo Press.

Stone, P. H. (1966). Frontogenesis by horizontal wind deformation fields. *Journal of the Atmospheric Sciences,* 23, 455–465.

Sundström, A. (1969). Stability theorems for the barotropic vorticity equation. *Monthly Weather Review,* 97, 340–345.

Taft, B. A., Lindstrom, E. J., Ebbesmeyer, C. C., Shen, C. Y. & McWilliams, J. C. (1986). Water mass structure during the POLYMODE local dynamics experiment. *Journal of Physical Oceanography,* 16, 403–426.

Talagrand, O. & Courtier, P. (1987). Variational assimilation of meteorological observations with the adjoint vorticity equation. I: Theory. *Quarterly Journal of the Royal Meteorological Society,* 113, 1311–1328.

Tanenbaum, A. (1990). *Structured Computer Organization,* 3rd ed. Englewood Cliffs: Prentice-Hall.

Tarantola, A. (1987). *Inverse Problem Theory. Methods for Data Fitting and Model Parameter Estimation.* Amsterdam: Elsevier Science Publishers.

Tarbell, S. A., Pennington, N. J. & Briscoe, M. G. (1984). *A Compilation of Moored Current Meter and Wind Recorder Data. Long-Term Upper Ocean Study (LOTUS) (Moorings 764, 765, 766,*

767, 770). Woods Hole Oceanographic Institution Technical Report, No. WHOI-84-36. Woods Hole: Woods Hole Oceanographic Institution.

Thacker, W. C. (1989). The role of the Hessian matrix in fitting models to measurements. *Journal of Geophysical Research*, 94, 6177–6196.

Thacker, W. C. & Long, R. B. (1988). Fitting dynamics to data. *Journal of Geophysical Research*, 93, 1227–1240.

Thiébaux, H. J. & Pedder, M. A. (1987). *Spatial Objective Analysis: With Applications in Atmospheric Science*. London: Academic Press.

Tziperman, E. & Thacker, W. C. (1989). An optimal-control/adjoint-equations approach to studying the oceanic general circulation. *Journal of Physical Oceanography*, 19, 1471–1485.

Wahba, G. (1990). *Spline Models for Observational Data*. Philadelphia: Society for Industrial and Applied Mathematics.

Wahba, G. & Wendelberger, J. (1980). Some new mathematical methods for variational objective analysis using splines and cross validation. *Monthly Weather Review*, 108, 1122–1143.

Weinert, H. L. (ed.) (1982). *Reproducing Kernel Hilbert Spaces, Applications in Statistical Signal Processing*. Stroudsburg: Hutchinson Ross.

WOCE International Planning Office (1988). *World Ocean Circulation Experiment Implementation Plan*, vol. 2 (Scientific Background). Wormley: World Meteorological Organization.

Wolibner, W. (1933). Un théorème sur l'existence du mouvement plan d'un fluide parfait, homogène, incompressible, pendant un temps infiniment longue. *Mathematische Zeitschrift*, 37, 698–726.

Worcester, P. F., Cornuelle, B. D., & Spindel, R. C. (1991). A review of ocean acoustic tomography: 1987–1990. *Reviews of Geophysics, Supplement*, 29, 557–570.

Worthington, L. V. (1976). *On the North Atlantic Circulation.* Baltimore: Johns Hopkins.

Wunsch, C. (1977). Determining the general circulation of the oceans: a preliminary discussion. *Science*, 196, 871–875.

Wunsch, C. (1978). The North Atlantic general circulation west of 50° W determined by inverse methods. *Reviews of Geophysics and Space Physics*, 16, 583–620.

Wunsch, C. (1984). An eclectic Atlantic Ocean circulation model. Part I: The meridional flux of heat. *Journal of Physical Oceanography*, 14, 1712–1733.

Wunsch, C. (1985). Can a tracer field be inverted for velocity? *Journal of Physical Oceanography*, 15, 1521–1531.

Wunsch, C. (1988a). Transient tracers as a problem in control theory. *Journal of Geophysical Research*, 93, 8099–8110.

Wunsch, C. (1988b). Eclectic modelling of the North Atlantic. II. Transient tracers and the ventilation of the Eastern Basin thermocline. *Philosophical Transactions of the Royal Society of London*, A325, 201–236.

Wunsch, C. (1996). *The Ocean Circulation Inverse Problem.* Cambridge University Press.

Wunsch, C. & Minster, J.-F. (1982). Methods for box models and ocean circulation tracers: mathematical programming and non-linear inverse theory. *Journal of Geophysical Research*, 87, 5647–5662.

Wüst, G. (1935). Schichtung und Zirkulation des Atlantischen Ozeans. Die Stratosphäre. In *Wissenschaftliche Ergebnisse der Deutschen Atlantischen Expedition auf dem Forschungs und— Vermessungsschiff "Meteor" 1925–1927*, 6, 1st Part, 2. (Emery, W. J. (ed.) (1978). *The Stratosphere of the Atlantic Ocean.* New Delhi: Amerind.)

Wyrtki, K., Constantine, K., Kilonsky, B. J., Mitchum, G., Miyamoto, B., Murphy, T., Nakahara, S. & Caldwell, P. (1988). *The Pacific*

Island Sea Level Network. JIMAR Contribution No. 88–0137–002. Honolulu: Joint Institute for Marine and Atmospheric Research, University of Hawaii.

Yaglom, A. M. (1962). *An Introduction to the Theory of Stationary Random Functions.* New York: Prentice-Hall.

Yakimiw, E. & Robert, A. (1990). Validation experiments for a nested grid-point regional forecast model. *Atmosphere-Ocean*, 28, 467–472.

Yeh, W. W-G (1986). Review of parameter identification procedures in groundwater hydrology: the inverse problem. *Water Resources Research*, 22, 95–108.

Yu, L. & O'Brien, J. J. (1991). Variational estimation of the wind stress drag coefficient and the oceanic eddy viscosity profile. *Journal of Physical Oceanography*, 21, 709–719.

SUBJECT INDEX

*, transposed conjugate,
91
^, inverse estimate, 277
β-plane, 2
 equatorial, 85, 163
χ^2-variable, 156, 199, 315
acoustic tomography, 49, 171, 173
adiabatic motion, 268
adjoint operator, 116, 167
adjoint representer, 119, 242
adjoint variable, 115, 132, 185
 discontinuities in, 130
advection-diffusion equation, 255
algorithm
 ideal minimization, 281
 RTS, 135
 Simplex, 39
 sweep, 129, 241, 279
altimetry, 71, 88, 104
anisotropy, 54, 85
annealing parameter, 177
annealing, simulated, 175, 177, 282,
 311
antenna element, principal, 152, 154,
 170
antenna mode
 residual, 152, 154
 solution, 152, 154, 173
approximation
 Boussinesq, 2, 273
 long-wave, 85

BLUE (best linear unbiased
 estimate), 48, 61, 65
behavior
 asymptotic, 93, 95–6, 133
 cuspoidal, 100
 high wavenumber, 93
 local spatial, 63
 singular, 96

binomial coefficients, 56
blurred observations, 19
bottom velocity, 8
boundary condition, 77, 80
 Charney, Fjörtoft & von Neumann,
 204, 207, 214
 classical, 204
 computational, 313
 Dirichlet, 203
 Euler-Lagrange, 279
 free-surface, 105
 interpolated 74
 Judovich, 216–7
 linear, 260
 long wave approximation, 85
 meridional, 85
 mixed, 203, 241
 Neumann, 203
 no-slip, 77
 number of, 260
 open, 203, 209, 275
 perturbation, 270
 persistent, 72
 radiation, 172
 rigid, 77, 172, 205, 261
 Robin, 203
 Sommerfeld, 204
 subcritical, 263–4
 supercritical, 263–4
 vanishing zonal transport, 85
boundary layer, 85, 95, 231, 241
boundary-value problem, two-point,
 131, 186
buffer-region, 267
buffer, 308

calculus of variations, 58, 115, 148,
 185, 298
 fundamental lemma, 299
canonical decomposition, 152

341